Logistics Systems

Hans-Christian Pfohl

Logistics Systems

Business Fundamentals

Springer

Hans-Christian Pfohl
Department of Law and Economics
TU Darmstadt
Darmstadt, Germany

ISBN 978-3-662-64348-8 ISBN 978-3-662-64349-5 (eBook)
https://doi.org/10.1007/978-3-662-64349-5

This book is a translation of the original German edition „Logistiksysteme" by Pfohl, Hans-Christian, published by Springer-Verlag GmbH, DE in 2018. The translation was done with the help of artificial intelligence (machine translation by the service DeepL.com). A subsequent human revision was done primarily in terms of content, so that the book will read stylistically differently from a conventional translation. Springer Nature works continuously to further the development of tools for the production of books and on the related technologies to support the authors.

This Springer imprint is published by the registered company Springer-Verlag GmbH, DE part of Springer Nature.
The registered company address is: Heidelberger Platz 3, 14197 Berlin, Germany

Preface to the 9th Edition

The new edition has again been extensively revised and updated. Where necessary, the latest available statistical studies were used, especially when discussing the macroeconomic aspects of logistics. Developments in information and communication technology (digitalization) relevant to logistics, e.g. RFID or cloud and blockchain technology, have been addressed primarily in connection with order processing and the infrastructure of the information flow. The list of literature has been brought up to date. The proven structure in five parts—formerly chapters A to D—with subdivision into sections has been retained. In accordance with the publisher's specifications, a bibliography has been added after each chapter. The previously common overall bibliography has been omitted.

For their support in revising and updating this edition, I would like to express my sincere thanks to my research associates Dr. Burak Yahsi and Tamer Kurnaz (M.Sc.). My special thanks go to Hendrik Bode (B.Sc.) for his tireless efforts in completing the artwork.

Darmstadt, Germany
September 2017

Hans-Christian Pfohl

Preface to the 8th Edition

The new edition has been comprehensively updated and expanded. This applies in particular to the statistical data, but also to significant technological developments that are relevant for logistics in various areas. These include RFID and scanner technology and the resulting possibilities for merchandise information systems or modern systems for the automation of intralogistics processes. Important aspects are also represented by current trends in logistics, which outline further developments at various points. In addition, many illustrations have been revised or newly presented and case studies have been updated. The literature sources have also been updated.

I would like to thank my research associates, Ms. Xin Shen (M.Sc.) and Mr. David Thomas (Dipl.-Wirtsch.-Inform.).

Darmstadt, Germany
September 2009

Hans-Christian Pfohl

Preface to the 7th Edition

This edition is a corrected reprint of the sixth edition. In terms of content, only adjustments to the reformed Commercial Code have been made. Otherwise, various printing errors that had unfortunately crept into the previous edition have been eliminated. In this regard, I would like to thank all my readers who have brought these to my attention.

I would like to thank my research associate, Oliver Boldt (Dipl.-Kfm.) for his support in revising this edition.

Darmstadt, Germany
August 2003

Hans-Christian Pfohl

Preface to the 6th Edition

The new edition has again been extensively revised and updated. This applies in particular to the statistical evaluations, but also to some recent developments that affect logistics, such as e-commerce, supply chain management or production networks.

I would like to express my sincere thanks to my research associate Alexander Koldau (Dipl.-Wirtsch.-Ing.) for his support in revising this edition, and to Ms. Tanja Schlag (stud.-wirtsch.-ing.).

Darmstadt, Germany
April 2000

Hans-Christian Pfohl

Preface to the 5th Edition

This edition has been extensively revised, updated and expanded. Recent developments in micro-, meta- and macrological systems have been taken into account by evaluating the relevant empirical studies conducted since the last edition. In order to characterize the conception of logistics, value- and benefit-oriented thinking was used in addition to systems thinking, referring back to the first version of this book, which was entitled *Marketing Logistics*.

The expansion of the new edition includes in particular the new chapter C. In it, the phase-specific subsystems of logistics are presented. Here, not only the classical subsystems of supply logistics, namely procurement, production and distribution logistics, are dealt with. Because of the often very specific problems of spare parts logistics, a special section is devoted to spare parts logistics. Since, in addition to the supply of the market, its disposal is becoming increasingly important, the reverse logistics, which is important for a circular economy, is also analysed as a new subsystem.

I received important support in editing the new edition from my former and current research associates, Ms. Birgit Ester (Dipl.-Kff.) as well as Hans Peter Buse (Dipl.-Wirtsch.-Ing.), Markus Engelke (Dipl.-Wirtsch.-Ing.), Michael Krings (Dipl.-Wirtsch.-Ing.), Dr. Rudolf Large, and Dr. Dirk Rohweder. I would like to thank them warmly for this. Special thanks are due to Christian Schäfer (Dipl.-Wirtsch.-Ing.) who, with the helpful support of Bernd Donabauer and Volker Kindermann (cand.-wirtsch.-ing.), took on the arduous task of preparing the final editorial version of this new edition.

Darmstadt, Germany
Summer 1995

Hans-Christian Pfohl

Preface to the 4th Edition

The increasing interest in the logistics conception requires a new edition of *Logistics Systems* after a relatively short time. This gave me the opportunity to correct some of the probably never avoidable printing errors. In addition, I have extended chapter D to include international aspects of logistics systems. In doing so, I am following the growing importance of international logistics systems in the context of the increasing Europeanization or worldwide globalization of corporate activities.

In addition to the *Fundamentals of Business Administration* presented here, the "technical fundamentals" of logistics systems have been published in this series under the title *Material Flow and Logistics* by my colleague Jünemann. I have refrained from making repeated cross-references to this basic volume in my explanations at the interfaces to technology. In the best logistical sense, both volumes offer the integrative fundamentals of logistics.

Darmstadt, Germany
Spring 1990

Hans-Christian Pfohl

Preface to the 3rd Edition

I have gladly accepted the publisher's invitation to prepare a third edition. First of all, I have of course corrected all printing errors of the second edition, as far as I have discovered them or have been made aware of them; furthermore, I have eliminated blurring in the wording of the text and the illustrations. In this context, I would like to thank all readers, especially my research associates and students, from whom I have received relevant advice. I would like to thank my research associate, Stephan Freichel (Dipl.-Wirtsch.-Ing.) for his intensive support in the editorial preparation of the third edition.

Due to the development trends in logistics systems that have occurred since the book was first published, some passages have been updated or expanded, and newly published literature has also been included.

Chapter D, which deals with macroeconomic aspects of logistics systems, is completely new. This is in no way intended to extend the basic principles of business administration. The macro-logistics system is only dealt with to the extent that it provides the framework for the micro- and metalogistics systems on a business management level.

Darmstadt, Germany
Spring 1988

Hans-Christian Pfohl

Preface to the 2nd Edition

Since the publication of the first version of this book entitled *Marketing Logistics. Design, Management and Control of the Flow of Goods in the Modern Market*, interest in the concept of logistics has increased considerably both in theory and in business practice. Some universities, technical colleges and universities of applied sciences now offer logistics courses. Job advertisements are increasingly looking for logisticians. In the corporate hierarchy, the logistics area is anchored at a higher management level.

This new edition attempts to do justice to this growing interest in the conception of logistics. It is a completely revised and expanded version of the first edition. It deals with the economic fundamentals not only of the marketing logistics system but also of all intra- and inter-organizational logistics systems. Technical aspects are only touched upon to the extent that they are absolutely necessary for a business understanding of logistics systems. The technical fundamentals of logistics systems will be dealt with by my colleague Jünemann in a further volume of this series.

The first chapter of this book deals with the basics of business logistics, whereby the approach underlying the logistics conception is discussed in detail. In the second chapter, the logistics subsystems of industrial and commercial enterprises are presented in order to show which problems are to be included in the functional analysis of logistics systems in terms of content. In the third chapter on institutional aspects of logistics systems, the organizational form of logistics systems in industrial and commercial enterprises is discussed first. Afterwards it is discussed which tasks can be taken over by logistics companies. Finally, interorganizational logistics systems are discussed, which are created by cooperation of different institutions in the logistics channel.

This book therefore aims to describe and explain the logistics systems with their business fundamentals that result from the logistics conception. The processes for designing such logistics systems are the subject of another book I am planning, which will be published in this series under the title *Logistics Management*.

I would like to take this opportunity to express my sincere thanks to all those who have contributed to the creation of this book. In particular, I would like to thank my former research associate, Dr. Klaus Wübbenhorst, for proofreading several versions of the manuscript, Stephan Freichel (cand.-wirtsch.-ing.) and Henrik Lewe for editing the manuscript on the word processing system, Holger Grotelüschen (cand.-wirtsch.-ing.) for the illustrations and Norbert Linn (Dipl.-Wirtsch.-Ing.) for the index. Last but not least, I would like to thank my wife and children for their understanding. They had to forego many weekends of leisure time with the author, which he spent in his study formulating logistics systems.

Darmstadt, Germany
Spring 1985

Hans-Christian Pfohl

Contents

Abbreviations

BVL	Bundesvereinigung Logistik e.V. (federal logistics association)
CSCMP	Council of Supply Chain Management Professionals
DBW	Die Betriebswirtschaft (business administration)
DIN	Deutsches Institut für Normung (German Institute for Standardization)
DVZ	Deutsche Verkehrszeitung (German Traffic)
ELA	European Logistics Association
GfL	Gesellschaft für Logistik e.V. (Society for Logistics) (now: DGfL)
GVB	Gesellschaft für Betriebswirtschaft und Logistik e.V. (Society for Business Administration and Logistics)
IJoPD	International Journal of Physical Distribution
MIR	Management International Review
NCPDM	National Council of Physical Distribution Management
TDM	Transportation and Distribution Management
VDI	Verein Deutscher Ingenieure e.V. (Association of German Engineers)
WSV	Wasserstraßen- und Schifffahrtsverwaltung des Bundes (Federal Waterways and Shipping Administration)
ZfB	Zeitschrift für Betriebswirtschaft (Journal for Business Administration)
ZfbF	Zeitschrift für betriebswirtschaftliche Forschung (Journal for Business Research)
ZfL	Zeitschrift für Logistik (Logistics Journal)

Part I
Basics of Business Logistics

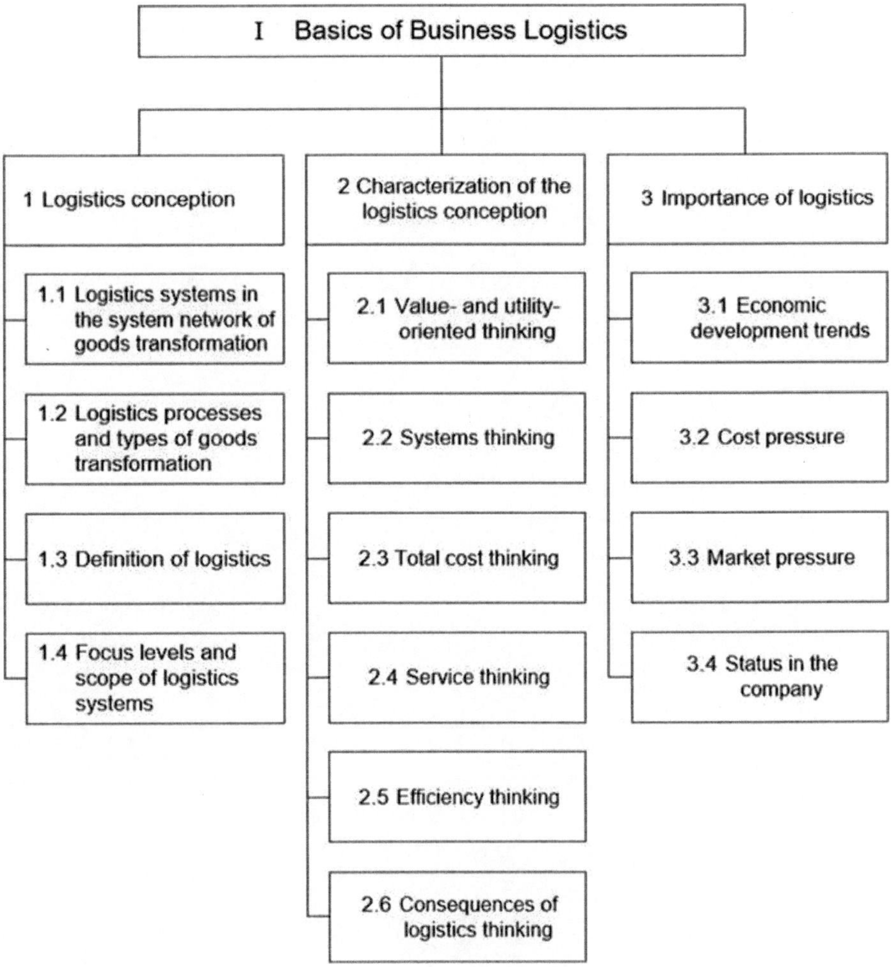

The presentation of the fundamentals of business logistics begins with a discussion of the conception of logistics. The definition of logistics is embedded in a substantive concretization of logistics systems and processes. This is followed by a characterization of the logistics conception, the logistical approach to problems in the company. The specific thinking in the analysis and design of logistics systems and processes is described with its various components. The consequences of this logistics thinking for the company are shown. Part I concludes with a detailed justification of the increasing importance of logistics as an instrument for rationalisation and gaining competitive advantages and the resulting high significance of logistics in the company.

1 Logistics Conception

1.1 Logistics Systems in the System of Goods Transformation

Systems of Goods Transformation

In an economy, the systems (sectors) of change of goods—in the sense of physical goods, tangible goods, real goods—listed in Fig. 1.1 can be distinguished.[1] The *provision of goods* takes place through production processes (extraction, processing and treatment processes) in industrial enterprises. The goods are qualitatively changed in this process. The goods are also qualitatively changed during the *use of goods*. Through consumption processes (using and consuming processes) in the broad sense, goods are consumed or worn out in households, but also in industrial, commercial or service companies. The link between the provision of goods and the use of goods is formed by the *distribution of goods*. It takes place through transformation processes (movement and storage processes), which do not change the goods qualitatively, but spatiotemporally. Systems for the spatiotemporal transformation of goods are *logistics systems*; the processes that take place in them are therefore *logistics processes*. They take place in so-called *logistics companies*. These are service companies whose purpose is to bridge space and time. However, they also take place in industrial, commercial or service companies in which the bridging of space and time is merely a partial task for fulfilling the actual purpose of the company.

Logistics processes cause the flow of goods that connects the systems of goods provision and goods use with each other. In all three systems, there are framework conditions that have a major influence on the flow of logistics processes. For example, the logistics processes will be very different for the production of loose or piece goods, for

[1] For the derivation of transfer or bridging needs and services from the division of labour, the dislocation and the time structures of the sectors of goods provision and goods use, cf. Ihde, 2001, pp. 1 f. and p. 57.

H.-C. Pfohl, *Logistics Systems*,
https://doi.org/10.1007/978-3-662-64349-5_1

Provision of goods	Distribution of goods[a]	Use of goods
System for the qualitative transformation of goods: Production processes	System of the spaciotemporal transformation of goods (logistics system): Transfer processes (logistics processes)	System for the qualitative transformation of goods: Consumption processes
in	in	in
Industrial companies	Logistics companies (bridging of time and space is the purpose of this type of service company)	Households
	or	or
	Industrial, commercial, or service companies (bridging of time and space is a partial task)	Industrial, commercial, or service companies
	Flow of goods →	
Production-oriented basic conditions for logistics processes: •Logistical product characteristics of technological and economical nature •Quantities of goods •Spatial structure of the provision of goods •Temporal structure of the provision of goods	Technological and institutional basic conditions for logistics processes: •Transport geography •Transport infrastructure •Transport means and auxiliary technology •Political and legal conditions •Stage of development of the logistics concept	Demand-oriented basic conditions for logistics processes: •Benefit of the bridging of time and space for the use of goods •Quantities of goods •Spatial structure of the use of goods •Temporal structure of the use of goods

[a] The legal transformation of goods in distribution of goods is disregarded in the figure.

Fig. 1.1 Systems of goods transformation

the distribution of goods in a country with a well or poorly developed road network, or for a use of goods for which rapid space bridging is of great or little importance.

Basic Structures of Logistics Systems

Logistics systems are characterized by the interplay of movement and storage processes. Graphically, the interaction of movement and storage processes—the latter can be more generally described as store processes—can be represented by a *network* in which nodes are connected by edges.[2] Objects are moved through this network. At the nodes, the objects are temporarily held (stored) or transferred to another path leading through the network. The different node connections (edges) represent the different ways an object can be moved through the network. Regardless of which objects (material goods, energy, information or people) flow through such a network, logistics systems exist.

However, this book deals only with logistics systems whose objects are physical goods. The information flows occurring in such logistics systems are not ends in themselves, but are derived from the physical flow of goods.

If we start from the *network idea*, we can distinguish the *basic structures of logistics systems* shown in Fig. 1.2.[3] In a *single-stage system*, space and time are bridged by a direct flow of goods between the delivery point at which the goods are made available, which can generally also be referred to as the source, and the receiving point at which the goods are used, which can generally also be referred to as the sink. Quite obviously, such a single-stage system has the advantage that the flow of goods between the delivery point and the receiving point is not interrupted, i.e. there are no additional storage processes and/or movement processes to transfer the goods to other routes.

In a *multi-stage system*, the space and time bridging is carried out by an indirect flow of goods between the delivery point and the receiving point. The flow of goods is therefore interrupted at least at one further point, at which additional storage and/or movement processes must then take place. The task of this interruption point is to break up or concentrate (bundle) the flow of goods. At a breakup point, goods arrive in large quantities from a delivery point and leave in small quantities (break-bulk point) to various receiving points. The breakup consists either in a pure reduction of the quantity units of a certain good in order to direct the flow of goods with regard to receiving points whose demand cannot be satisfied in large quantities but only in small quantities. However, the reduction can also consist of a sorting out. In this case, the flow of goods from a delivery point to the breakup point is not homogeneous, it does not consist of large quantities of a particular good, but is heterogeneous. In the heterogeneous system, the large quantities of goods are made up of different goods. The heterogeneous flow of goods is broken down at the breakup point into smaller homogeneous flows of goods to different receiving points. In a multilevel system, the breakpoint can also be a concentration point (consolidation point) where goods are bundled (collected or subjected to assortment). In the case of collection, a good arrives at the consolidation point in small quantities from different delivery points and is combined into larger homogeneous units. Assortment is another form of concentration. In assortment, different goods arrive at the concentration point from different delivery points and are grouped together to form assortments. In this case, the incoming goods flows at the concentration point are homogeneous, while the outgoing goods flows to the receiving points are heterogeneous.

[2] Cf. Ballou, 2004, pp. 41ff.

[3] Cf. Bowersox et al., 1968, pp. 120ff; Brauer/Krieger, 1982, p. 34; Bowersox/Closs, 1996, pp. 90ff.

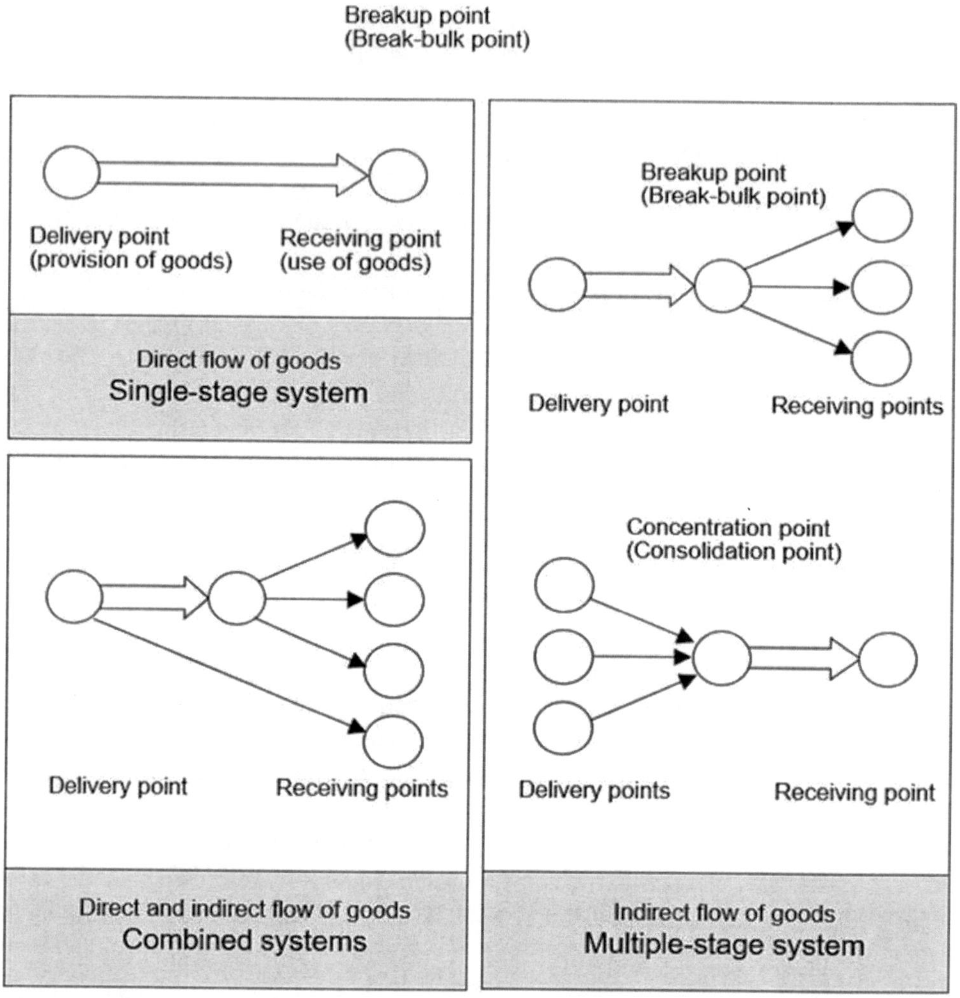

Fig. 1.2 Basic structures of logistics systems

One speaks of *combined systems when* direct and indirect flows of goods are possible side by side.

Single-stage systems have the advantage that the additional logistics processes required at the interruption points are avoided. The prerequisite, however, is that even in the case of large distances between the delivery point and the receiving point, the flow of goods is fast enough that the needs at the receiving point can be satisfied in time. If this is not possible, multi-stage systems are required in which, for example, the breakup point takes on the character of a distribution warehouse with which one moves as close as possible to a regional submarket in order to be able to quickly satisfy the needs arising with the customers of this market. However, the consideration that the economic efficiency of the flow of goods is generally directly related to the volume of this flow can also speak in favour of multi-stage systems.[4] The delivery warehouse then makes sense because it allows goods to flow from a production facility to a regional submarket in large units of volume. However, it must be taken into account that in a multi-stage system, additional logistics processes always arise at the interruption points, which can cancel out the advantages of a large-volume flow of goods between the delivery point and the interruption point or between the interruption point and the reception point. The logistics processes will be discussed in more detail below.

1.2 Logistics Processes and Types of Goods Transformation

Flow of Goods and Information

The basic function of logistics systems is the spatiotemporal change of goods. As can be seen from what has been said in the first section, the fulfilment of this basic function is often also associated with the function of changing the quantity and type of goods.[5] Ultimately, it is also part of the function of logistic systems to facilitate the aforementioned types of goods transformation. These functions are fulfilled by:

- Transport, handling and storage processes (core processes of the flow of goods),
- Packaging and signing processes (support processes in the flow of goods).

The flow of goods between the delivery point and the receiving point does not flow by itself, but requires the exchange of *information* between the two points. The information triggers the flow of goods in advance, accompanies it by clarifying it and follows it by confirming or not confirming it. Logistics processes therefore include not only those processes that manage the flow of goods, but also those that manage the corresponding flow of information. This information function of logistics systems is fulfilled by

- Order transmission and order processing processes (information flow).

The matrix in Fig. 1.3 shows a classification of logistics processes and the *types of goods transformation* they bring about. The term "handling" used to describe logistics processes is broadly defined. It refers to both the handling of goods, e.g. when goods are placed on a shelf, the grouping or breaking down of goods, e.g. in connection with pallets, and the sorting of goods in the context of order picking. The assignment of the goods transformation types to storage, transport and handling is obvious. Packaging facilitates or even enables the transport, handling and storage of many goods. Signing the packaging can provide important information for the type of transport, handling and storage that facilitates these processes. By transmitting and processing orders (*order processing*), a good goes

[4] Cf. Bowersox et al., 1968, p. 379.

[5] This function is also referred to as the structuring function, cf. Ihde, 2001, pp. 2 f.

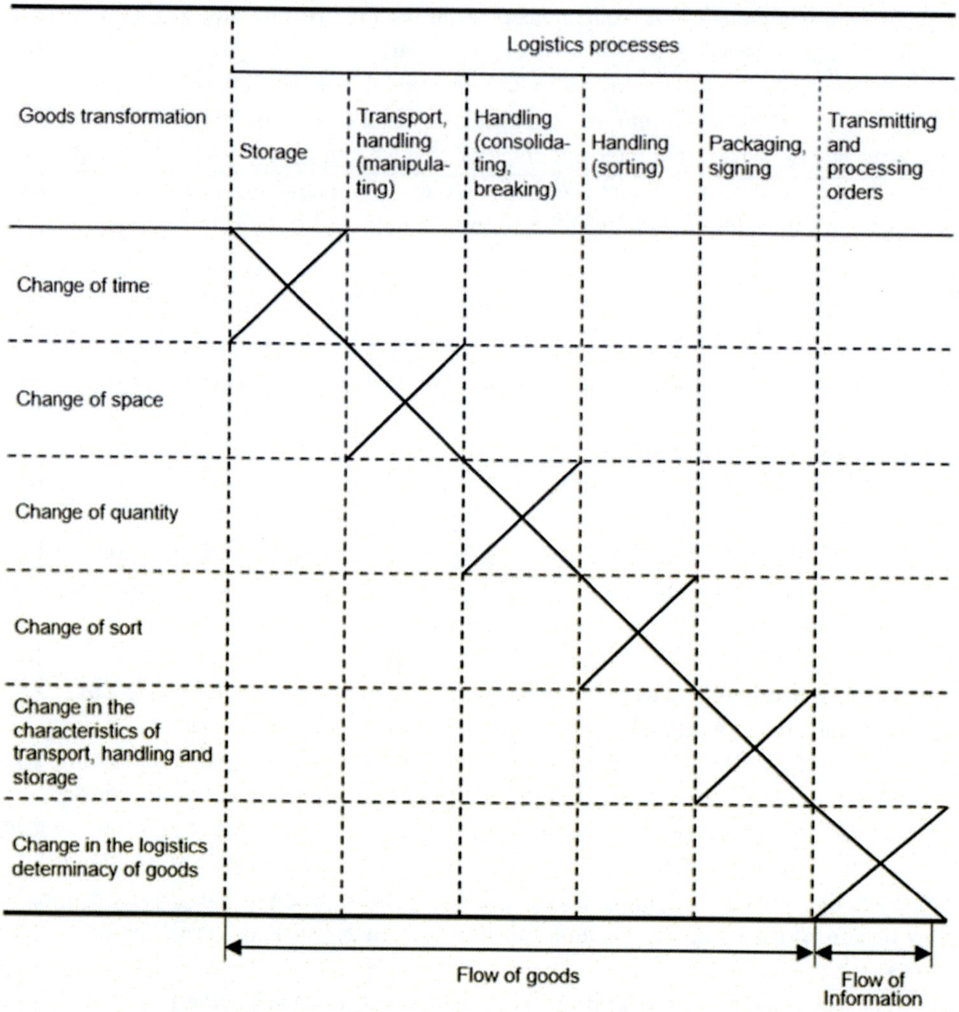

Fig. 1.3 Logistics processes and the transformation of goods they bring about (Source: Taken with modifications and additions from Jünemann, 1980, p. 2)

from being logistically indeterminate/undetermined to logistically determinate/determined. A good is more logistically determined the more extensive and precise information is provided regarding the nature of the flow of goods. If, for example, an order merely states that a good is to be delivered to the consignee in the 22nd week, it is much less logistically determined than if the order contained the information that the good is to be delivered to the consignee's loading bay 3 on Wednesday at 9:00 a.m. in the 22nd week.

The logistics processes mentioned are tasks whose execution realizes the flows of goods and information. In addition to these *realization tasks,* logistics tasks naturally include the *planning, implementation* and *control tasks* associated with them.

Order processing	• Form of order transmission • Form of order processing • Analysis of the order as information source • Forwarding of the order information
Inventory management	• Quantity of articles to store (selective inventory management, ABC-principle) • Order quantity and order point for replenishment of goods in stock • Safety stock • Inventory Level control • Short-term demand forecast
Warehouse	• Purchase or rent of warehouse and warehouse equipment • Number, site, capacities and delivery area of the warehouses • Internal (make) or external (buy) operation of the warehouses • Technical equipment for magazining and picking in the warehouse • Storage locations in the warehouse • Storage method (design of the stack) • Design of the loading ramp • Dispatch of transport means • Organisation of picking • Efficient deployment of warehouse personnel
Transport	• Types of means of transport • Internal or external operation of means of transport • Purchase or rent of the means of transport • Combination of means of transport • Organisation of transport processes (optimal transport routes, usage planning and loading of means of transport)
Package	• Fulfilment of the logistics functions of the packaging (protective, storage, transport, manipulative and informative functions) • Formation of logistics units (storage, loading, transport units etc.) as requirement for efficient transport chains

Fig. 1.4 Identification of logistical task areas

Logistical Task Areas

In order to get an impression of the *logistics tasks* to be performed in connection with the planning, implementation, realization and control of logistics processes, Fig. 1.4 shows important logistics decision elements. Based on the logistics processes, a distinction is made between logistics task areas. In the case of storage, the decision-making activities relating to inventories are grouped together under the term inventory management. The term warehouse includes the decision-making facts that determine where goods are stored and how they are to be received or delivered.

Decision facts regarding the handling processes can be found in the task area warehouse as well as in the task area transport. This is because handling processes combine various warehouse processes, various transport processes and warehouse and transport processes. The logistical task areas mentioned here are further substantiated in the second part.[6]

1.3 Origin and Definition of Logistics

Origin of the Word Logistics

After logistics systems and logistics processes have been presented in an introductory manner, the next section will deal with a definition of logistics. The word logistics brings with it various approaches to meaning in the German language. It therefore seems necessary to briefly draw attention to these different approaches.

In the science of *logic*, logistics is sometimes used synonymously with mathematical logic and symbolic logic. More recently, however, the use of the term logistics in this sense has receded in favor of the synonyms.[7]

Mathematical functions that can be defined as modified exponential functions are called logistic functions. Such logistic functions are used, for example, to describe the growth of the population or to represent the life cycle of a product from market launch to market saturation.

In the *military field*, logistics is used as a collective term for the tasks that serve to support the armed forces.[8] The term logistics in this case is derived from the French word "loger". Military logistics includes the transport, quartering and supply of troops as well as the transport, storage and maintenance of military goods.

From the military field, the term logistics has found its way into economic literature. In contrast to the military field, where logistics refers to troops and goods, the term logistics in the economic field is now primarily used to refer to goods. Moreover, in contrast to military logistics, maintenance—e.g. of production equipment—in companies is not generally included in logistics. Another essential difference is that logistical decisions in the military field are guided by objectives that are politically-militarily motivated, whereas logistical decisions in the economic field are made on the basis of technological, economical, ecological and social objectives.

Definition of Logistics

There are a variety of definitions for the term logistics or other terms used in its place. At this point it may suffice to present three definitions.[9]

[6] For a similar distinction of logistical task areas, see Stock/Lambert, 2001, p. 19ff. For a further overview of decision items in logistical systems, see also Kirsch et al., 1973, pp. 294ff; Krulis-Randa, 1977, pp. 200ff; Bowersox/Closs, 1996, pp. 25ff; Vahrenkamp 2007, pp. 9ff; Arnold et al., 2008.

[7] Cf. Behrendt, 1977, p. 21 and the literature cited there.

[8] On military logistics, cf. Krulis-Randa, 1977, pp. 39ff.; Ihde, 2001, pp. 23 f.

[9] Cf. Pfohl, 1972, pp. 17ff; Schary, 1984, pp. 7ff; Kummer, 1992, pp. 20ff; Isermann, 1998, pp. 21ff and the literature listed there.

The first definition approach can be described as a *flow-oriented definition of* logistics. It builds on the substantive concretization of the logistics term in the first two sections and reads as follows:

Logistics includes all activities through which the spatiotemporal transformation of goods and the associated transformations with regard to the quantities and types of goods, the goods handling characteristics and the logistical determinacy of the goods are planned, implemented, realized or controlled. Through the interaction of these activities, a flow of goods is to be set in motion that connects a delivery point with a receiving point as efficiently as possible.

Already at this point it can be pointed out what efficient means. For this purpose, we can refer to the *four R's*, which are mentioned to characterize the requirements for logistics[10]:

Logistics has to ensure that a receiving point according to its needs is supplied with the *right product* (in quantity and type), in the *right condition*, at *the right time*, at the *right place*, at the minimum cost, from a delivery point.

A flow-oriented definition also comes from the American logistics society Council of Supply Chain Management Professionals (CSCMP)—formerly the Council of Logistics Management (CLM)—and is widely used in the US. The definition reads:

> Logistics is the process of planning, implementing and controlling the efficient, cost-effective flow and storage of raw materials, semi-finished and finished goods and related information from the point of delivery to the point of receipt according to the customer's requirements.[11]

Also flow-oriented is the definition of the European umbrella organization of national logistics societies in Europe, the European Logistics Association (ELA). It reads:

Logistics is "the organization, planning, control, and execution of the goods flow from development and purchasing, through production and distributions to the final customer in order to satisfy the requirements of the market at minimum cost and minimum capital use."[12]

A second definition approach can be referred to as a *life cycle-oriented definition of* logistics. It builds on the life cycle of a product in the sense of its service life.[13] The concept of the life cycle is based on the idea that a product—more generally a system—is created through measures of planning, design and development and is finally decommissioned or scrapped after a period of operation. Life cycle phases are, for example, the initiation, planning, realization, operation and decommissioning phases. Logistics activities then relate to the support of transformation activities in the various life cycle phases. The

[10] Cf. Pfohl, 1972, pp. 28ff.

[11] Council of Supply Chain Management Professionals, n.d., p. 2.

[12] European Logistics Association, 1993, p. 1. This definition is used today by the ELA to define supply chain management.

[13] On this concept, cf. Pfohl/Wübbenhorst, 1983, pp. 144ff. See also Finkelstein/Guertin, 1988.

internationally active logistics society The International Society of Logistics (SOLE)—formerly the Society of Logistics Engineers—defines accordingly:

Logistics is "the supportive management that ensures more efficient use of resources and adequate performance of logistical elements during all phases of a product's life cycle, ensuring effective control of resource consumption through timely intervention in the system."[14]

Finally, a third definition approach can be called a *service-oriented definition of* logistics. It builds on the idea that a service can only be optimally provided to a customer if all activities for production are performed in a coordinated manner. The definition reads:

Logistics is "the process of coordinating all intangible activities that must be completed to fulfill a service in a cost and customer effective manner."[15] The focus of these activities is in the following three areas: Minimizing waiting time (the order fulfillment time), managing service capacity, and delivering the service through a distribution channel.[16]

The following remarks are based on the flow-oriented definition approach that is most widely used in science and practice. The life-cycle oriented definition can only prove useful when logistics is discussed in connection with the calculation, analysis and design of life-cycle costs. Life cycle costs are the total costs incurred by a system throughout its life. The service-oriented definition approach may prove useful in cases where logistics services are provided in close connection with other services. An example of this approach is the provision of a spare part at the customer's site at the time when the service technician performs maintenance activities on a machine at the customer's site.

After the content of the logistics term has been clarified, the following section will distinguish important logistics systems that occur in reality.

1.4 Focus Levels and Scope of Logistics Systems

A distinction between different logistics systems is necessary with regard to the differences in the problems that arise when designing a logistics system. A major influence on these problems will be the scope and the focus level (level of aggregation) of the defined logistics

[14] Coyle et al., 1992, p. 8.

[15] Arthur D. Little/The Pennsylvania State University, Center of Logistics Research, 1991, p. XXII.

[16] Cf. Arthur D. Little/The Pennsylvania State University, Center of Logistics Research, 1991, pp. 34ff. A different, more comprehensive service-oriented concept of logistics is represented by the banks, which understand logistics management as the creation of an expedient infrastructure with which the entire supply system is placed in the service of managing the business front. Logistics includes not only the performance of service functions but also the provision of assistance to all other areas. A bank's logistics management includes financial and operational accounting, IT and EDP, control and auditing, the bank's own research and development, real estate and security, as well as personnel management and training. On the concept of logistics in banks, cf. Lohmann, 1998, pp. 76ff.

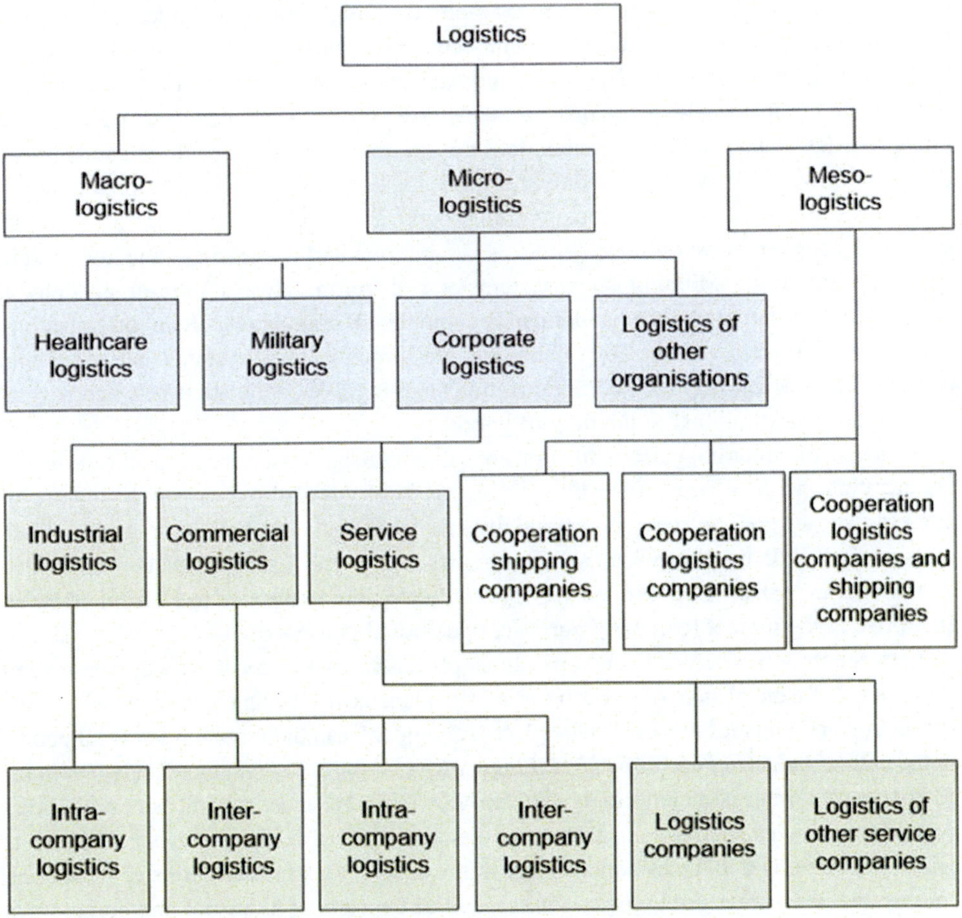

Fig. 1.5 Institutional delimitation of logistics systems

system. With regard to these two characteristics, logistics systems can be distinguished from both institutional and functional points of view. Institutionally, logistics systems differ according to the type and number of institutions considered in the system, functionally according to the type and number of functions considered in the system.

Institutional Delimitation of Logistics Systems

Figure 1.5 attempts an institutional delimitation of logistics systems.[17] Following the distinction between levels of aggregation commonly used in economics, *macro-, micro- and mesologistics* can be distinguished. *Macrologistics* systems are macroeconomic in

[17] Cf. Pfohl, 1974, pp. 73ff. and the literature cited there, as well as Felsner, 1980, p. 18; Endlicher, 1981, p. 29.

nature. A macrologistics system is, for example, the freight transport system in an economy. *Micrologistics* systems are of a microeconomic nature. Micrologistics include the logistical systems of individual public or private organizations, for example, the vehicle fleet of a company. *Mesologistics* systems are on a level between macro- and micrologistics. A mesologistics system does not, for example, include all the goods traffic in an economy, nor does it include only the goods traffic of a single organization, but rather the goods traffic of the organizations working together in a sales channel, for example, an industrial supplier, a wholesaler acting as an intermediary, a retailer, and the freight forwarder involved. Micrologistics systems are therefore always intra-organizational systems whose largest scope is defined by the legal boundaries of an organization. Mesologistics systems, on the other hand, are inter-organizational systems that go beyond the legal boundaries of individual organizations and involve cooperation between several organizations (institutions) in the flow of goods.

Systems of micrologistics can first be distinguished according to the type of organizations with different objectives.[18] In this book, only those organizations are of interest whose goal system is essentially characterized by economic goals. Such organizations will be referred to as enterprises. *Enterprise logistics* can be subdivided into *industrial, commercial and service logistics* according to the task to be fulfilled by an enterprise in the market (company purpose, operational purpose).

Between the terms company and operations/plant a distinction is often made: a company is the legal, financial unit of a business and a plant is the technical unit in which the production and logistics processes take place. A company can therefore have several plants. In the case of *industrial* and *commercial logistics*, it is useful to distinguish between intra-company and inter-company logistics. *Service logistics* must be further subdivided depending on whether the main task to be performed by a company on the market, its primary service, is a logistics service (e.g. in the case of freight forwarders or packaging companies) or whether the logistics services, as in the case of industrial and commercial companies, are merely secondary services that must be provided in connection with the performance of the actual market task (e.g. in the case of banks or insurance companies).[19] Companies whose main purpose, as in the first case, is the provision of logistical services, i.e. where logistical functions dominate,[20] are referred to as logistical business enterprises or *logistics companies,* or—since no distinction is often made between companies and businesses—as logistical businesses or *logistics service providers*.

Mesologistics systems can be differentiated according to which companies cooperate in the fulfilment of logistical tasks. Cooperation is possible among companies in the shipping industry. For example, shippers from different industries, but also from the same industry,

---

[18] Because of its great importance in a globally networked world, "*humanitarian logistics*" should be emphasized among the logistics of other organizations. Cf. Baumgarten et al., 2010.

[19] Cf. Frodl, 1998, p. 12ff.

[20] Cf. Kirsch et al., 1973, p. 84.

can use a common goods distribution system. Cooperation between logistics companies can take place, for example, between regionally specialized forwarding companies or between road transport companies and the railways. Cooperation between logistics companies and the shipping industry occurs, for example, when a shipper entrusts the delivery of its products to a logistics company.

Functional Delimitation of Logistics Systems

A first way to distinguish functional subsystems of logistics is to follow the different phases of a flow of goods from the procurement market through an industrial enterprise to the sales market and from there back to the procurement market. The *phase-specific subsystems of* logistics are then obtained. As can be seen in Fig. 1.6, the first phase of the flow of goods, consisting of raw and auxiliary materials, factory supplies, purchase parts and possibly merchandise and supplied spare parts, goes from the supplier on the procurement market to the procurement or receiving warehouse of an industrial enterprise. Between the procurement warehouse located directly at the production site and the procurement market, there may also be a supply warehouse which, for example, has to perform collection or sorting tasks. Of course, a flow of goods directly from the procurement market into the production process is also possible. The logistics system that deals with the first phase of the flow of goods is called *procurement logistics*, sometimes also supply logistics or physical supply system.[21]

In the second phase, raw and auxiliary materials, factory supplies as well as delivered spare parts and purchased parts flow from the procurement warehouse into the production process, where semi-finished products can be temporarily stored. Finished products and also semi-finished products as well as spare parts intended for customers flow from production to the sales warehouse. This logistics subsystem is called *production logistics*. Procurement logistics and production logistics together are sometimes also referred to as material logistics.

In the third phase, the flow of goods consists of finished goods, semi-finished goods—these then have the function of spare parts—and possibly also merchandise. Here, the flow of goods goes from the sales warehouse located at the production site via regional distribution warehouses to the customers in the sales market. Of course, direct delivery to customers from the sales warehouse or even from the production process is also possible in this case. The logistics in this third phase of the flow of goods is called *distribution logistics*. The term marketing logistics, which used to be used for this,[22] is nowadays more commonly used to designate the two market-linked logistics systems of procurement and

[21] On the latter, see Kirsch and others, 1973, p. 269.

[22] Cf. Pfohl, 1972.

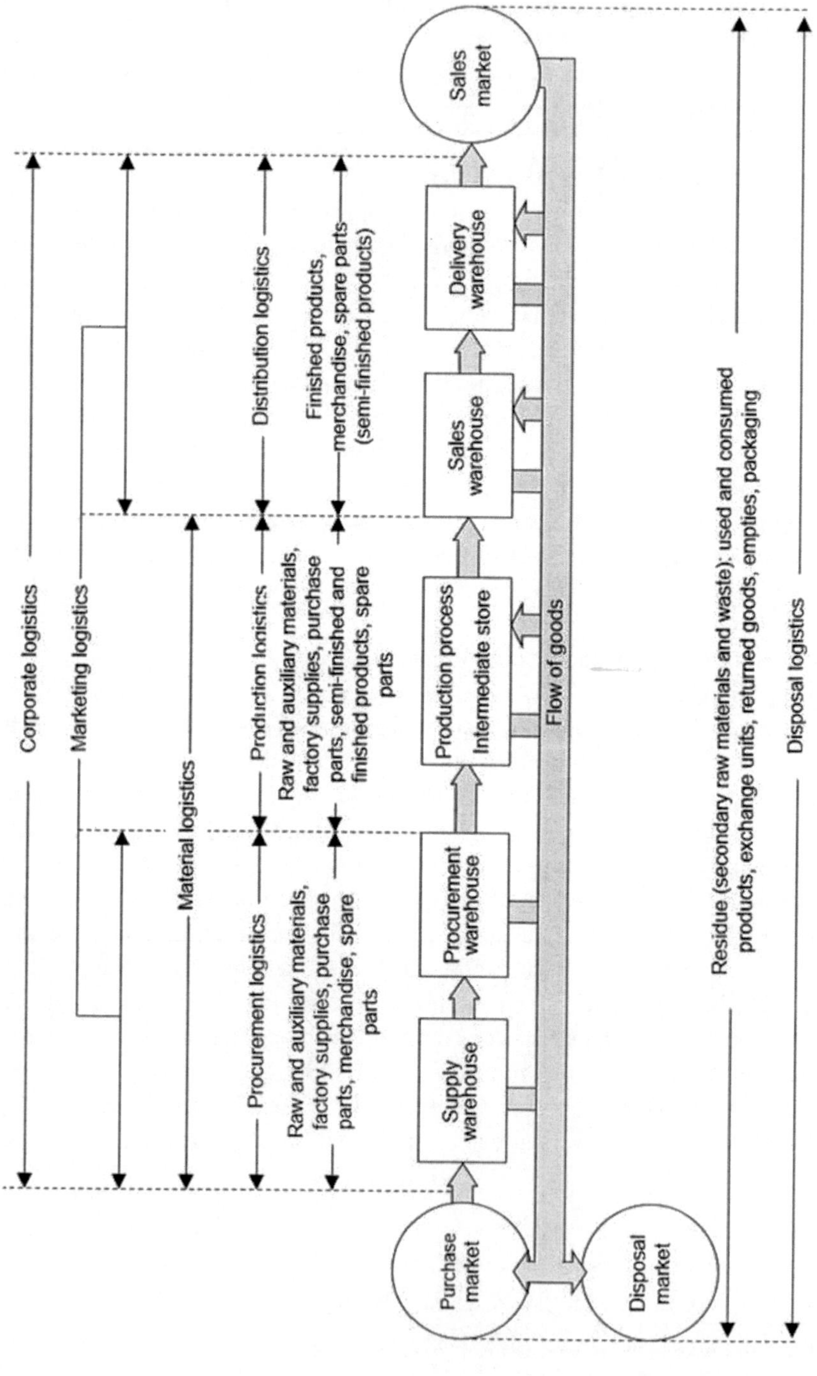

Fig. 1.6 Functional delimitation of logistics systems according to the phases of the flow of goods using the example of an industrial company

distribution logistics,[23] which used to be grouped together under the term physical distribution.[24]

Finally, in the fourth phase, the flow of goods flows in a reverse direction. It then consists of residues, which can be divided into secondary raw materials (valuable materials) and waste. While waste has to be disposed of, secondary raw materials are reused or recycled. Thus, damaged or incorrectly delivered goods returned by customers to suppliers (returns), empties to be returned, replacement units to be returned in the case of industrial and consumer goods, and used containers and packaging can also be counted as residues. This part of logistics can be referred to as *disposal logistics*. The term logistics in the post-purchase phase can also be found, although this also includes a part of *spare parts logistics* in addition to disposal logistics.[25] Spare parts logistics can be important for a company in connection with maintenance on the procurement side and customer service on the distribution side.

The logistics systems listed here can be summarized under the term corporate logistics. Corporate logistics is broken down in Fig. 1.6 using the example of an industrial company. In the case of a trading company, production logistics does not exist and the flow of goods consists only of merchandise and operating materials. Finally, in service companies there is only procurement logistics and the flow of goods consists only of operating materials.

For a further functional delimitation of logistics systems, reference can be made to Fig. 1.4. The logistics subsystems listed there are summarized again in Fig. 1.7. They reflect the activity-oriented content of the tasks to be performed in the logistics system. The result are the *activity-specific subsystems* of logistics. These are business subsystems like other business subsystems (we also speak of business functions such as sales, production, research and development, purchasing, financing, human resources, information), in which production factors are used for the purpose of creating and exploiting business services. The valuated input of production factors in logistics systems represents the *logistics costs*. Costs are only justified as system inputs from a business point of view if they are matched by corresponding activities as system outputs. The output of the logistics system can be characterized by the four R's of logistics mentioned in the previous section, namely to provide the right good, in the right condition, at the right time, at the right place. The output to be produced by a logistics system is also called a service. If raw materials, supplies or purchased parts have to be made available for the production process, then the *logistics service* is the supply service. If finished products, spare parts or merchandise have to be made available to the customer, the logistics service is referred to as the delivery service.

The arrows between the individual logistics subsystems are intended to make it clear that the interdependencies between these subsystems must be taken into account when

---

[23] Cf. Ihde, 1978, pp. 1 f.

[24] Cf. Pfohl, 1974, p. 77.

[25] Cf. Hallbauer/Knödel, 1980.

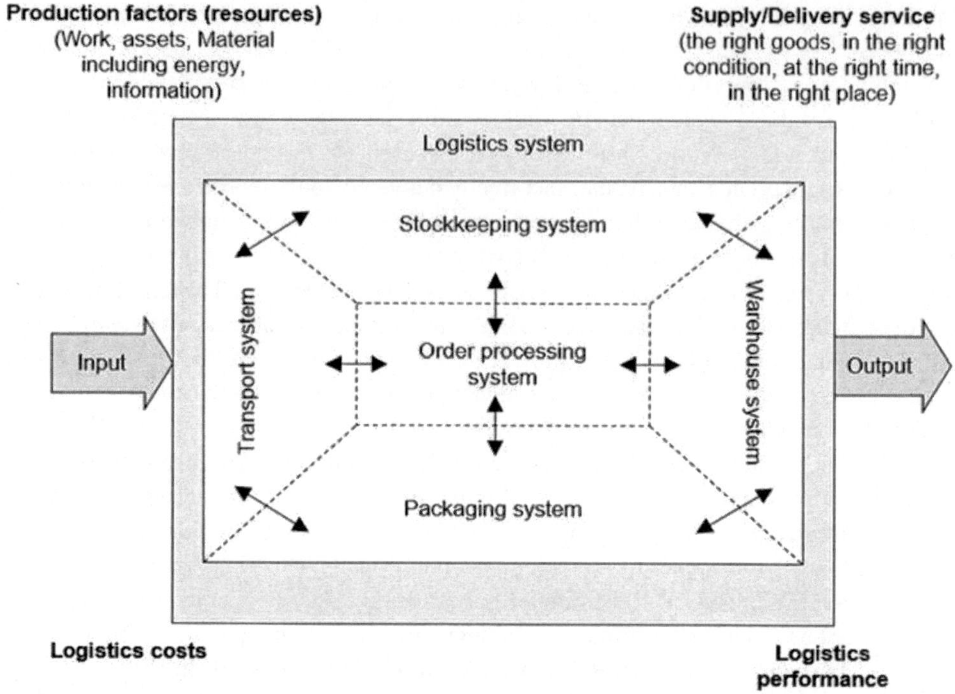

Fig. 1.7 Functional delimitation of logistics systems according to the content of logistics tasks

fulfilling logistics tasks. This already addresses the characterization of the logistics conception.

References

Arnold D u.a. (Hrsg) (2008) Handbuch Logistik. 3., neu bearb. Aufl. Berlin Heidelberg

Arthur D Little, The Pennsylvania State University, Centre of Logistics Research (1991) Logistics in Service Industries. Prepared for Council of Supply Chain Management Professionals (CSCMP). Oak Brook, IL

Ballou R H (2004) Business Logistics / Supply Chain Management. Planning, Organizing, and Controlling the Supply Chain. 5. Aufl. Upper Saddle River N.J.

Baumgarten H, Keßler M, Schwarz J (2010) Jenseits der kommerziellen Logistik. Die humanitäre Hilfe logistisch unterstützen. In: Schönberge R, Elbert R (Hrsg) Dimensionen der Logistik. Funktionen, Institutionen und Handlungsebenen. Wiesbaden, S.453-475

Behrendt W (1977) Die Logistik der multinationalen Unternehmung. Eine systemorientierte und verhaltenswissenschaftliche Analyse. Diss. Berlin

Bowersox D J, Closs D J (1996) Logistical Management. Überarb. Ausg. der 3. Aufl. New York u.a.

Bowersox D J, Smykay E W, LaLonde B J (1968) Physical Distribution Management. Logistics Problems in the Firm. Rev. ed. New York

Brauer K M, Krieger W (1982) Betriebswirtschaftliche Logistik. Berlin
Council of Supply Chain Management Professionals (CSCMP) (o J) What It's All About. Purpose, Objectives, Programs, Policies. Oak Brook, IL
Coyle J J, Bardi E J, Langley Jr C J (1992) The Management of Business Logistics. 5. Aufl. St. Paul, MN
Endlicher A (1981) Organisation der Logistik. Untersucht und dargestellt am Beispiel eines Unternehmens der chemischen Industrie mit Divisionalstruktur. Forschungsbe-richte zur Industriellen Logistik 18. Dortmund
European Logistics Association (ELA) (1993) What is ELA? Bern
Felsner J (1980) Kriterien zur Planung und Realisierung von Logistik-Konzeptionen in Industrieunternehmen. Bremen
Finkelstein W, Guertin J A R (1988) Integrated Logistics Support. The Design Engineering Link. Berlin u.a.
Frodl A (1998) Dienstleistungslogistik: Information, Kommunikation, Daten, Dokumente: Zur richtigen Zeit am richtigen Ort. München/Wien
Hallbauer A, Knödel W (1980) Logistische Prozesse in der Nachkaufphase. In: Poth L G (Hrsg) Marketing. Neuwied, Kennziffer 3.2.4.5, S. 1-47
Ihde G B (1978) Distributions-Logistik. Stuttgart/New York
Ihde G B (2001) Transport, Verkehr, Logistik. Gesamtwirtschaftliche Aspekte und einzelwirtschaftliche Handhabung. 3., völlig überarb. u. erw. Aufl. München
Isermann H (1998) Grundlagen eines systemorientierten Logistikmanagements. In: Isermann H (Hrsg) Logistik: Gestaltung von Logistiksystemen. 2., überarb. und erw. Aufl. Landsberg a. L., S. 21-60.
Jünemann R (1980) Bedeutung des Arbeitsschutzes in den Logistikbereichen der Wirtschaft. In: Institut für Logistik der GfL (Hrsg) Handbuch „Logistik und Arbeitsschutz". Dortmund, S. 112.
Kirsch W u.a. (1973) Betriebswirtschaftliche Logistik. Systeme, Entscheidungen, Methoden. Wiesbaden
Kummer S (1992) Logistik im Mittelstand. Stand und Kontextfaktoren der Logistik in mittelständischen Unternehmen. Stuttgart
Krulis-Randa J S (1977) Marketing-Logistik. Eine systemtheoretische Konzeption der betrieblichen Warenverteilung und Warenbeschaffung. Bern/Stuttgart
Lohmann L (1998) Banklogistik. Logistiksysteme und –prozesse in Banken. Berlin
Pfohl H-Chr (1972) Marketing-Logistik. Gestaltung, Steuerung und Kontrolle des Warenflusses im modernen Markt. Mainz
Pfohl H-Chr (1974) Die Logistik als Beispiel für Auswirkungen des Systemdenkens in der entscheidungsorientierten Betriebswirtschaftslehre. In: MIR 14 1, S. 67-85
Pfohl H-Chr, Wübbenhorst K L (1983) Lebenszykluskosten. Ursprung, Begriff und Gestaltungsvariablen. In: Journal für Betriebswirtschaft 33 3, S. 142-155
Schary Ph B (1984) Logistics Decisions. Text and Cases. Chicago u. a.
Stock J R, Lambert D M (2001) Strategic Logistics Management. 4. Aufl. Boston u.a.
Vahrenkamp R (2007) Logistik. Management und Strategien. 6., überarb. und erw. Aufl. München

2 Characterization of the Logistics Conception

2.1 Value- and Utility-Oriented Thinking

Value and Utility Types

The economic process serves to satisfy needs. As shown in Fig. 2.1, it can be broken down into three sub-areas, namely the provision (production), distribution and use of goods. Every economic activity must first be viewed from the perspective of the sub-area of use. Economic goods are ultimately used or consumed. This implies that they satisfy needs of people. This satisfaction of needs does not occur through the economic activity itself, but through the creation of utilities associated with it.

As Fig. 2.1 shows, five types of *utility* can be distinguished that arise with economic activity: Form utility, utility from the right to the good, information utility, place utility and time utility.[1] The form utility arises in the economic sub-sector of provision and relates to the form and quality of the economic good. The utility from the right to the good, information utility, place utility and time utility arise in the area of distribution. A machine produced in Stuttgart (form utility) can only satisfy a need in Munich if it is known in Munich that this machine exists in Stuttgart (information utility), if it is sent to Munich (place utility) and at the time when it is needed (time utility). In addition, the right to the good must be transferred to the user in an appropriate manner (utility from the right to the good through ownership, leasing or renting), on the basis of which he can dispose of the good. The location and time utilities as well as, to some extent, the information utility (through the order processing information that precedes the flow of goods) and the utility

[1] The four types of utility "form, possession, place and time" are distinguished by Converse, 1958, p. 116; Bowersox et al., 1968, p. 20; Langley/Holcomb, 1992, p. 1; Novack et al., 1992, p. 236. For a classification of utility under other aspects, see Corsten, 2007, p. 157 f. and the literature listed there. For a narrower concept of utility, cf. Large, 1995, pp. 34 f. The assignment of place and time utility to logistics can also be found in Morgenstern, 1955, p. 130.

H.-C. Pfohl, *Logistics Systems*,
https://doi.org/10.1007/978-3-662-64349-5_2

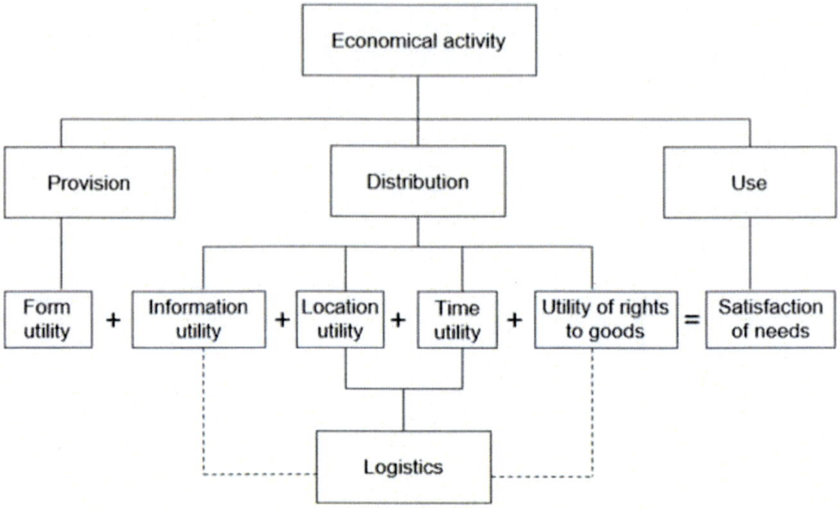

Fig. 2.1 Contribution of logistics to the satisfaction of needs

from the right to the goods (supplier is the owner of the stock in the customer's warehouse until the time the customer takes the goods) are generated in the logistics system.

The interest of economics is not directed towards the creation of the good itself with its natural properties,[2] but towards the creation of the value properties of goods—i.e. those properties that create utility. Accordingly, the purpose of companies is to create *value* by producing goods with those value characteristics that result in the satisfaction of the customer's needs (solution of problems).[3] The difference between the suitability and use value of a good, as shown in Fig. 2.2, is important for value creation. A good only has use value in the system of goods use if it is not only suitable for the satisfaction of a need, but is also available there. The availability of a good distinguishes the suitability value from the use value. Thus, in order to have value in the system of uses of goods, a good must have two properties: First, it must have the *suitability* generated in the system of goods provision for satisfying the customer's needs (solving problems). Second, it must have *availability* at the customer. The availability generated in the system of distribution has two dimensions, the actual availability and the legal availability. Actual availability is given when a good in the system of use of goods can be used at the desired time at the desired place. Legal availability is given when the customer has the right of disposal necessary for the specific use of the good. If the legal availability is given, but not the actual availability, then the promised use value of a good exists. The customer then has a right of disposal in the sense of

[2]This is the elementary interest of engineering.

[3]On this view of companies and on the value characteristics discussed below, cf. Large, 1995, pp. 3ff. and pp. 33ff; Large, 2013, pp.20ff. This view has an old tradition in business administration and has recently been "rediscovered" under the term value chain, cf. Porter, 2014, pp. 65ff.

	actual availability given	actual availability not given
Legal availability given	use value	promised use value
Legal availability not given	suitability value	suitability value

Fig. 2.2 Availability as a constituent property of use value (Source: Taken with minor modifications from Large, 1995, p. 27)

a claim to the production and provision of a good. The assured utility value in the sense of a promise of performance is particularly important for services, the production of which will be discussed in more detail below.

The explanations on value creation thinking show that logistics activities are necessary to generate the use value of a good. The use value can be increased not only by improving the suitability of the goods, but also by improving the availability of the goods. Actual availability can be improved by improving performance in logistics activities already provided by the supplier of a good or by the supplier taking on additional logistics activities from the customer. Figure 2.3 shows an example of the takeover of such value-adding activities previously exercised by the customer.

It makes sense for the offering company to take over such value-adding activities if the logistics services previously exercised by the customer can be offered at lower logistics costs or if higher services can be provided at the same logistics costs.[4]

Service Character of the Value Added

Logistical value creation activities have a service character, which has a great influence on the design of these activities.[5] Real services can be characterised by the features of *immateriality* and *integration of the external factor*. This is expressed in the following three proposed definitions.[6]

In terms of *potential-oriented*, service is understood as the ability and willingness to provide a service. In this view, service is a promise of performance and thus immaterial.

---

[4]See Fig. 1.7 in Sect. 1.4.

[5]Cf. Meffert, 1994, pp. 521 f. It should be noted that the following statements apply to the majority of services. However, there are also services which, because of their characteristics, are produced in a similar way to tangible goods. Likewise, there are tangible goods that are very similar to the services characterised here.

[6]Cf. Corsten, 1993, pp. 765 f.

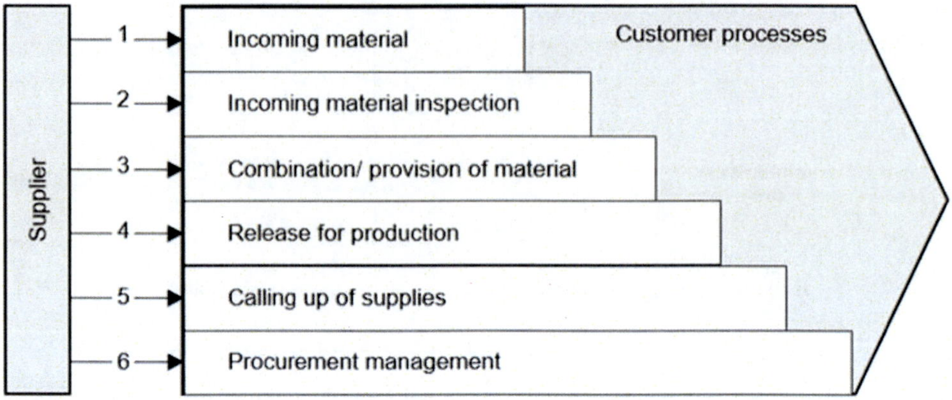

Stages of takeover:

Stage 1: Transport to the customer
Stage 2: Direct transport to the customer's incoming material inspection
Stage 3: Transport via loading ramp into warehouse
(Requirement: quality certification)
Stage 4: Direct transport to the customer's production line
Stage 5: Transfer to the customer's inventory management
Stage 6: Acquisition of other input factors for the customer – "one-stop-shop"

Fig. 2.3 Takeover of value-adding logistics activities of the customer by the supplier (Source: Based on Gopal/Cypress, 1993, p. 197)

This performance promise corresponds to the promised use utility value in Fig. 2.2. *Process-oriented* service is understood to be the service creation process in which a production factor is introduced by the service demand (either internal or external customer), which the service provider cannot dispose of himself without restriction and which is therefore referred to as an external factor. For example, if the service provider is a logistics company, the external factor includes the goods that are to be transported and stored for a customer. From a *result-oriented* perspective, service is understood to be the intangible result of this service creation process, which is concretized in the external factor. In the case of logistics, these are the types of goods transformation listed in Fig. 1.3.

Characteristic of service production is the distinction between *pre-combination and final combination of* production factors.[7] The aim of the *pre-combination* is to build up a performance potential, which is generally referred to as capacity, but in its situational availability is referred to as performance readiness. The aim of the *final combination* is to produce the demanded service by using the readiness to perform, other internal production factors and the external factor.

[7]Cf. Corsten, 1993, p. 767 f. For a characterisation of the logistical performance process as a two-stage combination process, see Isermann, 1998, pp. 26ff.

The planning of the service readiness has to take place in anticipation of the demand for the service (the final combination to be performed) and is complicated by the dependence on the external factor. In contrast to the production of material goods, fluctuations in demand cannot be compensated for by stock-keeping, which is why the service readiness must in principle be available before the service is offered on the market. The main problem associated with the need to maintain a certain level of service readiness is the associated fixed costs, which become idle costs in times of low demand. This problem is mitigated, on the one hand, by building capacity, if possible, with characteristics that facilitate quantitative, temporal and intensity adjustment to possible fluctuations in demand. On the other hand, the willingness to perform is also granted to provide a utility, which can be subdivided into the components of *use utility* and *provision utility*. "While the *use utility* arises via the use of the delivered service and is consequently tangible for the buyer, the perception of the *provision utility* represents a latent problem that the potential user often only becomes aware of through negative experiences, namely when the willingness to perform is not available or not available in sufficient quantity or quality at the time of demand."[8]

Due to the intangible nature of services and the fact that the act of production in the sense of the final combination and the act of consumption for the satisfaction of needs coincide, the customer cannot assess the quality of the service before purchase. This results in the *trust nature of* service and a change in the customer's perception of risk towards an *increase in perceived risk*. Both factors explain the overriding *importance of psychographic target variables* such as image and competence when offering services.[9]

2.2 Systems Thinking

Holistic Approach

The tasks summarized in the first section under the term logistics have, of course, always been performed in a company and not just since the logistics term came into existence. In this respect, the question arises as to whether logistics is merely a buzzword and whether logisticians are busy putting old wine into new bottles. Incidentally, this is a question that is always asked when new conceptions emerge in business administration, such as the marketing conception and the controlling conception. In answering this question, it must be assumed that it does not matter whether tasks have always been performed in the company or not, but only how these tasks are performed. A new conception involves a new way of looking at problems in companies and enables new solutions to problems.

[8] Corsten, 1993, p. 768.

[9] Cf. Meffert, 1994, pp. 525 f.

The systems theory approach, or *systems thinking for* short, is fundamental to the logistics conception.[10] Systems thinking has its origins in biology and was adopted from there into the economic sciences. A system is generally understood to be a set of interrelated elements. Characteristic for systems thinking is the holistic approach as well as the realization that for the explanation of the wholeness the explanation of its elements is not sufficient, but that in addition the explanation of the relations between the elements must step. Systems thinking is thinking in complex, interconnected relationships. In the coordination of the elements or subsystems, a distinction is made between the interaction models loose coupling, cooperation and unification as basic types.[11]

In the case of *loose coupling*, the subsystems acting with each other do not influence each other at all or only to a very small extent. They act largely autonomously, although they are dependent on each other. Communication between the subsystems is weak, which ultimately leads to a suboptimal use of resources and inconsistencies in the overall system. In *cooperation*, this weakness is compensated for by improved coordination of the resource and infrastructure requirements resulting from the individual goals of the subsystems involved. This is also achieved by improving information and communication behavior so that the individual interacting partners are each better informed about the requirements of the other subsystems. In the process of *unification*, the subsystems completely relinquish their autonomy. There is an alignment of long-term goals and visions. The aim is to make optimal use of common resources and infrastructure and to dispense with individual reserve capacities.

The relationships between the elements of a system can in principle be interpreted as *input-output relationships*, through which the relationship structure, e.g. of the network of a logistics system, is established. If one emphasizes the process character of these relationships in the exchange of objects between the system elements, then time as a system dimension takes on special significance. The dimension of time distinguishes the process structure from the relationship structure of the systems. However, KLAUS sees the *process approach*, which is already included in the system approach of logistics, as an extension of the thinking framework of logistics. In line with the process approach, he views logistics systems as a "web of flows and processes."[12] However, this flow principle has long been explicitly mentioned as the content of the logistics conception in addition to the principle of the holistic approach.[13]

The *coordination approach* assigned by WEBER[14] to the logistics concept as the actual new thing is also already contained in the systems approach. The coordinated design of flows of goods is the fundamental requirement of the theoretical logistics concept,

---

[10] Cf. Pfohl, 1974, pp. 70ff.; Krulis-Randa, 1977, pp. 34ff. and pp. 130ff.

[11] Cf. Merkel, 1995, pp. 75ff. and pp. 95ff.

[12] Klaus, 1998, pp. 66 f.

[13] See the characterization of the philosophy of logistics in Fey, 1989, pp. 32ff.

[14] Weber, 1992.

irrespective of its implementation in practice. The management of interdependencies, for example, is explicitly used to characterize logistics management, along with the management of material and information flows. However, logistics management cannot be limited to the management of interdependencies, since this management presupposes knowledge of the specific characteristics of logistical service processes.[15]

The application of systems thinking puts the treatment of logistical problems on a new footing, which may be a major reason why today, after a long period of neglect, increasing attention is being paid to these problems both in science and in practice. This is because new statements of a terminological (definitional), descriptive (depicting), theoretical (explanatory) and praxeological (shaping) nature are becoming possible as a result.

Performance of Systems Thinking

Definitional statements: On the basis of systems thinking, terms can be formed and defined that allow an exact recording of logistical problems. The core of such a concept formation and distinction is the logistics definition given in the first section with reference to the system concept. The definition of various logistical subsystems based on this definition is important, for example, for the realization of organizational task analyses. However, the system-oriented logistics definition is also fundamental for carrying out cost analyses, since it determines which costs are to be regarded as logistics costs.

Descriptive statements: For the description of real flows of goods, systems thinking offers two advantages: The first advantage is that the logistics definition based on it makes it possible to describe the various logistics systems in a uniform terminology. This offers the opportunity to identify previously unseen commonalities of problems and problem solving in, for example, military logistics and corporate logistics or procurement and distribution logistics. The second advantage is that when describing logistics systems, one is forced to grasp the complex logistical system interrelationships. For example, one will no longer describe order processing, warehousing, transportation, etc. in isolation, but rather their interaction in realizing the flow of goods. Even if one focuses on the description of a logistical subsystem, attention will be drawn to the description of the interfaces with the other subsystems.

Explanatory statements: Systems thinking makes it possible to recognize connections that would otherwise have been very difficult or impossible to recognize at all. Because consistently applied, this approach to thinking should result in the destruction of generally accepted ways of thinking and in some completely new ways of thinking.[16] For a long time, a lack of systems thinking prevented logistical systems from being viewed as an actual whole and the relationships between the individual system elements from being grasped.

Since systems thinking emphasizes recognizing the relationships between the individual system elements, the decision regarding an element will only be made from the perspective

[15] Cf. Fey, 1989, pp. 111ff.

[16] Cf. Churchman, 1970, p. 20.

of its contribution to the performance of the entire system. Failure to consider this aspect can lead to poor decisions. For example, an isolated decision regarding the change from one means of transport to another means of transport with different technical characteristics or transport speeds may result in unforeseen packaging requirements or a considerable increase in stock. Systems thinking reveals these interrelationships. This is because it forces one to consider the effects of a change in one logistical subsystem on the other logistical subsystems. Similarly, when problems occur in one logistics subsystem, one will look for the causes of the problem not only in that subsystem, but also in all other subsystems.

Shaping statements: Shaping statements are intended to give the decision-makers instructions for their actions. Logistical decisions concern the structure of logistics systems and the course of logistics processes. Based on the definitional, descriptive and explanatory statements, models based on the systems approach can be developed to support this decision, taking into account the objectives to be aimed at. The decision support provided by these models takes into account logistical interdependencies to a much greater extent than would be possible with other models. This provides a better basis for evaluating the available decision alternatives, thus facilitating the rational selection of the optimal alternatives.

Resource and Process Interdependencies

In summary, it can be said that systems thinking helps to avoid suboptimal isolated solutions in logistics decisions by taking into account resource and process interdependencies[17] and to strive for optimal overall solutions.

If the factual interrelationships between different logistics subsystems are captured by analyzing the relationship structure of a system, the result is that logistics decisions can be made taking into account resource interdependencies, i.e. existing *bottlenecks* (scarce logistics capacities of a quantitative and qualitative nature) or free *potentials* (free logistics capacities of a quantitative and qualitative nature). Systems thinking thus makes it possible to include bottleneck and synergy effects in the decisions.

If the temporal relationships between the various sections of the logistics chain of the flow of goods are captured by analyzing the process structure of the system, the result is that logistics decisions are made taking into account the process interdependencies, i.e. existing *autonomy costs* (buffering of the subsections of the logistical chain by stocks creates scheduling freedom in these sections) and *coordination costs* (coupling of the subsections of the logistical chain by information exchange—communication relationships—promotes consideration of the dependencies between the sections due to service interdependencies). *Flow* or *process thinking*, by substituting autonomy costs with coordination costs, enables short lead times of goods in the logistics chain and thus flexible reactions to delivery service requirements. Flow thinking as a manifestation of systems

[17] See the distinction between resource and process strategies in Fey, 1989, pp. 8ff.

thinking emphasizes the dimension of time over the dimension of capacity in the logistics system.

Closely related to systems thinking are *total cost thinking* and *service thinking*. As can be seen from Fig. 1.7, logistics costs can be seen as system inputs and service as system outputs. Logistics decisions must be made with regard to the input and output effects.

2.3 Overall or Total Cost Thinking

Overall or Total Costs

The same interdependence that exists between the elements of the logistics system is also present in the costs caused by these elements. Reducing costs in one logistics subsystem may lead to an increase in costs in other subsystems and, if the cost reduction is less than the cost increase, to a cost increase for the entire logistics system. For example, a reduction in transportation costs without taking into account a possible associated increase in packaging or warehousing costs may result in an increase in delivery costs. Total cost thinking therefore requires that all logistics costs relevant to a logistics decision be recorded. The term total costs is used here only to indicate the requirement to record all relevant logistics costs and should not be confused with its common use in cost accounting in the sense of prime costs or full costs. The costs that are relevant for a decision are those that are only incurred if a decision alternative is realized and are not incurred if this decision alternative is not realized.

Figure 2.4 provides an overview of the costs to be taken into account on the basis of the logistical overall or total cost concept. First of all, these are the logistics system costs arising in the functional logistics subsystems—as compiled and identified in Figs. 1.4 and 1.7.[18] They are caused by the use of production factors in the logistics subsystems. The production factors can be referred to as primary cost elements and the logistics subsystem costs as secondary cost elements. In addition to these logistics system costs, however, costs must also be taken into account that are directly related to the logistics system costs. On the one hand, these include service level costs, which are caused by a lower service level. If the delivery service level is too low, these include, for example, the lost orders and customers recorded in the form of shortage costs or the costs incurred in processing complaints. In the case of supply service levels that are too low, these are the costs incurred as a result of business interruptions or changeovers. On the other hand, it includes the lot costs, which vary with the number of lots to be produced by production or to be supplied by the supplier. In the case of production lots, these are the fixed setup costs incurred in production. In the case of order lots, it is the portion of the order costs attributable to purchasing—and not to logistics order processing.

[18] Another classification of logistics system costs would be the distinction between autonomy and coordination costs mentioned in the previous section.

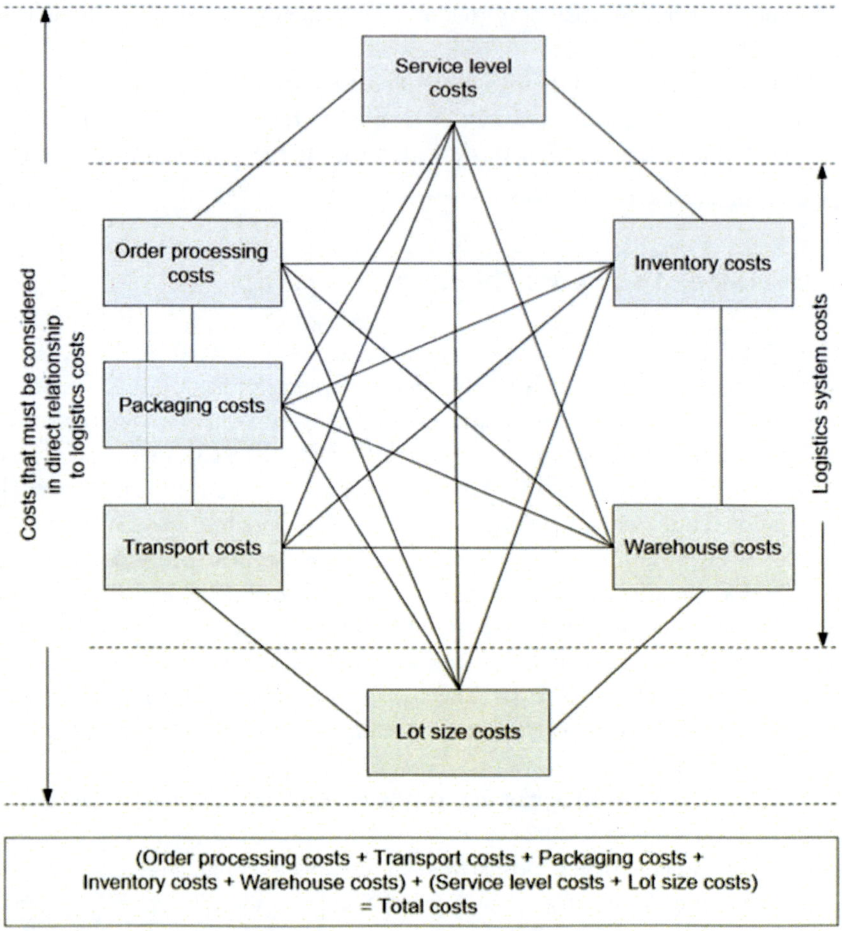

Fig. 2.4 Composition of total costs in the logistics process (See also Stock/Lambert, 2001, p. 29)

Conflict of Objectives

Total cost thinking is of great importance for logistics decisions because logistics systems are characterized by a multitude of *cost conflicts.* Cost reductions in one subsystem therefore often cause cost increases in another subsystem. For example, lower inventory costs lead to higher transportation costs because deliveries have to be made more frequently. Further examples of conflicting goals are[19]:

- packaging costs and costs due to transport damage,
- order processing costs and transport costs due to unfavourable route planning,

[19] Cf. Poruks/Sitta, 1970.

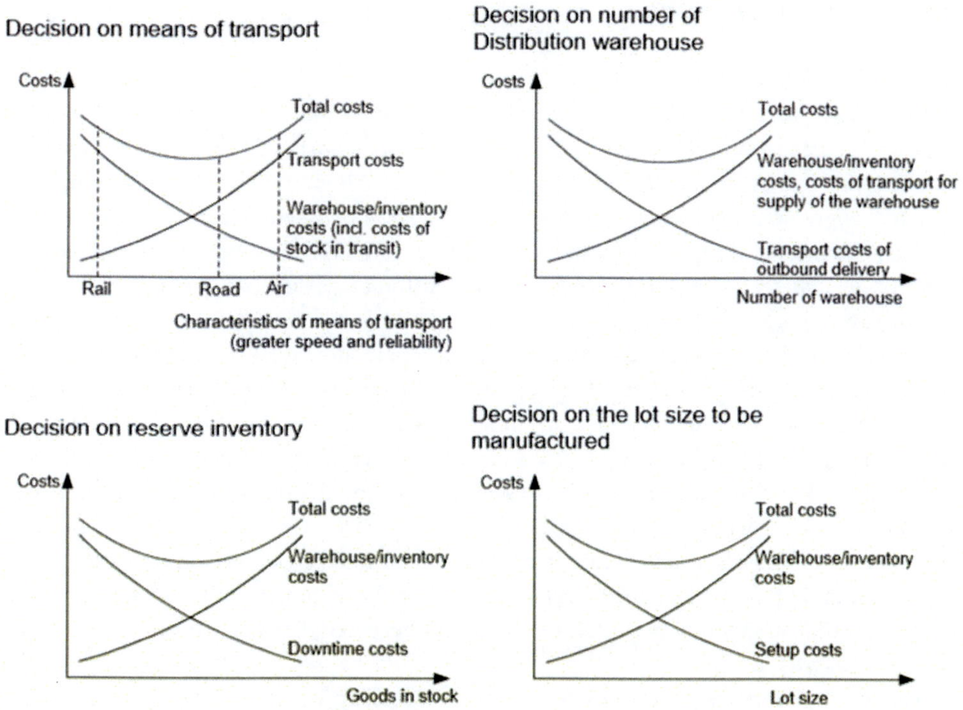

Fig. 2.5 Typical cost patterns in the logistics system (Source: See also Ballou, 2004, p. 46)

- inventory costs and production costs due to small lot sizes,
- inventory costs and purchasing costs due to more frequent and smaller orders.

Figure 2.5 shows some examples of cost curves with conflicting objectives in logistics systems. Logistical thinking presupposes knowledge of the cost conflicts prevailing in a system. If one has a basic idea of such tendential cost progressions, then one knows which costs are to be recorded in a detailed cost analysis in a concrete decision situation.

Total or total cost thinking has probably been applied for the first time to logistical problems related to air freight.[20] If, for example, freight costs alone are considered, the use of air freight is certainly only justified for very few goods. However, if the effect of air freight on total logistics costs is taken into account, a much more favourable picture emerges for its use.

However, as can be seen from Fig. 1.7, logistics thinking is never just cost thinking, but also performance thinking. The incurrence of logistics costs is only justified if they are caused by corresponding logistics services. The demands for low logistics costs are contrasted by demands for high logistics performance.

[20]Cf. Lewis et al., 1956, pp. 64ff; Heskett et al., 1973, p. 530.

2.4 Service Thinking

Customer Orientation

In the 1990s, *customer orientation*, which is actually an old buzzword, moved into the focus of business research and teaching. Due to constantly increasing customer requirements in many markets, it became more and more important for offering companies to increasingly orientate themselves towards the wishes and demands of customers when offering products and services. In addition to price, formerly the most important decision criterion for buyers, the fulfillment of complex customer wishes has become an essential competitive criterion.[21]

Customer satisfaction has become a corporate goal in almost all companies, as it can be used to achieve lasting *customer loyalty.* Satisfied regular customers bring the company more profits in the long run than new customers who have to be won over and over again.[22] An essential means of retaining such long-term or regular customers for one's own customer base is *customer proximity*. This is achieved by maintaining close contact with the customer in order to quickly identify changes in customer requirements. In meeting *customer requirements, a* distinction is made between three areas of requirements[23]:

It is essential that the *basic requirements are* met. Their fulfilment is taken for granted by the customer. Accordingly, failure to meet these requirements results in extreme customer dissatisfaction.

The *performance requirements* are usually explicitly agreed. They go beyond the requirements that are customary in the industry and will also usually not be able to be fulfilled by all suppliers. In contrast to the basic requirements, the fulfilment of the performance requirements is usually perceived and positively evaluated by the customer.

Enthusiasm requirements are not requirements in the strict sense. They are never explicitly expected, nor are they missed if they are not fulfilled. However, a fulfillment of the enthusiasm requirements is evaluated extremely positively by the customer.

At a time when many products are barely distinguishable from competing products, logistical services that go beyond basic requirements are becoming increasingly important. They represent a way to differentiate from competitors and ensure customer satisfaction.

Service: Supply and Delivery Service

Logistics services as the output of logistics systems are services provided in connection with supplying a company with materials (supply service) or delivering goods to customers (delivery service). From the point of view of the vendor-customer relationship, the supply service and the delivery service are two sides of the same coin. This is because the supplier's delivery service must meet the customer's supply service requirements. From

[21] Cf. Pfohl, 1998, pp. 3ff.

[22] Cf. Simon/Homburg, 1995, p. 18; Hinterhuber et al., 2003, p. 7.

[23] Cf. Pfohl, 1998, p. 14.

the point of view of the flow of goods through a company, the *delivery service* refers to the flow of goods to the customer, while the *supply service* refers to the flow of materials from the supplier to the procurement warehouse (in the case of a trading or service company) or into the production process (in the case of an industrial company). In *both cases*, the service can be characterized by the four R's mentioned in the definition of logistics. In each case, it is about ensuring the availability of the good.

The service requirements depend on the one hand on the needs of the customer, and on the other hand on the needs of the company's own production. In the following, it is sufficient to discuss service thinking using the example of the delivery service. The problem of the supply service is essentially the same. It is only necessary to think of one's own production as an internal customer instead of an external customer. The service is ultimately the result of the logistical transformation of goods, about which Fig. 1.3 provides information.

Delivery Service as Primary and Secondary Output

Delivery service is one of the services additionally offered by industrial and commercial enterprises with the sale of in-kind outputs.[24] The in-kind outputs can be described as *main* or *primary outputs* and the additional services as *secondary outputs*. The latter can be characterized by the fact that the supplier, in addition to offering his products, takes over functions that can also be fulfilled by the customer in the process of creating the service. The basic feature of the secondary service is therefore the takeover of additional functions and thus usually of costs that affect the process of producing the output at the customer. For the definition of the secondary output it is irrelevant whether the supplier tries to cover the secondary service costs via a separate price demand or via the product price, whereby the additional revenue necessary to cover the costs in the second case could be achieved both via a higher product price and via an increased sales volume.

The *delivery service* comprises the secondary output through which functions are taken over by the supplier's distribution logistics that would otherwise have to be fulfilled by the customer's procurement or material logistics. A logistics company can also be involved between the vendor and the customer to perform these services. The task of the logistics company is to provide the vendor (sender) with the required delivery service or the customer (recipient) with the required supply service. This service is then the primary output of the logistics company. To systematize the possible manifestations of this service, it seems useful to distinguish between four service components.

[24] Cf. Pfohl, 1977, p. 241 and the literature cited there.

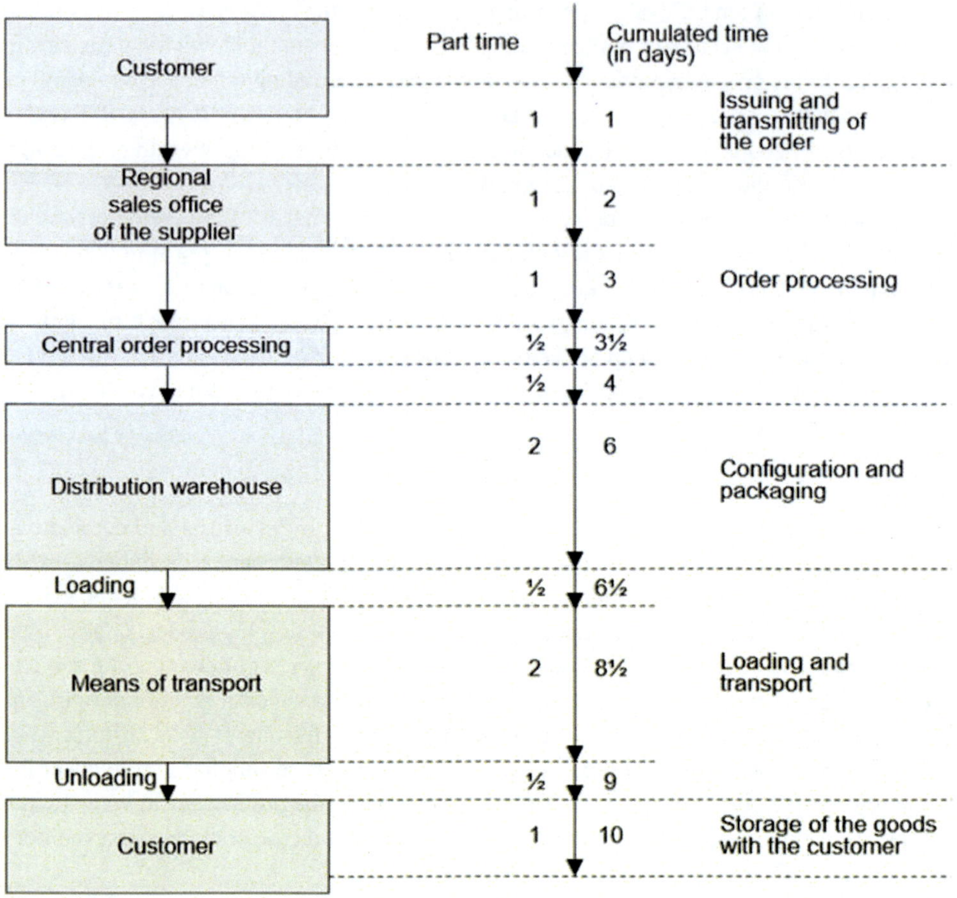

Fig. 2.6 Example of the time course of a 10-day delivery period

Service Components

The delivery service is essentially made up of delivery time, delivery reliability, delivery quality and delivery flexibility.[25]

Delivery time (order period) is the period of time between the issuance of the order by the customer and the receipt of the goods. It is important for the customer because shorter delivery times enable him to have lower stocks and to plan at shorter notice. Figure 2.6 gives an example of the composition of the delivery time.[26] The supplier can influence all the partial times given there—including the times for activities that are not carried out by

[25]Cf. Pfohl, 1972, pp. 177ff. and 1977, pp. 241 f.; Bender, 1976; LaLonde/Zinszer, 1976, p. 148; Havighorst 1980, pp. 58ff.; Stock/Lambert, 2001, pp. 117 f.

[26]Cf. Heskett et al., 1973, pp. 246 f.

the supplier or a third party, but by the customer himself. He can, for example, influence the time for the completion of the order by the customer by means of suitable order forms. The time for the storage of the goods at the customer can be influenced by the supplier insofar as the coordination of the transport, packaging and storage requirements between supplier and customer can greatly facilitate the acceptance and inspection of the goods, the transport to the customer's warehouse and the storage in the customer's warehouse.

The term delivery time is used here to describe the output of logistics systems, i.e. in the sense of a delivery time that is solely dependent on logistics. A distinction must be made between the delivery time for which logistics alone is not responsible, but for which other areas of the company, such as production, are also responsible.[27]

Delivery reliability (delivery dependability, adherence to delivery dates) is the reliability (probability) with which the delivery time is met. It is important for the customer because it enables lower inventories and helps to avoid disruptions in the operating process. Delivery reliability depends on the following two influencing factors:

- Reliability of the workflow,
- Readiness for delivery.

Compliance with the promised delivery time is initially determined by how *reliably the* partial times that make it up are adhered to. The work processes occurring in the individual phases of the delivery time must be realized according to plan in the time allotted for them. During order processing, for example, it can happen that incoming orders are left unprocessed, or during transportation it is possible that a forwarder does not keep the promised transportation times. The reliability of the delivery time is therefore determined by its most unreliable phase!

On the other hand, delivery reliability in meeting the delivery time will depend quite significantly on the *readiness for delivery*. It indicates the extent to which the supplier is able to deliver from stock. If an order encounters shortages in a delivery warehouse, it will generally not be possible to meet the normal delivery time. Unless, in this case, it is possible to supply the customer directly from the central warehouse or another distribution warehouse using faster means of transport.

Readiness for delivery is usually measured by percentages, but these are based on very different definitions. Figure 2.7 provides an overview of frequently used measures of readiness for delivery. From the possible definitions, each supplier must select the one that is appropriate for his situation. For example, it may be appropriate to include only the frequency of shortages in the definition, but not the size of the shortages. In other cases, however, it may be appropriate to use the size of the shortage as the basis for defining the readiness for delivery. It is then not considered to be relevant how often the demand cannot be satisfied by the warehouse, but it is essential what percentage of the demand cannot be

[27] See Wagner, 1978; delivery time is also discussed in Part III, Sect. 10.3 on subsystems of production logistics.

1 based on number of orders		
1.1 Service level (%)	=	$\frac{\text{fulfilled orders} \cdot 100}{\text{received orders}}$
1.2 Service level (%)	=	$\frac{\text{ex store fulfilled orders} \cdot 100}{\text{fulfilled orders}}$
1.3 Service level (%)	=	$\frac{\text{ex store fulfilled orders} \cdot 100}{\text{received orders}}$
1.4 Service level (%)	=	$\frac{\text{lost orders} \cdot 100}{\text{fulfilled orders}}$
1.5 Measurement of service level	=	Lost orders per time period
1.6 Measurement of service level	=	Set aside orders per time period
1.7 Measurement of service level	=	Lost and set aside orders per time period
2 based on demand		
2.1 Service level (%)	=	$\frac{\text{delivered quantity} \cdot 100}{\text{inquired quantitiy}}$
2.2 to 2.7 analogous to 1.2 to 1.7		(instead of quantities also values could be compared)
3 based on time measurement (e.g. days)		
3.1 Service level (%)	=	$\frac{\text{Share of periods with available inventory} \cdot 100}{\text{total period}} \text{ e.g. } \frac{330 \cdot 100}{365} = 90{,}5\%$
3.2 Service level (%)	=	$\frac{\text{Share of periods with out lost or set aside orders} \cdot 100}{\text{total period}}$
4 based on monitoring intervals		
4.1 Service level (%)	=	$\frac{\text{Intervals with goods in stock above safety stock} \cdot 100}{\text{total number of intervals}}$
4.2 Service level (%)	=	$\frac{\text{number of intervals without goods instock above safety stock} \cdot 100}{\text{total number of intervals}}$
4.3 Service level (%)	=	$\frac{\text{Intervals without lost orders} \cdot 100}{\text{total number of intervals}}$
4.4 Measurement of service level	=	Frequency of using the safety stock in a period
4.5 Measurement of service level	=	Frequency of shortages in a period
4.6 Measurement of service level	=	Frequency of lost orders in a period
5 based on delivery delays		
5.1 Measurement of service level	=	$\frac{\text{Number of days with delivery delay}}{\text{fulfilled orders}}$
5.2 Measurement of service level	=	$\frac{\sum \text{Days with delivery delay} \cdot \text{ordered quantities}}{\text{fulfilled orders}}$

Fig. 2.7 Formulas for calculating readiness for delivery (Source: Steinbrüchel, 1971, p. 27)

met by the warehouse. Demand can be recorded in terms of quantity or value. However, quantity data is generally only useful if the value of the various items in a distribution warehouse is approximately the same. Each company will ultimately have to choose the definition that takes into account the occurrence of shortages in the way that actually corresponds to their effect on sales.[28]

[28] Cf. Pfohl, 1972, pp. 181ff.

The *Delivery quality* determines the extent to which the delivery itself gives the customer cause for complaint. This in turn depends on two factors:

- Delivery accuracy,
- Delivery condition.

Delivery accuracy indicates the extent to which the ordered products will be delivered in the desired type and quantity. If the supplier cannot deliver the ordered product, he should only deliver another product to the customer as a substitute if he has obtained the customer's consent beforehand. Otherwise, the supplier risks upsetting the customer and possibly losing him. He may also incur additional costs in dealing with the customer's complaint and taking back the goods. The customer must also receive the quantity he ordered. This is because if larger quantities than those ordered are delivered, the customer's storage costs increase. If smaller quantities are delivered, this may result in shortages in the customer's warehouse.

The *delivery condition* essentially depends on the extent to which the packaging fulfils its protective function when the goods are delivered. Damage to the goods during delivery leads to customer complaints and/or additional costs due to returns or price reductions to be granted.

Delivery quality is measured by how often customers complain about deliveries and can thus also be measured by percentages.

Delivery flexibility means whether the supplier's delivery system allows the customer's special needs to be met or whether the customer's procurement logistics have to comply with rigidly prescribed rules of the supplier's distribution logistics. Delivery flexibility essentially depends on the following three influencing factors:

- Order modalities,
- Delivery terms,
- Information of the customer.

The *order modalities* determine the order size, the purchase quantity, the time of order placement as well as the type of order creation and order transmission. The less the customer's freedom of choice regarding these order modalities is restricted, the higher the level of this delivery service component. However, diversity in order modalities places a burden on the distribution logistics system at the supplier and generally results in higher logistics costs. The definition of minimum order sizes, minimum purchase quantities and predefined times by which an order must be placed so that the promised delivery time can be met, as well as standardization of order creation and transmission, are ways of fixing the order modalities.

If the order modalities refer to the information flow between supplier and customer, the *delivery modalities* refer to the flow of goods. The delivery modalities determine in particular the type of packaging, the transport variant to be used and the possibility of delivery on call. The transport variant also includes the arrangement for self-collection.

Finally, delivery flexibility also includes accurate and prompt *information to the customer* about delivery possibilities, the status of order processing, foreseeable delivery delays and the handling of complaints about inadequate delivery.

In contrast to the first three service components, the level of delivery flexibility can only be quantified to a limited extent. However, this component also has an impact on logistics costs at the supplier and on customer satisfaction. As with the other service components, there will be a trade-off between seeking to minimize logistics costs and seeking to maximize delivery service. This trade-off must be resolved based on logistical efficiency thinking.

2.5 Logistical Efficiency Thinking

Efficiency Thinking

Logistics systems are efficient if their design takes into account logistics costs (input) and logistics services (output) as design goals. In solving logistics problems, efficiency thinking requires neither a one-sided orientation towards the goal of minimizing costs nor a one-sided orientation towards the goal of maximizing services, but rather a compromise between these goals.

The efficiency target initially corresponds to the familiar *productivity target*, which is measured by the output/input ratio (e.g. number of pallets handled/working hour). Thus, efficiency thinking is applicable to the *technological dimension of* the logistics system, which requires thinking in terms of quantities and qualities.[29] This thinking deals with problems of performance (quantitative and qualitative capacity as well as operational elasticity) and with problems of performance readiness (susceptibility to failure and user-friendliness) of logistics systems. Performance readiness in the broad sense also includes the time required for planning and implementing the logistics system. It is often better to have systems with satisfactory performance in use in good time than to have systems with maximum performance in use only after a long time. Flexibility in the sense of the *adaptability of* systems is often highlighted as a particular problem. For example, a distinction is made between short-term efficiency as productivity under constant conditions and adaptability as productivity under changing conditions.[30] Adaptability is of great importance for logistics systems insofar as changes in the level, composition and geographical distribution of demand lead to flows of goods that place very different demands on logistics systems.

[29] Cf. in the following Pfohl/Stölzle, 1997, pp. 84 f.

[30] Cf. Ulrich/Fluri, 1995, p. 164.

However, efficiency thinking can also refer to the *economic dimension of* the company. This requires thinking in terms of values. At its heart are problems of prices and of sales and costs. In addition to the particular problem of allocating logistics costs to logistics services, which remains unresolved in many companies because of the one-sided orientation of cost and activity accounting, there is also the problem of recording the effect of the delivery service on sales. In Fig. 2.8, typical cost and market response functions are plotted against the level of service. A detailed discussion of these curves is given in the Logistics Management volume of this series. Suffice it to point out here that the strongly progressive increase in costs with increasing service levels is fundamentally characteristic of logistics systems. Improving an already very good level of service by a few percent causes a far disproportionate increase in costs—if one disregards a shift in the cost curve due to process innovation. No similarly empirically validated statements can be made for the revenue curve. However, in some cases the curve in Fig. 2.8 can be demonstrated. Of particular interest here is the saturation phenomenon that can also be observed with other market reaction functions. Increasing an already good delivery service by a few percent leads only to a strongly disproportionate increase in sales. It follows from this that the greatest profit contribution of the delivery service is by no means at a maximum service level. On the economic level, therefore, efficiency thinking corresponds to the familiar *profitability target*, which is measured by the profit/capital ratio. The input variable in this case is the capital tied up in logistics systems, the output variable is the profit contribution made by logistics systems.

Technical-Economical Thinking

When explaining efficiency thinking, it became clear that efficiency thinking is both technically and economically oriented. However, this is not to express that the *social* and *environmental dimension* of the company would not play a role in the logistics sector. Social goals, such as employee satisfaction or the long-term preservation of employee health, have the same significance for the logistics sector as for the other areas of the company. This also applies to the ecological goals of protecting the environment. In this respect, the logistics division does not differ from the other divisions in terms of social and ecological thinking. On the other hand, the special combination of technical and economic thinking that is characteristic of logistics is not always typical of the other corporate divisions in the same way.

Logistics provides an excellent example of an area of business where technical and economic problems overlap. A logistics manager must be able to assess and exploit the possibilities that technical progress in packaging, transport and warehousing opens up for the flow of goods. But he or she must also be able to balance cost and service and work with procurement, production, and sales managers who are often trained only in either economic or technical thinking. Management positions in the logistics sector therefore require a particular combination of economic and technical thinking. However, this also addresses the consequences of logistics thinking for the company.

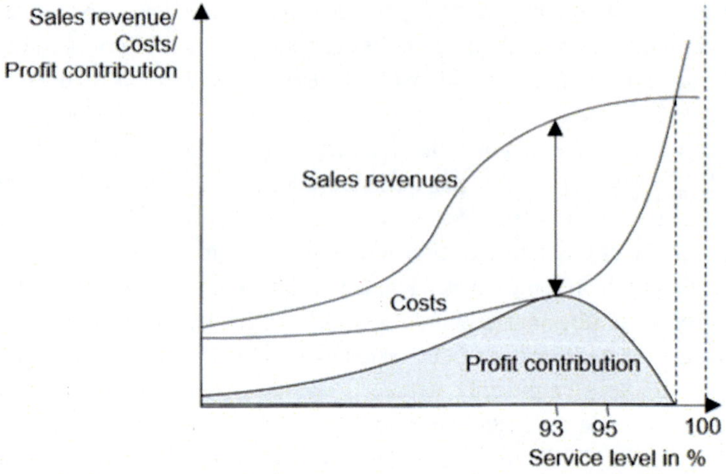

Fig. 2.8 Effect of service level on profit (Source: Taken with modifications from Buxton, 1975, p. 35)

2.6 Consequences of Logistics Thinking

In order to analyse the possible consequences of logistics thinking, it is useful to distinguish between a functional, an instrumental and an institutional dimension of the logistics conception.[31] Functional consequences arise if logistics is seen as a conceptually definable complex of tasks in the company. In this case, it would be necessary to examine whether logistics should be regarded as a new business management function. Instrumental consequences refer to the use of software and hardware technologies. The question here would be about possible changes in the technologies to be used to support information processing and to facilitate the flow of goods. Institutional consequences concern the organization of a company and the organization of cooperation between companies. Possible changes in the organizational structure of a company (intra-organizational consequences) or in the cooperation between companies (inter-organizational consequences) as a result of logistics thinking should be identified here.

Functional Consequences

The system thinking characteristic of the logistics conception requires that the logistics tasks be seen in the overall context, and that logistics be seen as a self-contained task area of the company. In business administration, such a task area is often described as a business function that must be performed in every company. Logistics is thus seen as a function in addition to the other functions to be performed in the company, such as procurement or

[31] Cf. Pfohl, 1980, pp. 1201 f.

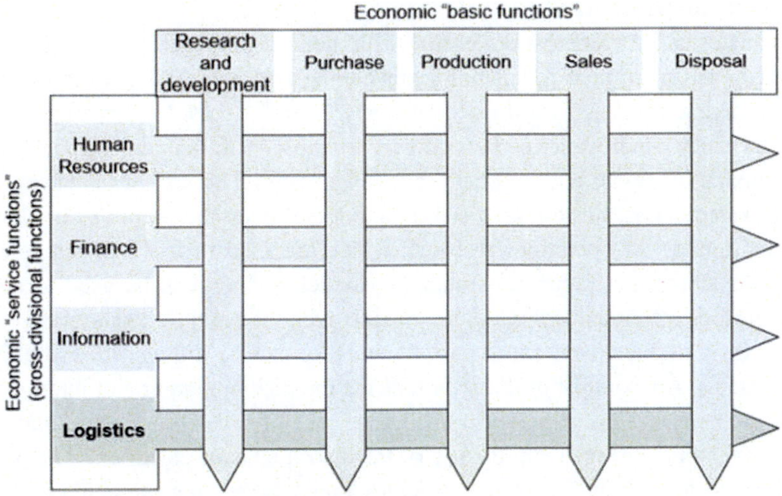

Fig. 2.9 Logistics as a cross-sectional business function

financing. The prevailing opinion is that the logistics function has a cross-sectional or overarching character.[32]

If one considers the functions of research and development, procurement, production and sales, which can be derived directly from the market task, as *basic*[33] *business functions*, then a series of cross-sectional business functions (services) can be derived from them, which must necessarily be performed in addition to these basic functions. In addition to the logistics function, these include in particular the functions listed in Fig. 2.9, which deal with personnel, finance and information. If these functions are referred to as *services*, their service character is emphasized in relation to the basic business functions. If, on the other hand, one speaks of *cross-sectional functions, this emphasizes the* fact that they permeate the basic business functions. It should be emphasized at this point that the term "services" does not imply any valuation of these functions in relation to a basic business function. Which functions are of greater importance for the success of a company depends on the existing competitive and cost situation.

If logistics is seen as a business management function, it follows that a special sub-discipline of business management has to deal with the specific logistical decision facts in a functional classification of business management. As with the other business functions, however, it does not necessarily follow that logistics must be institutionalized in an organizational unit within the company.[34]

[32] Cf. Pfohl, 1983, p. 726 and the literature cited there.

[33] Waste disposal is also regarded as a basic business function, cf. Pfohl, 1993, pp. 214ff.

[34] Cf. Kirsch/Esser, 1976.

Instrumental Consequences

If one associates an instrumental dimension with the logistics conception, logistics is first understood as an instrument for planning, managing and controlling the flow of goods. Logistics thinking can have an impact on many areas of these information processing instruments, which can be described as *software technology*, both in the support of logistics decision-making processes and in order processing, material control or inventory management. For example, logistics thinking suggests that the focus should not be on developing optimization models of operations research, which are used to find isolated solutions in logistics subsystems, but rather on simulation models, which can be used to map more complex logistics interrelationships. Such models make it possible to show the effects of changes in the variables that cannot be influenced or can be influenced in the design of logistics systems. An example of this would be the interactive planning of internal material flow with the help of simulation software. In the case of order processing, material flow control or inventory management, the use of the instruments of electronic data processing then has an effect. Without IT support, complex logistics systems can neither be planned, nor controlled, nor monitored.

Consequences regarding the use of methods can be shown using the example of cost and performance accounting. This must be methodically converted in such a way that it allows cost types to be allocated to logistically relevant reference values on the basis of detailed cost type recording. On the one hand, these can be cost centers that allow meaningful control of the logistics managers. On the other hand, they can be orders that take on the functions of cost objects or delivery service components to which source-related logistics costs are to be allocated.

However, instrumental consequences of logistics thinking are also possible in the area of *hardware technology*, which is to be understood as transport, handling, storage and packaging technology. This is expressed in particular when increased attention is paid to the interfaces between these areas of technology (e.g. between transport and handling technology). These interfaces are typically weak points in the technology of logistics systems. The technical development in the various subsystems must be coordinated with each other. The technical systems must be compatible with each other. This realization has also resulted, for example, in the increasing demand for "everything from a single source" when purchasing technical systems. Capital goods manufacturers are increasingly meeting this need of their customers through the corresponding systems selling.

Institutional Consequences

The use of new instruments generally initially requires only process-organizational changes in the company. However, it is not always possible to implement the logistics conception without institutional change, i.e. without *intra-organizational* change in the existing distribution of tasks, competencies, responsibility and power in micrologistics systems. This is because when logistical tasks are fragmented in the organizational structure, firstly, logistical system interrelationships are more difficult to identify. Secondly, the pursuit of logistics objectives is made more difficult by the conflicting interests of the various organizational units that perform logistics tasks. Obviously, then, an organizational grouping of logistical tasks can facilitate the realization of the logistics conception.

One characteristic of systems thinking is the realization that system behavior is strongly dependent on the system environment. Systems thinking therefore suggests that parts of this environment should not be accepted as variables that cannot be influenced (data), but should be turned into variables that can be influenced by extending the system boundaries. This means that the logistics conception can also lead to changed forms of *inter-organizational cooperation* between different companies in mesologistics systems. This then leads to a changed distribution of tasks, competencies, responsibility and power between different companies. For example, it may make sense for two industrial companies to use a joint logistics system, at least in part, or for one industrial company to transfer its distribution logistics to a specialized logistics company.

Obstacles to the Realization of the Logistics Conception

There are many examples of an increase in the efficiency of the company as a result of the realization of the logistics conception in functional, instrumental and institutional terms.[35] *Increases in efficiency* are achieved by reducing logistics costs, accelerating the turnover of capital, increasing the level of service, improving decision-making processes, reducing conflicts in the company organization, and improving cooperation between different companies.

Despite these obvious opportunities to increase efficiency, there are a variety of obstacles to realizing logistics design. In particular, the following *main obstacles* to a realization of the logistics conception can be named:

- The lack of assertive logistics-oriented managers (e.g. lack of assertiveness against a strong sales force),
- the lack of decision-relevant information (e.g. missing information from cost accounting),
- the existence of intra-organizational boundaries in the company (e.g. lack of agreement in solving logistical problems on the procurement side and on the sales side of a company),
- the existence of inter-organizational boundaries between companies (e.g. poor communication between consignor, forwarding agent, receiving agent and consignee).

The following section discusses the importance of logistics for the company, on which it ultimately depends whether it is worth overcoming the obstacles identified.

[35] Cf. Pfohl, 1983, p. 721; Pfohl, 2016, pp. 50ff. and the literature listed there.

References

Ballou R H (2004) Business Logistics / Supply Chain Management. Planning, Organizing, and Controlling the Supply Chain. 5. Aufl. Upper Saddle River N.J.

Bender P S (1976) Design and Operation of Customer Service Systems. New York

Bowersox D J, Smykay E W, LaLonde B J (1968) Physical Distribution Management. Logistics Problems in the Firm. Rev. ed. New York

Buxton G (1975) Effective Marketing Logistics. The Analysis, Planning and Control of Distribution Operations. New York

Churchman C W (1970) Einführung in die Systemanalyse. München

Converse P D (1958) The Other Half of Marketing. In: Seelye A L (Hrsg) Marketing in Transition. New York, S. 114ff

Corsten H (1993) Stichwort “Dienstleistungsproduktion”. In: Wittmann W u.a. (Hrsg) Handwörterbuch der Betriebswirtschaft. 5. Aufl. Stuttgart, Sp. 765-776

Corsten H (2007) Produktionswirtschaft. Einführung in das industrielle Produktionsmanagement. 11., vollst. überarb. Aufl. München/Wien

Fey P (1989) Logistik-Management und Integrierte Unternehmensplanung. München

Gopal C, Cypress H (1993) Integrated Distribution Management. Burr Ridge, NY

Havighorst D (1980) Konzept und Leistungspotential der Marketing-Logistik. Weinheim

Heskett J L, Glaskowsky N A, Ivie R M (1973) Business Logistics. Physical Distribution and Materials Management. 2. Aufl. New York

Hinterhuber H H, Handlbauer G, Matzler K (2003) Kundenzufriedenheit durch Kernkompetenzen. Eigene Potentiale erkennen, entwickeln, umsetzen. 2., überarb. Aufl. Wiesbaden

Isermann H (1998) Grundlagen eines systemorientierten Logistikmanagements. In: Isermann H (Hrsg) Logistik: Gestaltung von Logistiksystemen. 2., überarb. und erw. Aufl. Landsberg a. L., S. 21-60

Kirsch W, Esser W M (1976) Plädoyer für eine betriebswirtschaftliche Logistik. In: Journal für Betriebswirtschaft 26 4, S. 208-218

Klaus P (1998) Jenseits einer Funktionenlogistik: Der Prozeßansatz. In: Isermann H (Hrsg) Logistik: Gestaltung von Logistiksystemen. 2., überarb. und erw. Aufl. Landsberg a. L., S. 61-78

Krulis-Randa J S (1977) Marketing-Logistik. Eine systemtheoretische Konzeption der betrieblichen Warenverteilung und Warenbeschaffung. Bern/Stuttgart

LaLonde B J, Zinszer P H (1976) Customer Service. Meaning and Measurement. Chicago, IL 1976

Langley C J Jr, Holcomb M C (1992) Creating Logistics Customer Value. In: Journal of Business Logistics 13 2, S. 1-17

Large R (1995) Unternehmerische Steuerung von Ressourceneignern: Ein verstehender Ansatz zur Theorie der Unternehmung. Wiesbaden

Large R (2013) Strategisches Beschaffungsmanagement. Eine praxisorientierte Einführung mit Fallstudien. 5., vollst. überarb. Aufl. Wiesbaden

Lewis H T, Culliton J W, Steele J D (1956) The Role of Air Freight in Physical Distri-bution. Boston

Meffert H (1994) Marktorientierte Führung von Dienstleistungsunternehmen – neuere Entwicklungen in Theorie und Praxis. In: DBW 54 4, S. 519-541

Merkel H (1995) Logistik-Management-Systeme: Grundlagen und informationstechnische Umsetzung. München/Wien

Morgenstern O (1955) Note on the Formulation of the Theory of Logistics. In: Naval Research Logistics Quarterly 2 o. Nr., S. 129-136

Novack R A, Rinehart L M, Wells M V (1992) Rethinking Concept Foundations in Logistics Management. In: Journal of Business Logistics 13 2, S. 233-267

Pfohl H-Chr (1972) Marketing-Logistik. Gestaltung, Steuerung und Kontrolle des Warenflusses im modernen Markt. Mainz

Pfohl H-Chr (1974) Die Logistik als Beispiel für Auswirkungen des Systemdenkens in der entscheidungsorientierten Betriebswirtschaftslehre. In: MIR 14 1, S. 67-85

Pfohl H-Chr (1977) Zur Formulierung einer Lieferservicepolitik. Theoretische Aussagen zum Angebot von Sekundärleistungen als absatzpolitisches Instrument. In: ZfbF 29 5, S. 239-255

Pfohl H-Chr (1980) Aufbauorganisation der betriebswirtschaftlichen Logistik. In: ZfB 50 11-12, S. 1201-1220

Pfohl H-Chr (1983) Logistik als Überlebenshilfe in den Achtziger Jahren. In: ZfB 53 8, S. 719-734

Pfohl H-Chr (1993) Die Bedeutung der Entsorgung für die Unternehmenslogistik. In: Pfohl H-Chr (Hrsg) Ökologische Herausforderungen an die Logistik in den 90er Jahren. Umweltschutz in der Logistikkette bei Ver- und Entsorgung. Berlin, S. 211-257

Pfohl H-Chr (1998) Kundennähe. Bedeutung für die Logistik. In: Pfohl H-Chr (Hrsg) Kundennahe Logistik: Wertschöpfend – Agil – Beziehungsorientiert. Berlin, S. 1-45

Pfohl H-Chr (2016) Logistikmanagement. Funktionen und Instrumente. 3., neu bearb. und aktual. Aufl. Berlin/Heidelberg

Pfohl H-Chr, Stölzle W (1997) Planung und Kontrolle. 2., neu bearb. Aufl. München

Porter M E (2014) Wettbewerbsvorteile: Spitzenleistungen erreichen und behaupten. 8., durchgesehene Aufl. Frankfurt a. M.

Poruks A, Sitta F (1970) Bedeutung und Hauptproblemkreise der Physical Distribution. In: Wirtschaftspraxis. B.-Organisationspraxis, Lieferung 263, 3 Blätter

Simon H, Homburg C (1995) Kundenzufriedenheit als strategischer Erfolgsfaktor – Einführende Überlegungen. In: Simon H, Homburg C (Hrsg) Kundenzufriedenheit, Wiesbaden, S. 15-27

Steinbrüchel M (1971) Die Materialwirtschaft der Unternehmung. Bern/Stuttgart

Stock J R, Lambert D M (2001) Strategic Logistics Management. 4. Aufl. Boston u. a.

Ulrich P, Fluri E (1995) Management: Eine konzentrierte Einführung. 7., verb. Aufl. Bern

Wagner G R (1978) Lieferzeitpolitik. 2., überarb. Aufl. von "Die Lieferzeitpolitik der Unternehmen". Wiesbaden

Weber J (1992) Logistik als Koordinationsfunktion. Zur theoretischen Fundierung der Logistik. In: ZfB 62 8, S. 877-895

3 Importance of Logistics

3.1 Business Development Trends

The business development trends are to be described on the basis of three target variables that are of great importance for business decisions due to the globalization of business activities and the dynamics in technology development. These are changes in costs, in the market and in risk.[1]

Target Development

As a result of the globalization of business activity and the dynamics of technological development, *cost pressure* is increasing. Every company must therefore be able to achieve Economies of Arbitrage (price advantages), Economies of Scale (size advantages), Economies of Scope (bundling advantages), Economies of Speed (time or speed advantages) and Economies of Structure (change advantages) in order to achieve a favourable cost position. Should a company be unable to meet these cost requirements, it has lost its "ticket" to play in the marketplace. The resource or production factor orientation associated with cost pressure is therefore of essential consequence for logistics.

However, globalization and technological development also result in *market pressure*. Customers are becoming increasingly demanding and quickly learn to transfer positive experiences from submarkets to other submarkets. This requires an extreme orientation of the company to the needs of the customer. A company can differentiate itself from competitors primarily through differentiated problem-solving offers tailored to market segments. These problem-solving offers require not only product innovations but also increasingly process innovations.

[1]For other proposed systematizations of development trends, see Bowersox et al., 1999, pp. 172 ff.

H.-C. Pfohl, *Logistics Systems*,
https://doi.org/10.1007/978-3-662-64349-5_3

In future, greater attention will have to be paid to the *risk* (uncertainty) of entrepreneurial decisions. This is because the gap is widening between the demands on companies' willingness to take risks and their ability to do so. This is primarily a consequence of the increasing complexity and dynamism of markets, which is associated with poorer forecasting possibilities for entrepreneurial decisions, and the declining profitability of companies. It demands that the greatest attention be paid to the financial impact of entrepreneurial decisions. From this, two new focal points in management can essentially be identified.

Management Consequences

The first focus arises from the declining profitability of companies. This makes it necessary to give special weight to the profitability goal among the corporate goals.[2] The sales thinking that prevailed in the 1970s, especially in large companies, has given way to profitability thinking. The focus is now on the quality rather than the quantity of the markets.[3] Potential increases in sales are examined to see whether the cost increases they cause do not create problems rather than solve them. Marketing-oriented management must be supplemented by production factor- or resource-oriented management.[4] This is because profit will be more positively influenced by the sales side or the cost side depending on the growth of a market.

Production factor orientation in management requires that equal attention be paid to all production factors when combining them and that their respective cost development be constantly monitored. In the past, the production factors information and material were neglected in this respect. While information, in addition to the classic production factors of labor, operating resources and material, was for the most part not seen as such at all, in the case of the production factor of material the focus was one-sidedly on the cost-influencing factor of price. In contrast, the cost-influencing factor quantity, which is expressed in inventories, was neglected. The production factor orientation has therefore resulted in (in particular) a new information management and a new inventory management.[5] The object of *information management* is the economic use of the production factor information, which also includes the substitution of other production factors by information if information costs develop accordingly. The object of *inventory management* is an integrated view of all inventories available in the company with the aim of reducing the capital commitment and thus increasing profitability through a greater capital turnover rate.

---

[2]This change can be seen externally through an increasing focus on shareholder value, which is also communicated externally by most companies. This is to be increased by increasing cash flow. On the influence of logistics on return on investment, see Pfohl, 2016, pp. 55ff.

[3]Cf. LaLonde, 1979, pp. 16ff.

[4]Cf. Bender, 1983, pp. 27 f.

[5]Cf. LaLonde, 1979, pp. 16ff.

Fundamental to inventory management is the consideration that capital tied up in current assets is just as much an investment as capital tied up in fixed assets.

The second management focus results from the dynamics of the markets and the associated *shorter reaction times* for entrepreneurial decisions. It first requires an expansion of the company's *planning and control systems*. Only in this way can the dangers and opportunities arising in the company's environment be identified in good time. The instruments of controlling can be used to limit the risk in the long term.

However, highly developed planning and control systems cannot completely eliminate the uncertainty of entrepreneurial decisions. For this reason, the company's *flexibility* is becoming increasingly important, enabling it to react to changes in the environment at short notice. To implement the demand for flexibility, companies can introduce flexible organizational structures or flexibly align production and logistics systems. Flexibility of the organizational structure is achieved, for example, by reducing bureaucracy. Flexible *organizational structures* enable decision-making processes with which the company can react to unforeseen situations. Flexibility in *production* and *logistics is* achieved by bringing production and logistics decisions closer to the time of purchasing decisions in the sales market.[6] This means postponing production until there is clarity about the products desired by the buyer and implies modular product design, smaller production batches, lower changeover costs, and more universally deployable inputs and employees. It also means deferring inventory until the emerging demand in the sales markets becomes clear and requires centralized warehousing—combined with fast and reliable delivery.

Flexibility is often further increased by specialization of the companies. This is done partly within the companies by creating specific modules for the respective tasks,[7] partly through a cross-company division of labour. Instead of taking over the most extensive parts of the logistical chain and production processes, only certain tasks are performed. Other tasks are handed over to other companies that are specialized in the respective areas. This specialization allows companies to concentrate on their tasks and to react more flexibly to the tasks assigned to them. This gives rise to flexible logistics chains (also known as supply chains) and networks that provide services jointly.[8]

The two new focal points in management highlighted from the business development trends show the fundamental importance of logistics for the company. This is because the logistics conception can make both a contribution to production factor-oriented management and a contribution to the flexibility of the company. This can be seen from a study of new production and logistics strategies in German automotive companies in the 1980s.[9] It showed that innovative logistics strategies, in contrast to new production strategies, take

---

[6]This concept is called postponement, cf. Bowersox/Closs, 1996, pp. 471 f.; Pfohl/Pfohl, 2000, pp. 40 ff; Pfohl, 2016, pp. 110ff.

[7]Cf. Picot et al., 2003, pp. 230ff.

[8]See Part IV, Sect. 17.3.

[9]For this study conducted over the period from 1981 to 1990, see Graumann, 1993.

effect more quickly in the company and do not have a negative impact on profitability even in the short term, since they are apparently associated with lower monetary inputs and costs and risks are shared with suppliers. The need to exploit the potentials associated with logistics design for the company follows from the cost and market pressures faced by many companies today. It will therefore be addressed in the next two sections.

3.2 Cost Pressure

Initial Situation

The exact *amount of logistics costs* in a company is very difficult to determine. One reason for this is that the definition of logistics costs varies from company to company, from industry to industry and from country to country, and that different cost accounting systems are often used to determine the costs.[10] In a study published in 2009, which compared companies from different industries in Europe, logistics costs were found to be 3.5–8.7% of sales (see Fig. 3.1). Figure 3.2 shows the importance of logistics costs by estimating their share in the gross national product of various countries. In terms of comparability, however, the geographical conditions and the existing infrastructure must be taken into account. Various studies have shown that the proportionate logistics costs have declined over many years, but have been rising again in recent years, although they are estimated at different levels.[11]

Although new cost accounting methods have been developed in the meantime,[12] with which costs can be allocated to the cost causers more precisely than in the past, logistics cost accounting presents a difficulty in many companies.[13] The amount of logistics costs is often underestimated because not all costs caused by logistics processes are recognized as logistics costs. They then either remain hidden in overhead surcharges, e.g. in procurement, production and sales costs, or the entire operational logistics system is not seen, but only a logistics subsystem. For a very long time, the focus was solely on distribution logistics. Later, procurement logistics and production logistics also gained the attention they deserved. Moreover, the more or less completely recorded logistics costs are not allocated to the logistics services (the service) according to their cause, so that the actual price of these services is not known, which leads to exaggerated service requirements on the part of production and marketing in the company.

---

[10] On the problems of determining logistics costs, see Pfohl, 1996 and the literature cited there.

[11] Cf. Straube/Pfohl, 2008, p. 47; European Logistics Association/A. T. Kearney, 2009a, p. 13 and 2009b, p. 12. In the USA, the lowest value for the share of logistics costs in gross national product was 7.37% in 2009, after which they rose again to around 7.9% in the years 2011 to 2015. For the first time, they fell to 7.5% in 2016. Cf. Solomon/Gooley, 2017.

[12] Cf. Weber, 2002; Pfohl, 2016, pp. 214ff.

[13] Cf. Straube/Pfohl, 2008, p. 48.

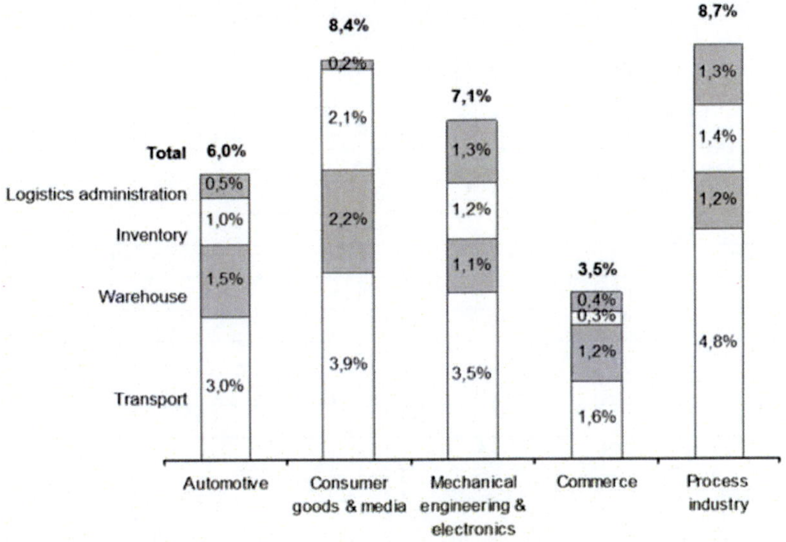

Fig. 3.1 Breakdown of logistics costs of various logistics sub-functions and industries as a percentage of sales. Results of a study conducted in 18 European countries in 2008/2009 (Source: European Logistics Association/A. T. Kearney, 2009a, p. 14 and 2009b, p. 13)

However, it is not only decisions about logistics services that are often made without sound knowledge of the logistics costs they cause. Procurement, production and sales decisions also have effects on logistics costs that are not taken into account. Examples of such effects are shown in Part III.

Future Trend

The share of logistics costs in gross national product and the sometimes considerable differences between the various countries (cf. Fig. 3.2) show that the total volume of logistics expenditure is substantial. Studies from 2008 show that logistics costs are primarily driven up by rising energy, fuel and transport prices as well as high personnel expenses.[14] It should also be taken into account that certain logistics costs will continue to rise as a result of the increasing tendencies towards globalization, environmental and resource protection, and security requirements (cf. Fig. 3.3).

Although cost pressure has been overtaken in recent years by information technology and supply chain management in the ranking of the main factors influencing the development of logistics in the regular surveys of logistics managers in the USA,[15] logistics costs must by no means be neglected. In many markets, logistics costs are not a sufficient

[14] Cf. Straube/Pfohl, 2008, pp. 48ff.

[15] Cf. the evaluation of various studies in Pfohl, 1999, p. 205; LaLonde et al., 2007, p. 15; LaLonde/Ginter, 2008, pp. 15 f.

Country*	Gross national product*	Estimated logistics costs*	Share in %
Belgium	331	31.9	9.6
Denmark	228	14.2	6.2
Germany	2,424	205.0	8.5
Finland	180	22.6	12.6
France	1,892	113.2	6.0
Greece	229	19.2	3.1
United Kingdom	2,019	108.3	8.4
Ireland	186	12.9	6.9
Italy	1,536	82.6	5.4
Luxembourg	36	5.2	14.4
Netherlands	560	46.1	8.2
Norway	284	22.4	7.9
Austria	271	16.9	6.2
Portugal	163	9.1	5.6
Sweden	332	28.2	8.5
Switzerland	310	16.0	5.2
Spain	1,050	82.7	7.9
Total	**12,028**	**836.5**	**7.0**
USA	9,699	914	9.4

* in billions of euros

Fig. 3.2 Comparison of national logistics costs and the gross national product of selected countries. As of 2007 for Europe and as of 2008 for the USA (The logistics expenses are extrapolated on the basis of the determinations of the freight transport volume. Klaus based his calculations on transport costs. Based on the typical distribution of logistics costs in companies, the costs for warehousing, order processing, administration and capital tied up in inventories were estimated, cf. Klaus/Kille, 2008, pp. 153ff. Wilson uses a similar approach in her annual studies.). * in billions of euros (Source: Klaus/Kille, 2008, p. 153ff; Wilson, 2009, p. 2ff)

condition for gaining competitive advantage. But they remain a necessary condition in all markets. They are the entry ticket for being able to play a role in a competition at all.

The *potential for savings in* the area of logistics costs is assessed quite differently. Logistics managers still see cost reduction potential in inventory costs. Potential savings in transport costs could still result from better utilisation of existing transport capacities, for example through more cooperation on the part of both suppliers and demanders on the transport market. There is also potential for savings in the increased use of information as a production factor, which substitutes other production factors that generate higher costs, e.g. material stocks or warehouses. This applies above all to the improved support of logistical planning and control processes. However, there is still potential to be tapped in

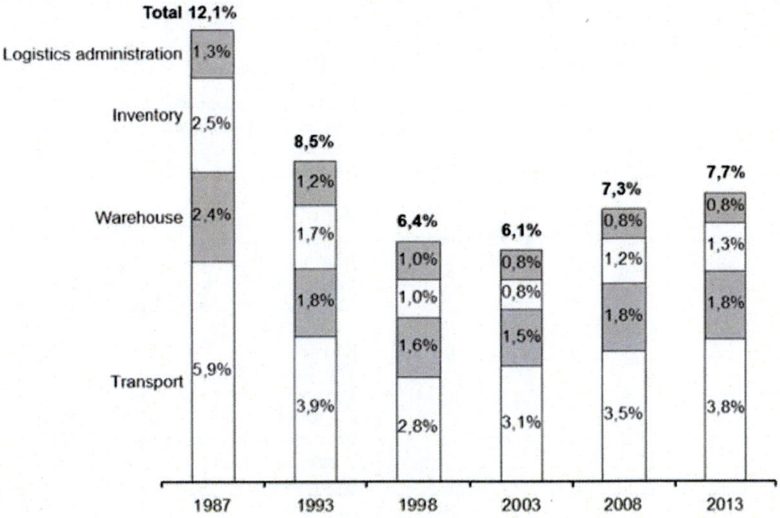

Fig. 3.3 Share of logistics costs as a percentage of annual sales. Results of a study conducted in 18 European countries in 2008/2009 (Source: European Logistics Association/A. T. Kearney, 2009a, p. 13 and 2009b, p. 13)

the more comprehensive, cross-company optimization of interorganizational logistics chains.[16]

3.3 Market Pressure

Initial Situation

Competition is fought out in buyers' markets with all the instruments of marketing policy. As empirical studies in the 1980s and 1990s have already[17] consistently shown, delivery service is of great importance in many sales markets of industrial and trading companies. Together with product quality, it is often the most important factor influencing purchasing decisions (supplier selection).[18] Evaluation criteria for supplier selection include not only the object to be procured, but also the overall performance of the supplier and also the logistics services are an important component of the criteria (cf. Fig. 3.4). On the one hand, this follows from a purchasing behavior characterized by the desire to *shift inventory to* the

[16]Cf. Pfohl, 2004, pp. 3ff.

[17]Cf. A. T. Kearney, 1993, pp. 21 f.; Marr, 1984, p. 34.

[18]Cf. Baumgarten/Wolff, 1999, pp. 46 f. In some sectors, manufacturers or suppliers are required by their customers to provide certain delivery service components that must be fulfilled in order to place an order, e.g. in the case of just-in-time delivery (cf. Part III, Sects. 9.2 and 10.3) or cross-docking (cf. Part II, Sect. 6.1).

Criteria for assessing the economic situation of the supplier	• Legal form • Image • Capital base • Market position • Quality of management • Quality of the employees • Cost structure • Earnings • Organization • Research and development activities
Criteria for assessing the basic suitability as a supplier	• Customer distance • Delivery options • Possibilities for just-in-time connection • Flexibility with regard to possible changes at a later date • Service • Warranty/ Goodwill • Recycling opportunities • Coordination and integration of IT systems • Possibility of joint investments • Possibility of joint production planning and control • Research and development activities
Criteria for the evaluation of the supplier with regard to the procurement object	• Quality • Price • Delivery conditions • Terms of payment • Delivery dates

Fig. 3.4 Evaluation criteria for supplier selection (Source: based on Ehrmann, 2008, p.288)

supplier. This is associated with smaller order sizes and shorter order intervals as well as the demand for special delivery conditions. On the other hand, it is a consequence of the *substitutability of products* found in many markets. Competitive advantages can then only be gained by heterogenising the supply with services, e.g. logistical services. This possibility is used by suppliers to differentiate themselves from competitors, but is now also increasingly demanded by customers. In such markets, the need for goods to be available at all times (e.g. presence of the goods on the retailer's shelves) is obvious. This development is also evident from the shift in the selection criteria for outsourcing logistics services (cf. Fig. 3.5). Factors such as costs and reliability are becoming relatively less important because they are regarded as necessary prerequisites. In contrast, the factors of innovative capability and flexibility/customer orientation are gaining in importance as differentiating features.

The importance of delivery services in industry and commerce is reflected in the importance of logistics companies as sales supporters to industrial and commercial enterprises. Many industrial and trading companies are dependent on these sales supporters because they are unable to provide the necessary logistics services themselves, or can do so only with great difficulty. The reasons for this are the lack of qualified logistics specialists, the prevalence of resistance to the organizational changes associated with the logistics concept and—especially in small and medium-sized enterprises—the lack of experience in the use of logistics tools (e.g. IT support). As the empirical study presented in Fig. 3.6

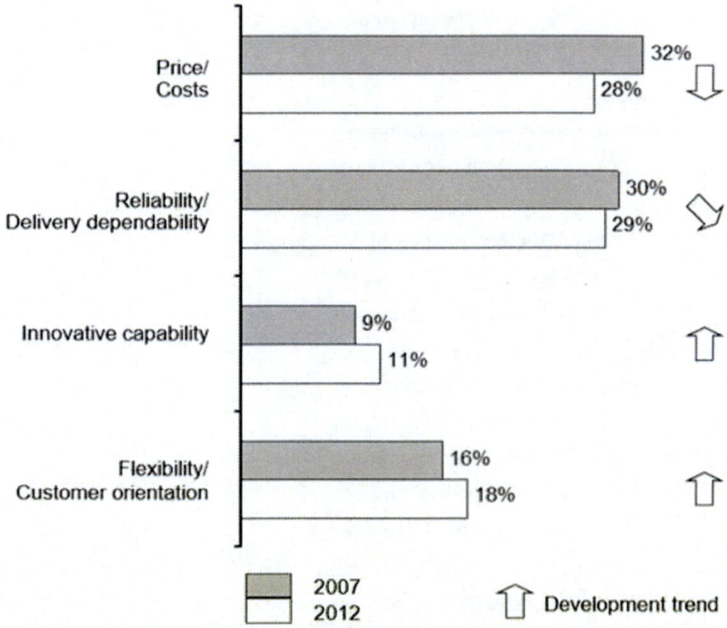

Fig. 3.5 Four selected criteria for the outsourcing of services by shippers to logistics service providers (Source: European Logistics Association/Arthur D. Little, 2007, p. 12)

shows, transport costs are by no means the only decisive factor in the choice of logistics companies. Of greater importance is usually the *quality of* the *logistics services offered.*[19]

Future Trend

The rather still increasing substitutability of products and the increasing spread of knowledge about the advantages resulting from the logistics conception among both customers and competitors will further increase the importance of the delivery service as an instrument for asserting oneself in competition in many markets. However, the disproportionate increase in logistics costs generally associated with an increase in the delivery service level forces a *differentiated delivery service policy.* Beyond a basic service level that is as high as possible and offered equally to all customers, it is necessary to offer an individual service tailored to the specific requirements of certain customers or customer groups and thus to bind the customer to the company in the long term. In this regard, the segmentation of customers according to their requirements and their importance to the company is particularly necessary. This will lead to a *differentiated marketing strategy* with regard to

[19] Cf. Large et al., 2000, pp. 41ff; European Logistics Association/Authur D. Little, 2007, p. 12.

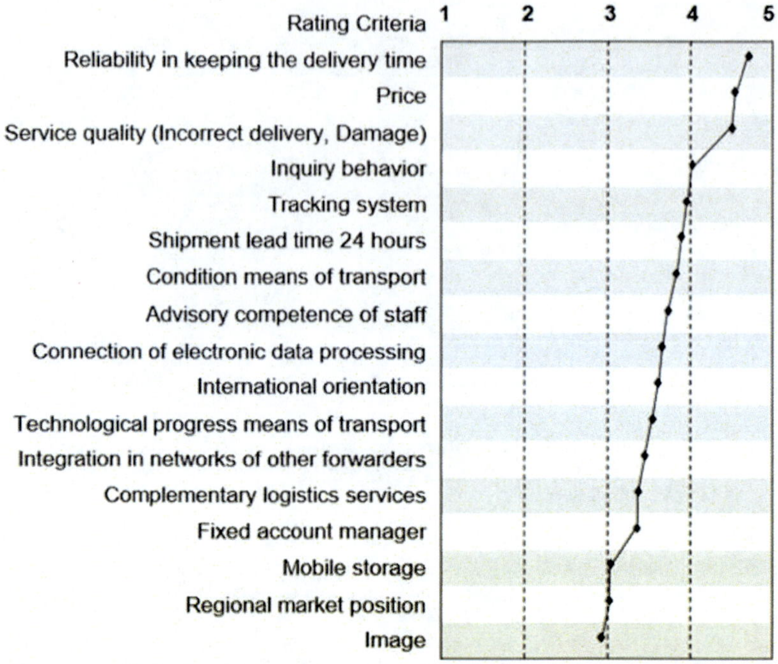

Fig. 3.6 Importance of assessment criteria in the selection of forwarding or transport companies for transport realisation in the procurement sector (1 = no importance, 5 = very great importance). Results of a survey conducted in 1997 among 533 companies in Germany (Source: Pfohl/Schäfer, 1998, p. 86)

complementary logistics services, as has long been customary for the use of other marketing instruments (e.g. quality differentiation, price differentiation).[20]

This is significantly influenced by the continuing trend within a company to concentrate on core competencies and to transfer other services to specialized companies (outsourcing, cf. Part IV, Sect. 17.3). This leads to an increase in the number of *interfaces in logistics chains.* At the same time, efforts are being made to improve coordination between the companies within the logistics chains, which ultimately leads to a constant increase in the demands placed on companies and logistics service providers in terms of delivery services as well as information and communication skills.

Logistics companies will be confronted with three main developments when it comes to offering logistics services.[21] Firstly, fierce competition in the transport markets will initially do little to change the *low level of earnings* appreciably in favor of transport

[20]Cf. Pfohl, 1998, pp. 31ff. Studies have shown that companies that offer customer-specific supplementary services are more successful than average, cf. ibid., pp. 29 f.; Straube et al., 2005, pp. 20 f.; Doch, 2009, pp. 1 f.

[21]On the development of the transport industry in general, see Aberle, 2009, pp. 40 ff.

companies.[22] Regulatory measures (increase in mineral oil tax, levying of a distance-related motorway toll) have had an overall negative impact on the earnings situation of road haulage companies.[23] The eastward enlargement of the EU has contributed to simplified market access for even more competitors, who are subject to more favourable framework conditions in terms of costs. The ability to provide customer-specific logistics services will emerge as an important competitive advantage. On the one hand, this allows higher prices to be achieved; on the other hand, if the range of logistics services is extended beyond transport services, transport tariffs will only play a subordinate role in pricing.[24]

This addresses a second development in the transport markets: The *qualitative change in demand*.[25] In line with the logistics conception, the shipping industry increasingly sees transport only in the overall logistics context, which means that demand is shifting from pure transport services to integrated logistics services. For transport companies, this means that transport is never an end in itself, but always only a means to the end of overcoming space and time. A transport company must therefore not limit itself to a specific means of transport from the outset, but must first include all possibilities for fulfilling the transport function (e.g. combined transport in addition to road transport) in its supply strategies. This is because the trend towards a higher quality profile is also becoming apparent in the demand for transport itself,[26] with the result that it is necessary to refine the means of transport on offer.

This is related to *technical progress* as the third development in transport markets. Technical progress within existing transport systems in terms of routes (e.g. railways for faster trains), vehicles (e.g. new forms of propulsion) and stations (e.g. new transhipment technologies) or goods flow control (e.g. RFID technology[27]) forms the basis for the industrialisation of the production of transport services. It manifests itself in increased mechanisation and automation, in the creation of mass production and in greater capital intensity.

3.4 Position in the Company

Logistics is not of equal importance for all companies, but its significance will depend in particular on the importance of the delivery service for marketing and the supply service for production, as well as the importance of logistics costs. These two factors influencing the

[22] For the current competitive and price situation, see regular reports: Market Observation Freight Transport by Federal Office for Freight Transport.

[23] Cf. Federal Office for Goods Transport, 2006a, p. 14.

[24] On the competitive situation after the EU's eastward enlargement, cf. Federal Office for Goods Transport, 2006b, pp. 1 ff.

[25] Cf. the development of logistical demand broken down by economic sector in Stabenau, 1994, p. 19.

[26] Cf. Aberle, 2009, p. 239ff.

[27] RFID: Radio Frequency Identification.

significance of logistics in the company are presented below. Then, in conclusion to this basic chapter, statements are made on the increasing importance of the significance of logistics in the company.

Hypotheses on the Importance of Service

The importance of delivery service as an instrument of marketing policy can be represented by five hypotheses that seem plausible[28]:

The higher the degree of *substitutability* for a product, the greater the importance of the delivery service. This hypothesis is plausible because with easily substitutable products—they are generally characterized by the same or similar functional performance, quality and prices—the customer can easily switch suppliers.

The higher the demands on the performance of transport, packaging and warehousing that emanate from the *physical product characteristics*, the greater the importance of the delivery service. This hypothesis is plausible because such requirements imply a high level of delivery service, which becomes a prerequisite for the saleability of a product. One example is the requirements that perishable goods place on delivery.

The higher the *delivery service level of the competition,* the greater the importance of the delivery service. This hypothesis is plausible because the customer's service expectations are also shaped by the delivery service level of the competition.

The more the *customer's location is close* to conurbations, the greater the importance of the delivery service. This hypothesis is plausible because delivery warehouses are generally located near population centres and the transport connections and the range of transport services are good. Since the customer knows this, his delivery service expectations are also shaped by this.

The greater the *customer's dependence* on a product due to its production, storage or sales conditions, the greater the importance of the delivery service. This hypothesis is plausible because the customer may incur large costs or sales losses due to poor delivery service when such dependence exists. This is true, for example, if the customer has relatively low inventory and therefore depends on fast and reliable delivery.

The second and fifth hypotheses can also be applied to the supply service. Thus, the greater the demands made by the physical product properties of the material and the greater the dependence of production on a feedstock, the more important the supply service is for a company.

Hypotheses on the Importance of Logistics Costs

With regard to the importance of logistics costs, four plausible hypotheses can be formulated depending on the distance of the procurement and sales markets from the company and the type of product, from which in turn indications of the importance of logistics in the company can be derived. The cost trends are shown in Figs. 3.7 and 3.8.

[28] The following closely based on Pfohl, 1977, p. 248.

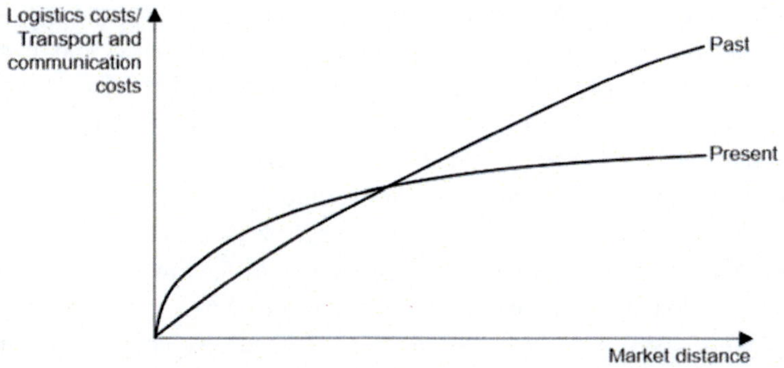

Fig. 3.7 Trend in logistics costs—in particular transport and communication costs—per unit of goods as a function of market distance

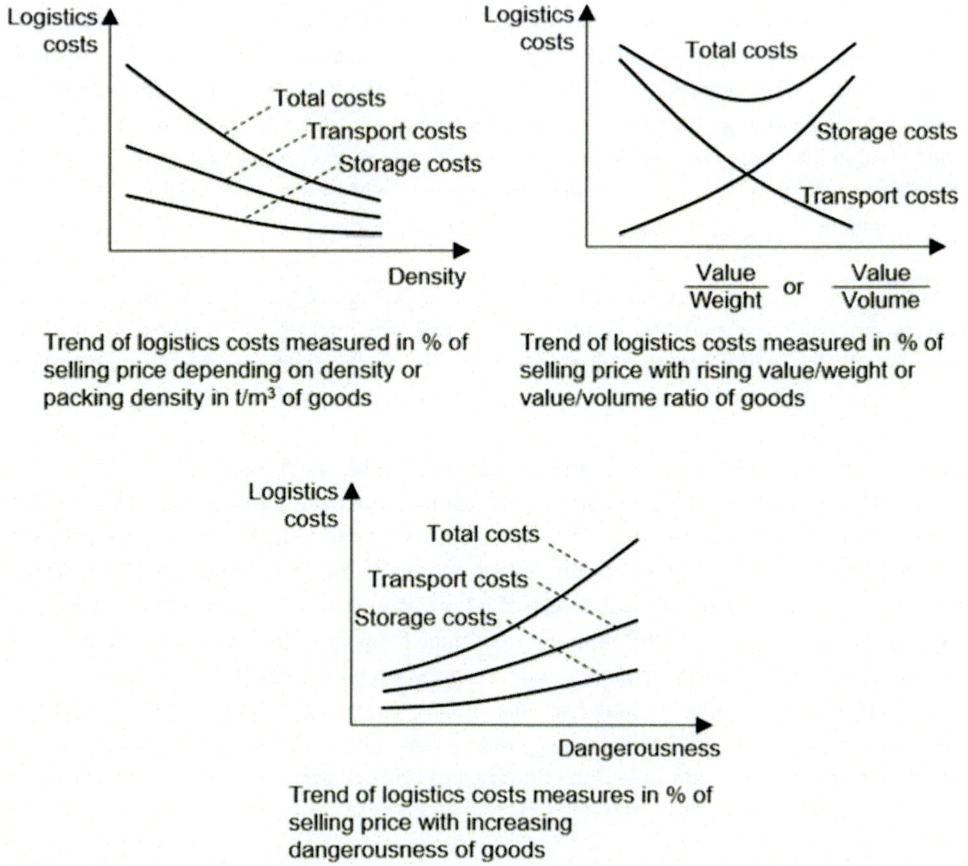

Fig. 3.8 Trend in logistics costs depending on the type of product (Source: Cf. Ballou, 2004, p. 72ff.; Heskett et al., 1973, p. 45ff)

The greater the *distance of the markets* from which the goods are procured or to which the goods are sold, the greater the share of logistics costs in sales. This hypothesis is plausible because as distance increases, so do the transport and communication costs of bridging space. However, these costs increase only degressively over long distances due to the exploitation of economies of scale (size effects), as Fig. 3.7 shows. Newer developments in transport and communications technology are also leading to a growing degression in distance costs, so that market distance is playing an increasingly minor role.

The greater the *density* or *packing density of the goods,* the lower the share of logistics costs in sales. This hypothesis is plausible because storage space can be better utilized by goods with high density, e.g. steel blocks or books, than by goods with low density, e.g. pillows or balls. This tends to decrease relative storage costs. Since transport capacities can be better utilized for the same reason, relative transport costs also tend to fall.

The higher the *value/weight ratio* or the *value/volume ratio of* the goods, the lower the relative share of logistics costs in sales initially. Above a certain ratio, however, the share increases. This tendency of total costs results from the different course of warehousing and transportation costs. Relative inventory costs increase as the value/weight or value/volume ratio increases. This hypothesis is plausible because goods with a high value/weight or value/volume ratio, e.g. electronic items or diamonds, tie up more capital in storage than goods with a low value/weight or value/volume ratio, e.g. vegetables or timber. The result tends to be the opposite if one considers the relative level of transport costs.

The greater the *dangerousness* or *sensitivity of* the goods, the higher the relative share of logistics costs in sales. This hypothesis is plausible because toxic, explosive, radioactive, sensitive or perishable goods tend to incur higher storage and transportation costs than non-hazardous or less sensitive goods. For example, fresh fruit and frozen foods must be transported in special transport and stored in special warehouses. The possibility of spoilage also limits the length of time these goods can remain in the logistics system.

Importance of Logistics in the Company Depending on the Sector

From what has been said about the importance of logistics costs and service, it is clear that logistics is not of equal importance to all industries.[29] The importance of logistics depends on what goods are flowing through the logistics system. It is easy to see that the logistics problems involved in a one-time delivery of a high-value capital good made to order for a particular customer are quite different in importance from those involved in the continuous supply of relatively cheap consumer goods to an anonymous market.[30]

An analysis of logistics costs and service shows, for example, that logistics must be of great importance for *everyday consumer goods.* These are goods that are relatively cheap and that are purchased without lengthy quality and price comparisons because there are no

---

[29] Cf. Pfohl, 1972, pp. 82 f. For a discussion of the industry dependence of logistics, cf. Schumacher, 1988, pp. 96ff.; Kummer, 1992, pp. 40 f.

[30] Cf. Hallbauer/Knödel, 1977, pp. 16ff.

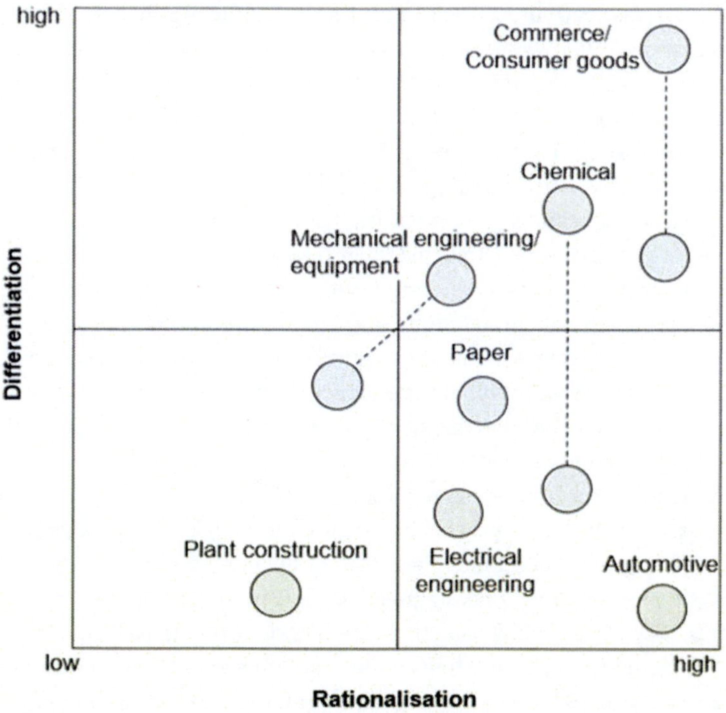

Fig. 3.9 Importance of logistics as an instrument of rationalisation and of delivery services as an instrument of differentiation for some sectors (Source: based on Kowalski, 1992, p. 130)

significant differences in competing products. This applies, for example, to large areas of the food industry. It is therefore not surprising that companies in the food industry in particular must be counted among the pioneers in the field of logistics.

For many products in the *chemical industry,* the share of logistics costs in sales is also very high and there are hardly any price and quality differences between competing products, so that logistics is also very important for these goods. Of course, the importance of logistics for an industry can also be high if the share of logistics costs is relatively low but service is very important. This is likely to be the case for *office machinery* and *household appliances*, including the related spare parts business. In Fig. 3.9, an attempt is made to position various industries in a matrix according to the importance of logistics as an instrument of rationalisation and delivery service as an instrument of marketing policy for differentiation.

To ensure its place in the company, logistics must be implemented in two ways, firstly as a cross-sectional function and secondly as a corporate principle.

Logistics as a Cross-Sectional Function and as a Corporate Principle

The implementation of *logistics as a cross-sectional function* in companies[31] leads, on the one hand, to the exploitation of the experience curve and synergy effects associated with functional specialization, provided that the appropriate instrumental and institutional safeguards are in place.[32] The division egoism of vertically compartmentalized functions, which is promoted by corresponding incentive mechanisms and organizational task, competence and responsibility delimitations, is replaced by cross-sectional functions with a uniform objective that penetrate the other functions horizontally. Special importance is attached to the intra-organizational design of the interface tasks between logistics on the one hand and procurement, production, sales and disposal on the other. The vertical opposition in an organization of mistrust is replaced by horizontal cooperation in an organization of trust in which mutual dependencies are also accepted. Across companies, this statement can be transferred to the inter organizational design of the interfaces between the company and its suppliers and customers.

The implementation of *logistics as a corporate principle*[33] (management concept) leads to the consideration of the approach inherent in the logistics conception in problem definition and solution in other functional areas of the company as well. The coordination of decisions is therefore already taking place in the heads of the employees when dealing with problems. The distinction between function and corporate principle is also found in marketing. On the one hand, marketing can be implemented as a specialized function. On the other hand, implementation as a corporate principle means that the other functional areas of the company should also act in a customer-oriented manner.[34]

In some cases, it is even argued that logistics, with its traditional feature of process orientation, offers the potential for a logistics-oriented organisational or management theory[35] or can be[36] understood as a special manifestation of a (meta-)management theory in which all process-related approaches to corporate management are to be integrated. Certainly, one does not need to go that far to do justice to the importance of logistics! However, it has a great influence on important sub-areas of business administration, which will be dealt with in conclusion in this basic chapter[37]:

---

[31] See Fig. 2.9.

[32] Cf. Pfohl, 1991, pp. 4 f. Weber, 1996 distinguishes between logistics as a functional specialization and logistics as a cross-company coordination function and two development stages of logistics. According to the degree of flow orientation, he further details this distinction into functional specialization, coordination function of logistics, flow or process orientation and supply chain management. Cf. Weber/Dehler, 2000; Pfohl, 2016, p. 25.

[33] Discussed in detail in Pfohl, 2016, pp. 36ff.

[34] On this difference, cf. Meffert et al., 2012, pp. 13 f.

[35] Cf. Klaus, 1998, pp. 74ff.

[36] Cf. Weber, 1996.

[37] Cf. Pfohl, 1991, pp. 5ff. and the literature listed there.

In *traffic management* as a special business administration in accordance with the institutional structure of business administration, the logistics conception has led to an expansion of the experience and knowledge object. Today, objects of experience are all companies that offer logistical services on the market. The terms logistics companies or logistics service providers have become established to describe them. In gaining new knowledge, traffic management is increasingly dependent on the research results of neighbouring technical sciences and computer science, whereby the view of the complex interfaces to the logistics systems of industry and trade is being opened up. This development is also expressed in the fact that one of the sponsors of the logistics idea is the Gesellschaft für Verkehrsbetriebswirtschaft und Logistik e. V. (GVB). (GVB).

In *materials management* as a business sub-discipline according to the functional structure of business administration, the logistics conception has led to a clear profiling of the two different materials sub-functions of purchasing and materials logistics, both of which contribute to the fulfillment of the materials management objective of supplying the company with materials in the broad sense. Characteristic for purchasing is the marketing orientation for active procurement market research and procurement market influence with regard to securing the supply capacity. Characteristic for material logistics is the logistics orientation in the design of the supply chain from the supplier to production (connection of supply capacities with production capacities) at the industrial company or to the sales points (sales capacities) at the trading company. Logistical evaluation criteria are increasingly being taken into account in the development of purchasing strategies and the selection of suppliers. In the area of materials management, there is also an association as a carrier of the logistics idea, namely the Bundesverband für Materialwirtschaft, Einkauf und Logistik e. V. (Federal Association for Materials Management, Purchasing and Logistics) (BME).

In *production management* as a sub-discipline of business administration according to the functional structure of business administration, the logistics conception has led to a new weighting of the goals pursued in production control and thus to new ways of handling the conflict of goals known as the dilemma of production control. Compared to the goal of the highest possible capacity utilization, the goal of the shortest possible lead time has gained in importance. In addition, logistical issues relating to assembly are becoming increasingly important as the vertical range of manufacture tends to decline.

Since the origins of *business research* (operations research), quantitatively formulated models have played a major role in supporting decisions in logistics subsystems (e.g. optimal order quantity, optimal locations or optimal routes). Under the influence of logistics design, the focus shifted away from optimization models for solving logistical subproblems towards the formulation of complex simulation models with which logistical system interrelationships can be mapped and design alternatives can be tested in computer experiments.

In *computer science*, under the influence of the logistics conception on the basis of relational database systems, cross-functional, cross-divisional and cross-company information systems are developed for order processing as well as management and control of the flow of goods in and between companies. In addition, the logistics idea is applied to the design of information systems themselves. We then speak of information logistics, whose task is to provide the right information, in the right state, at the right time, at the right place, at the minimum cost.

References

A T Kearney (1993) Logistics Excellence in Europe. A Study Report Prepared by A T Kearney on Behalf of the European Logistics Association (ELA). o. O. 1993

Aberle G (2009) Transportwirtschaft. Einzelwirtschaftliche und gesamtwirtschaftliche Grundlagen. 5., überarb. und ergänzte Aufl. München

Ballou R H (2004) Business Logistics / Supply Chain Management. Planning, Organizing, and Controlling the Supply Chain. 5. Aufl. Upper Saddle River N J

Baumgarten H, Wolff S (1999) The Next Wave of Logistics - Global Supply Chain efficiency. Berlin/ Boston

Bender P S (1983) Resource Management. An Alternative View of the Management Process. New York u.a.

Bowersox D J, Closs D J (1996) Logistical Management. Überarb. Ausg. der 3. Aufl. New York u.a. 1996

Bowersox D J, Closs D J, Stank Th (1999) 21st Century Logistics: Making Supply Chain Integration a Reality. Oak Brook, IL

Bundesamt für Güterverkehr (2006a) Marktbeobachtung Güterverkehr: Sonderbericht: Eineinhalb Jahre streckenbezogene Lkw-Maut – Auswirkungen auf das deutsche Güterverkehrsgewerbe. Köln

Bundesamt für Güterverkehr (2006b) Marktbeobachtung Güterverkehr: Sonderbericht: Zwei Jahre EU-Osterweiterung – Auswirkungen auf das deutsche Güterverkehrsgewerbe. Köln

Doch S A (2009) Logistische Leistungsdifferenzierung im Supply Chain Management: Theoretische und empirische Entwicklung eines Gestaltungsansatzes für die Differenzierung der logistischen Leistungserstellung produzierender Unternehmen zur Erfüllung individueller Kundenwünsche. Berlin

Ehrmann H (2008) Logistik. 6. Aufl. Ludwigshafen

European Logistics Association (ELA), A T Kearney (2009a) Supply-Chain-Excellence in der globalen Wirtschaftskrise

European Logistics Association (ELA), A T Kearney (2009b) Supply Chain Excellence amidst the Global Economic Crisis

European Logistics Association, Arthur D Little (2007) Innovation Excellence in Logistics. Value Creation by Innovation. Brussel

Graumann M (1993) Die Anwendung neuerer Entwicklungen in Produktion und Logistik bei deutschen Herstellern von Personenkraftwagen 1981-1990. In: ZfB 63 5, S. 443-470

Hallbauer A, Knödel W (1977) Produktprofile und logistische Leistungen. In: Poth L G (Hrsg) Marketing. Neuwied, Kennziffer 3.2.4.4. S. 1-57

Heskett J L, Glaskowsky N A, Ivie R M (1973) Business Logistics. Physical Distribution and Materials Management. 2. Aufl. New York 1973

Klaus P (1998) Jenseits einer Funktionenlogistik: Der Prozeßansatz. In: Isermann H (Hrsg) Logistik: Gestaltung von Logistiksystemen. 2. überarb. und erw. Aufl. Landsberg a. L., S. 61-78

Klaus P, Kille C (2008) Die Top 100 der Logistik – Marktgrößen, Marktsegmente und Marktführer in der Logistikdienstleistungswirtschaft, Ausgabe 2008/2009, Hamburg

Kowalski M (1992) Qualität in der Logistik. In: Athur D Little (Hrsg.): Management von Spitzenqualität. Wiesbaden, S. 128-136

Kummer S (1992) Logistik im Mittelstand. Stand und Kontextfaktoren der Logistik in mittelständischen Unternehmen. Stuttgart

LaLonde B J (1979) A New Management Style for Distribution in the 1980s. In: Lusch R F, Zinszer P H (Hrsg) Contemporary Issues in Marketing Channels. The University of Oklahoma. Norman, OK, S. 15-18

LaLonde B J, Ginter J L (2008) The Ohio State University 2008 Survey of Career Patterns in Logistics

LaLonde B J, Ginter J L, Stock J R (2007) The Ohio State University 2007 Survey of Career Patterns in Logistics

Large R, Kovács Z, Lichtenberger I (2000) Einkauf von Logistikdienstleistungen. In: Beschaffung aktuell o.Jg. 2, S. 41-44

Marr N E (1984) The Impact of Customer Services in International Markets. In: International Journal of Physical Distribution & Materials Management 14 1, S. 33-40

Meffert H, Burmann Chr, Kirchgeorg M (2012) Marketing: Grundlagen marktorientierter Unternehmensführung: Konzepte – Instrumente – Praxisbeispiele. 11., überarb. und erw. Aufl. Wiesbaden

Pfohl H-Chr (1972) Marketing-Logistik. Gestaltung, Steuerung und Kontrolle des Warenflusses im modernen Markt. Mainz

Pfohl H-Chr (1977) Zur Formulierung einer Lieferservicepolitik. Theoretische Aussagen zum Angebot von Sekundärleistungen als absatzpolitisches Instrument. In: ZfbF 29 5, S. 239-255

Pfohl H-Chr (1991) Unternehmensführungstrends und Logistik. In: Pfohl H-Chr (Hrsg) Logistiktrends '91: Unternehmensführung – Marketing – Technologie – Infrastruktur – Logistische Spitzenleistungen. Berlin, S. 1-32

Pfohl H-Chr (1996) Stichwort "Logistikkosten und –leistungen". In: Kern W, Schröder H-H, Weber J (Hrsg) Handwörterbuch der Produktionswirtschaft. 2. Aufl. Stuttgart, Sp. 1129-1141

Pfohl H-Chr (1998) Kundennähe. Bedeutung für die Logistik. In: Pfohl H-Chr (Hrsg) Kundennahe Logistik: Wertschöpfend – Agil – Beziehungsorientiert. Berlin, S. 1-45

Pfohl H-Chr (1999) Ständiger Wandel. Änderung der Organisations- und Personalstruktur als Voraussetzung zur Logistikexzellenz. In: Pfohl H-Chr (Hrsg) Logistik 2000plus: Visionen – Märkte - Ressourcen. Berlin, S. 167-220

Pfohl H-Chr (2004) Grundlagen der Kooperation in logistischen Netzwerken. In: Pfohl H-Chr (Hrsg) Erfolgsfaktor Kooperation in der Logistik. Berlin, S. 1-36

Pfohl H-Chr (2016) Logistikmanagement. Funktionen und Instrumente. 3., neu bearb. und aktual. Aufl. Berlin u.a.

Pfohl H-Chr, Pfohl P (2000) Postponement in der Supply Chain. In: Hossner, R. (Hrsg.): Jahrbuch der Logistik. S. 40-45

Pfohl H-Chr, Schäfer Chr (1998) Analyse des Beschaffungsverhaltens von Industrie- und Handelsunternehmen zur Aufdeckung von Zeitpuffern im Beschaffungsentscheidungsprozeß – Ergebnisse einer Unternehmensbefragung. Arbeitspapiere zur Unternehmensführung und Logistik Nr. 24. Fachgebiet Unternehmensführung, Institut für Betriebswirtschaftslehre, Technische Universität Darmstadt. Darmstadt

Picot A, Reichwald R, Wigand R (2003) Die grenzenlose Unternehmung: Information, Organisation und Management. 5., aktual. Aufl. Wiesbaden

Schumacher W (1988) Die Entwicklung der betriebswirtschaftlichen Logistik und ihr Einfluß auf das zukünftige Leistungsbild des deutschen Speditions- und Lagereigewerbes. Köln

Solomon M B, Gooley T (2017): 28th annual „State of Logistics Report". In: Supply Chain Quarterly 11 Special Issue, S. 10-13

Stabenau H (1994) Verkehrsbetriebslehre. Betriebswirtschaftliche Grundlagen für eine langfristig orientierte Unternehmenspolitik in strukturell sich ändernden Verkehrsmärkten. 3. Aufl. Düsseldorf

Straube F u.a. (2005) Trends und Strategien in der Logistik. Ein Blick auf die Agenda des Logistik-Managements 2010. Hamburg

Straube F, Pfohl H-Chr (2008) Trends und Strategien in der Logistik - Globale Netzwerke im Wandel. Umwelt, Sicherheit, Internationalisierung, Menschen. Hamburg

Weber J (1996) Stichwort "Logistik". In: Kern W, Schröder H-H, Weber J (Hrsg) Handwörterbuch der Produktionswirtschaft. 2. Aufl. Stuttgart, S. 1096-1109

Weber J (2002) Logistikkostenrechnung: Kosten,- Leistungs- und Erlösinformationen zur erfolgsorientierten Steuerung der Logistik. 2., gänzl. überarb. u. erw. Aufl. Berlin u.a.

Weber J, Dehler M (2000) Entwicklungsstand der Logistik. In: Pfohl H-Chr (Hrsg) Supply Chain Management: Logistik plus? Logistikkette – Marketingkette – Finanzkette. Berlin, S. 45-68

Wilson R (2009) 20th annual "State of Logistics Report". Council of Supply Chain Management Professionals (CSCMP), Washington D C

Part II
Activity-Specific Subsystems of Logistics

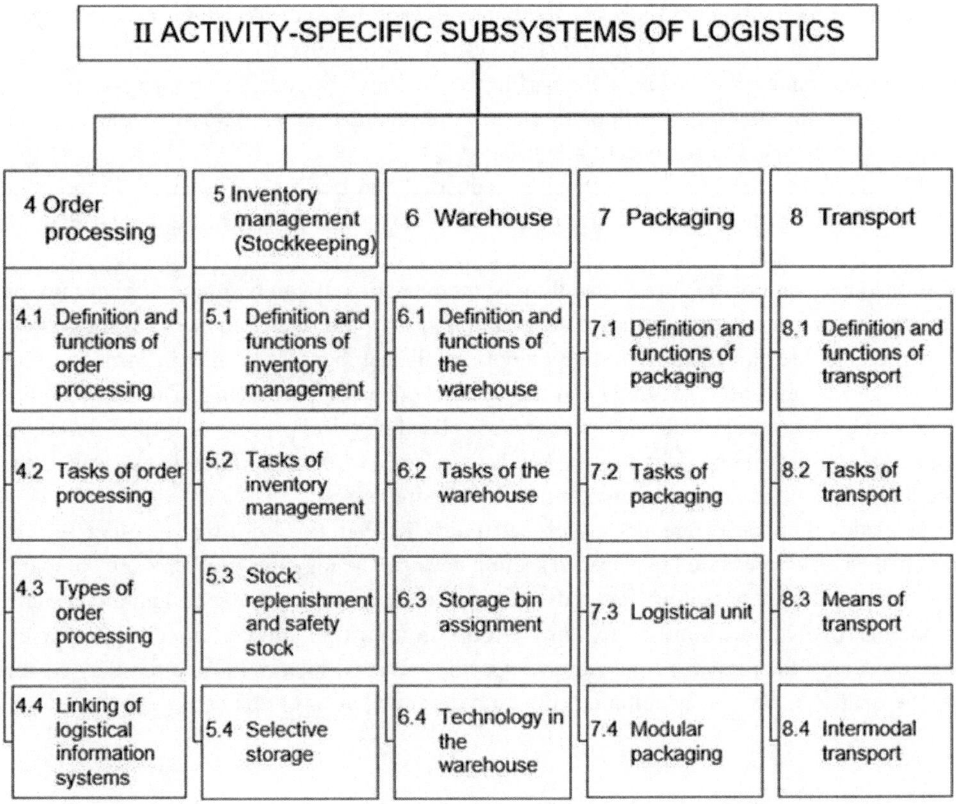

Each system can be classified into a superordinate system and broken down into a number of subsystems. A breakdown of the overall logistics system into subsystems enables the decision-making problems characteristic of logistics to be specified. For this purpose, the overall logistics system is broken down into the following *activity-specific subsystems* in accordance with the functional delineation made in Part I. The subsystems are as follows:

- Order processing,
- Inventory management,
- Warehouse,
- Packaging,
- Transport.

Within the framework of this book, it is not possible to present the entire breadth and depth of the task-specific subsystems of logistics; instead, it is necessary to limit the presentation to fundamental aspects. To this end, each task-specific subsystem is defined and its functions in the overall logistics system are discussed. This is followed by a characterization of the tasks to be performed in the task-specific subsystem. Finally, specific decision-making situations that are to be regarded as particularly typical for the respective task-specific subsystem are discussed.

It is limited to the logistics systems in industrial and trading companies.[1] In the trade literature, the logistics system is also discussed as part of the merchandise management of the trading company.[2] Merchandise management is understood to mean the physical, administrative and dispositional handling of merchandise. It can be broken down into the functional subsystems merchandise information system and merchandise process system. The merchandise information system comprises all merchandise-related information processes and is presented separately in the section on order processing. The merchandise processing system comprises all merchandise-related physical processes. This includes the logistical sub-processes of the flow of goods to be dealt with in the following as well as the modification of goods by normal commercial manipulation.

Logistics companies are dealt with separately in Part IV. Logistics in other service companies is not presented in a special section because the logistics problems are comparatively minor if only operating materials are considered as logistical objects and the logistics problems involved are similar in nature to those in industrial and commercial companies. The transfer of the logistics conception to specific services, such as those offered by a bank on the market, cannot be accomplished within the framework of this book.

---

[1] Specifically on trading companies, see Prümper, 1979; Toporowski, 1996; Schnedlitz/Teller, 1999; Arnold et al., 2008, pp. 525ff; Seeck et al., 2014. Specifically on logistics in the construction industry, see Günthner/Zimmermann, 2008; Ebel, 2012.

[2] Cf. Ebert, 1986, pp. 49ff. and pp. 70ff.

Order Processing

4

4.1 Definition and Functions of Order Processing

Definition

The *order* is the basis of the information flow in the logistics system. Figure 4.1 shows what information should definitely be included in a fully completed order form. The completeness and correctness of the information are so important because they are the input for the logistics system. Errors in this information can generally no longer be detected by checks in the system, but only become apparent when the goods arrive at the receiving point. In addition, the order is an important source of information for other areas of the company.

As an *external* order (customer order), it is an important link for the supplier's distribution logistics and the customer's procurement logistics. As an *internal order*, it is the link between intra-organizational logistics systems, such as between production logistics and procurement logistics or central warehouse and external warehouse. Order processing is defined below using the external order as an example. However, these definitions can be applied analogously to internal orders with different receiving and delivery points. When defining order processing, a distinction can be made between two perspectives—one from the area of distribution logistics and one from the area of production logistics.

Definitions from the *field of distribution logistics* emphasize the functional dimension of order processing. Accordingly, order processing can be defined as the transmission and "data processing and control of orders from the time the order is placed with the customer until the shipment documents and invoices arrive at the customer's premises".[1] Order processing thus focuses on the flow of forms for the completion of an order. It is part of the order cycle along with picking, packing and transportation.[2] Differences between the

[1] Klee/Türks, 1970, p. 69. Cf. also Türks, 1972, pp. 67 f.; Specht/Fritz, 2005, pp. 159 f.
[2] Cf. Stock/Lambert, 2001, pp. 146ff.

H.-C. Pfohl, *Logistics Systems*,
https://doi.org/10.1007/978-3-662-64349-5_4

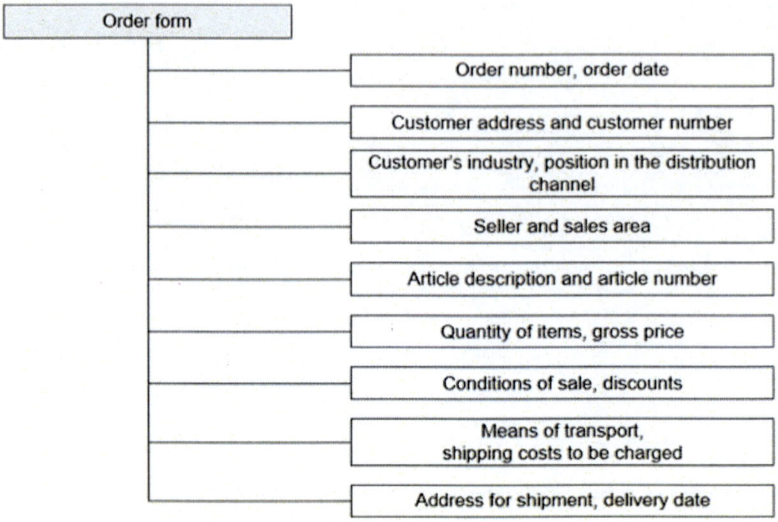

Fig. 4.1 The order as a source of information

various definitions of this view are mainly based on the inclusion of material flow tasks and the degree to which the financial flow is included (for example, customer accounts receivable).

In a broader definition from the *field of production logistics* compared to this narrow distribution logistics view, order processing is defined as the market-driven control of material and information flows from the raw material supplier to the end customer. It thus becomes a comprehensive concept of all functional areas involved in the order flow.[3] Order processing can be defined as the central task area for fulfilling the performance obligation within industrial production. It forms "the link between external information and its internal implementation by converting market-induced inquiries and orders into concrete internal specifications and courses of action. As a result, order processing touches almost all areas of the company that are either involved in the direct production process or for which, as adjacent administrative areas, corresponding data and control information must be provided."[4]

This view is typical for customer-specific assembly, production and construction, whereby order processing is equated with production planning and control in many cases. These definitions also frequently differentiate between technical and commercial order processing. While *technical* order processing includes sales-related quotation writing, design, work preparation, procurement, parts production, assembly and shipping, *commercial* order processing includes costing, purchasing and financial accounting.

[3]On this concept, see Rohweder, 1995; Knolmayer/Mertens/Zeier, 2000, p. 2 and pp. 25ff. See also Part III, Sect. 10.3.

[4]Wildemann, n.d., p. 49.

The time required for order processing is an essential component of the *delivery time*. This is because it does not only occur during the physical movement of the goods between the point of delivery and the point of receipt. It also includes the time required for the communication processes that take place beforehand and for the necessary processing of the order documents. To make clear the importance of order processing for the delivery time, the term *order period* is also used instead of delivery time. Very often, the share of the order transmission time in the length of the delivery period is underestimated. For example, an analysis of a company's delivery time showed that the time for order transmission was 20% of the total delivery time when the delivery time was short; on the other hand, it was 75% when the delivery time was long.[5] Thus, in this case, the cause of long delivery time was mainly due to slow order transmission rather than the other elements of delivery time, which are usually considered to have a much greater influence on the length of delivery time than order transmission. The analysis of the delivery time will reveal opportunities for rationalisation, particularly in order transmission and also in order processing, which can lead to cost reductions and a considerable shortening of the delivery time.

Order transmission and order processing are very closely related. Activities that are part of order processing in one company can be carried out during order transmission in another. If, for example, the customer has access to the supplier's Internet-based standard order forms, the completion of these electronic order forms by the customer is part of the order transmission. If, on the other hand, the customer sends the order to the supplier in paper form, then in many companies it is necessary to transfer this order into their IT system, which is then part of the order processing and often causes errors.

Functions

Order processing has *three functions*: Ensuring an information flow that precedes the flow of goods, ensuring an information flow that accompanies the flow of goods, and ensuring an information flow that follows the flow of goods. Based on these three information flows, it is possible to plan, manage and control the flow of goods. The flow of goods is thus logistically determined.

The *information flow that precedes the flow of goods* is intended to inform all offices involved in the flow of goods in good time about the incoming goods. In this way, they receive the necessary planning and disposition leeway, which is a prerequisite for an optimal realization of the flow of goods from a cost and service point of view. For example, unnecessary waiting times of delivering trucks at the unloading ramp can be avoided.

The *information flow accompanying* the flow of goods should provide all parties involved in the flow of goods with the information required for the operational execution of transport, handling and storage activities on site. This includes, for example, the correct handling of dangerous goods. In addition, the flow of information accompanying the flow of goods should enable the flow of goods to be tracked through the logistics network. The

[5] Cf. Johnson/Parker, 1961, p. 44.

flow of goods must be monitored until the goods have arrived at the receiving point. Only if one is informed about the stage of clearance of each order, can clearance be accelerated or delayed if necessary. Only on the basis of this information can it be ensured that the goods arrive at the receiving point at the desired time.

The *information flow that follows* the flow of goods consists of information that can only flow after the flow of goods has been realized. This can be, for example, the invoice in the case of a corresponding organizational form of invoicing. However, information can also flow in the opposite direction to the flow of goods. This includes information that, based on the accompanying information, is used to confirm the status of the processing of an order. An example of this is the information back to the delivery point about the times at which certain critical stations in the transport chain, such as border crossings in international transport, have been passed by the flow of goods. However, the downstream information flow also includes information from the recipient to the supplier about the quality of the delivery service, e.g. in the form of complaints. Finally, the information flow that follows the flow of goods includes information that can be taken from the order in the context of order evaluation and that is of interest to other areas of the company (such as market research).

In addition to the control of information flows accompanying the flow of goods, the term *information* or *office logistics* has recently been gaining increasing importance. Many companies are increasingly concentrating on their core competencies and leaving the area of receiving, sending, processing, providing, archiving and destroying large amounts of data to other service companies.[6] The aim of information logistics is, in particular, to provide employees with the right information at the right time and in the right place.[7] The business field of office logistics as a subarea of information logistics in turn includes mailroom logistics, archive logistics as well as the destruction of files and data media. A typical service provided by an office logistics service provider within the scope of mailroom logistics is, for example, the media-independent, i.e. both physical and digital, sorting and distribution of incoming and internal in-house mail. The task of archive logistics, on the other hand, is the handling of file inventories. The service provider often takes over the respective files from its customers, records them in terms of information technology and provides them with an identification number. The files are then stored in a security warehouse, which is often chaotically organized. The advantages of outsourcing these tasks to a service provider lie primarily in the variabilization of fixed costs, the achievement of economies of scale in investments, and the increase in transparency within the company through clean and accurate process documentation.[8]

[6] Cf. Peters, 2003.

[7] Cf. Lienemann, 2001, pp. 13ff.

[8] Cf. Peters, 2003; Klotz, 2003, p. 25.

4.2 Tasks of Order Processing

Regardless of the type of order processing, certain basic tasks can always be distinguished that must be performed. These tasks are most easily identified by following the path of order information in order processing shown in Fig. 4.2. It becomes clear that order processing is closely linked to the other task-specific subsystems of logistics.[9]

Transmission

The starting point for the order information is the order creation at the customer. The method of transmission of the order largely determines the *form of order creation*. Orders can be transmitted by the customer by letter, telex, fax, telephone or electronic data processing to a sales representative, to a decentralised sales office or directly to the supplier's head office. The sales representative can also take the order from the customer himself and then forward it in turn. Each of these options requires a different solution to the problem of controlling the creation of the order, which becomes particularly urgent when the order is no longer processed manually but by using electronic data processing.

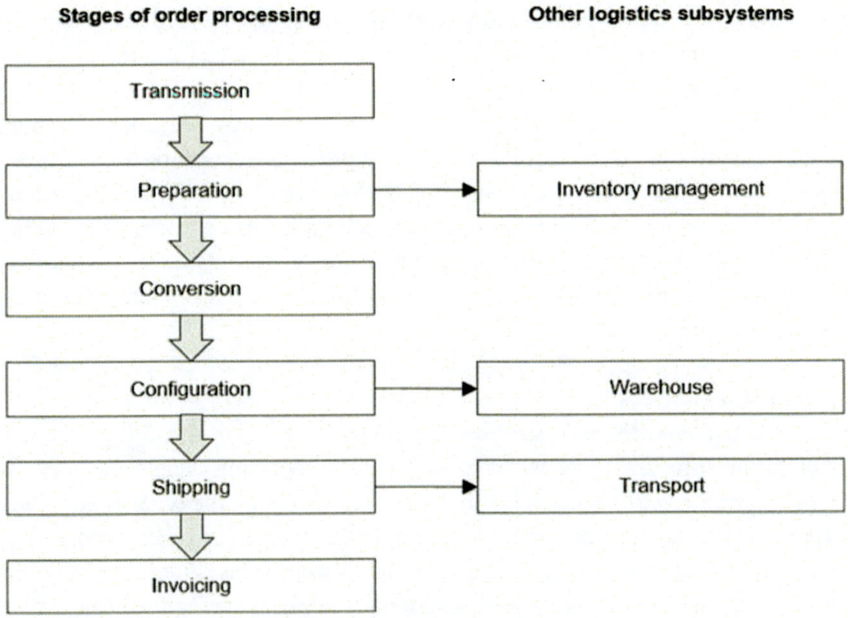

Fig. 4.2 The path of order information in order processing and the connection to the other subsystems of logistics (Source: Closely based on Türks, 1972, p. 69)

[9]Cf. for the following Pfohl, 1972, pp. 89ff.; Türks, 1972, pp. 68ff.; Arnold et al., 2008, pp. 191ff. and pp. 406 f.

A classic method of controlling order creation is the *use of sales representatives* who themselves fill out a pre-printed order form at the customer's premises. It is also possible to provide the customer with order forms on which all items are already pre-printed (order sets) and in which only the desired quantity has to be inserted, with the request for exclusive use. If the customer is allowed to use his own order forms, the supplier still has to prepare the order information after receipt of the order. This incurs additional costs and introduces additional sources of error. Cost considerations and the reduction of the likelihood of errors occurring that can be attributed to order processing argue in favour of leaving the responsibility for filling in the order form used during order processing to the customer as far as possible. However, it should also be borne in mind that the customer should find it as easy as possible to place an order and that, from the point of view of delivery service, the supplier should be flexible in responding to the customer's wishes in this respect.

In order to meet the high customer requirements with regard to the availability of ordered goods, short delivery times (which often depend significantly on the order processing time) and information on the status of order processing (so-called status information) up to the automated notification of delivery dates, *electronic data processing*[10] is used in order processing. In this way, long order transmission times and high-cost errors in order creation and transmission can be avoided by using modern systems of order data entry. It is possible to import the order data from the delivery point directly into the computer at the receiving point.[11]

The right choice of transmission method for the order can only be made if the impact on the overall delivery time is taken into account. Higher costs resulting from faster order transmission can be more than compensated by the benefits of a shorter delivery time. Care must also be taken in the choice of transmission method to ensure that it does not become the cause of bumpy orders. Efforts must be made to utilise the logistics system as evenly as possible and therefore to eliminate, as far as possible, factors that cause an accumulation of orders at certain times.

Preparation and Conversion

The order is transmitted to an order receiving point of the company, which prepares the orders for further processing in the company. This processing adapts the order to the company's *internal requirements.* This includes first of all that the order is supplemented with possibly still missing information. In addition, the order must be checked with regard to price conditions, delivery modalities and the creditworthiness of the customer. Finally, the order must be planned into the logistics system. A prerequisite for this is the availability of the desired product in the warehouse. It can therefore be seen that there is a flow of information from the preparation of the order to inventory management (warehousing).

[10]Cf. Stock/Lambert, 2001, pp. 170 f.

[11]Cf. Pfohl, 1997, pp. 5ff.; Stock/Lambert, 2001, pp. 149ff.

Order processing can initiate measures in the area of inventory planning or production planning if the desired goods are not in stock.

After the *orders* have been prepared, they are usually *converted* (rewritten) either manually, mechanically or electronically into order confirmations and into internal processing documents, such as the delivery notice for processing the order in the warehouse, including all shipping documents.[12] However, due to increasing EDP integration, the conversion is automated and the flow of information can accordingly be paperless.

Cofiguration and Shipping

Based on the prepared and converted orders, the *goods are assembled (picked)* in the warehouse and shipped. This involves further information processing tasks. For example, the warehouse documents must be planned on the basis of their content (e.g. order size, rush order, orders to be delivered together) and on the basis of the organization of picking. After picking, the papers may be supplemented with data on weight, item type, packaging, and staging date. At this stage, order processing provides information for the warehouse and warehousing, for example, for the control of warehouse operating devices or for warehouse accounting.

After configuration, the *shipping documents are* completed, which may need to be supplemented with freight, transport and time data. If there are options, the optimum means of transport and transport route for the delivery of the goods are determined in this phase. Close information relationships therefore exist with the transport system, because the loading and transport of the goods are triggered by the information processing in this phase.

Invoicing

Billing (invoicing) of orders can take place after shipping scheduling or before or even in parallel with the compilation and shipping phases. In the first case, one speaks of *post-billing*, in the second case of *pre-billing*. The main difference between these two types of billing is that in pre-billing, the invoice is created during the preparation and conversion phase, that is, before the goods are physically picked. With post-billing, on the other hand, the picking process must be completed before the invoice is created. Pre-invoicing is based on the idea that all paperwork should be completed in one phase as far as possible, so that further paperwork is as unnecessary as possible. The prerequisite for this is either that there are always sufficient stocks available or that there is always an up-to-date stock update that provides information about the available stocks. In addition, the time required to complete this paperwork must not be too long, as otherwise picking would be delayed for too long. However, this would contradict the requirement that orders should flow into the warehouse as continuously and quickly as possible.

Post-invoicing puts these thoughts of the fast flow of information into the warehouse and the picking act in the foreground. If the invoice is to be sent together with the goods,

[12] Cf. Stock/Lambert, 2001, pp. 149ff.

however, care must be taken to ensure that the invoice is created quickly. Otherwise, the time advantage gained by post-invoicing may be lost. If the invoice is sent separately, additional postage costs are incurred, which are not negligible for a large number of orders.

The basic considerations for pre- and post-invoicing can of course also be transferred to the other order processing activities. In principle, it is possible to do as little preparatory work as possible before picking in order to allow the orders to run into the warehouse without long delays. The other option is to complete as much paperwork as possible in one operation before picking.

4.3 Types of Order Processing

Different types or systems of order processing can be distinguished according to the *tools* used *to manage the flow of forms or documents*. In all types of order processing, the aim is to avoid manual work when writing, to carry out the same information processing activities once and to ensure a rapid flow of information to the points at which the flow of goods is triggered. In business practice, this objective is referred to as "operational excellence" measures.

Manual Types

Before the introduction of electronic data processing, manual types of order processing were optimized. The main aim here was to keep the amount of paperwork and the effort required to implement the order as low as possible and to prepare the information available in the order for further use in production, in the warehouse and in transport in a single operation if possible.

However, due to the easy access to electronic data processing systems and the need to realize ever shorter delivery times, almost exclusively machine-based types of order processing are now in use.

Machine Types

Order processing lends itself to the *use of electronic data processing* because it generally consists of a large number of routine, yet time-consuming activities and the information to be taken from the order form has to be evaluated in a variety of ways (e.g. the preparation of article turnover statistics or supplier turnover statistics). The level of computer support is lower if the turnover of individual orders is low or the number of orders or articles is low. Computer support is less extensive if order processing must be very flexible because many exceptions must be made when processing orders.

The reason for this is that the achievable increase in efficiency and the associated process-related savings potential will not exceed the IT implementation and maintenance costs incurred as well as the administrative costs (for example, due to the regular need to create supplier master data). This discourages small and medium-sized companies in particular from investing in IT systems.

Today, however, there are so-called "Supplier Information Management" systems (SIM) that enable inter-company master data and order maintenance for faster transaction processing.[13] The SIM system is real-time and helps to improve the completeness and accuracy of existing data. By implementing it in the inter-company order process, even smaller suppliers can speed up the processing of their orders with a (relatively) small contribution to sales. They maintain their own order information on the inter-company IT platform (the "cloud" technology is explained in a later part of this paper) and usually have direct access to contacts at the buying company through so-called "chat" functions. By using these SIM systems, inefficiencies are reduced and thus the overall costs of order processing are reduced.[14]

In principle, however, all the decision-making, checking, writing and transmission tasks listed in Fig. 4.3 can be performed by electronic data processing systems. Depending on the stage of order processing in which electronic data processing is used, a distinction can be made between single-stage and multi-stage types. Another way of distinguishing between the different types of order processing is based on the type of information processing. Accordingly, a distinction is made between batch processing and real time processing.

Different approaches to the use of electronic data processing exist in the order conversion and invoicing phases. In *single-step types of order processing*, electronic data processing is used in only one of these phases. In *multi-stage types, electronic data processing* is used in both phases and possibly in other phases as well.

Restricting to a single-step system with a machine order conversion, for example, lends itself to companies that use the services of a factoring bank for invoicing or perform contract billing, so that invoicing is infrequent while orders are often processed via call-off. However, specific technology companies today are enabling a digital solution for "factoring" either before or after successful delivery of contracted products to the buyer company. A direct technological link between internal IT systems and the external IT platform solutions also makes multi-stage types viable for companies wishing to use such a "factoring" concept. A single-stage system with automated invoicing is also advantageous in companies that are able to work with the manual stock replacement method due to their warehouse assortment.

Multi-stage systems are particularly advantageous if the following conditions are met[15]:

- immediate output of order confirmations,
- preliminary check of the available stock,
- comprehensive check of creditworthiness,
- comprehensive calculation of order conditions,
- continuous control of the order backlog,

[13] See further on logistics platforms Hausladen, 2011, pp. 94ff, 168–171.

[14] Cf. Yahsi, 2017, pp. 239ff.

[15] Cf. Türks, 1972, p. 78.

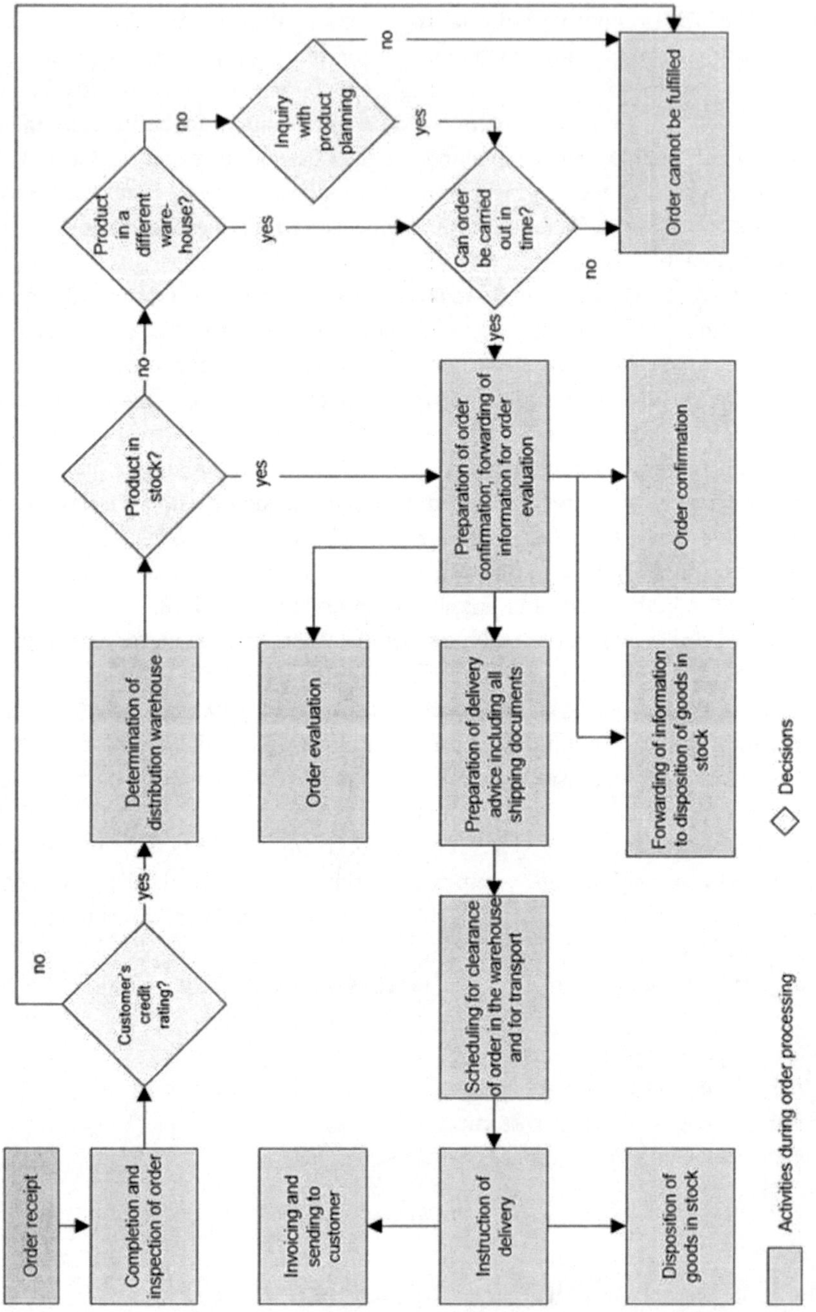

Fig. 4.3 Flow chart of order processing for the use of electronic data processing

- secure control of uninvoiced deliveries,
- equally high order and invoice throughput,
- unsuitable for factoring, unless there is an IT interface to platform solutions (so-called "cloud" solutions) of technology companies.

When using multi-stage automated types of order processing, the question arises in particular as to whether the data should or must be processed in batch or dialog mode (in real time).

If electronic data processing is used in *batch mode*, the orders are stored for a certain period of time (e.g. for one day) and the order volume accumulated in the queue is processed in one burst (e.g. during the night). The accumulation of orders can take place both at the order clerk and in the computer itself.

In *real-time operation*, the computer is constantly ready to take over order processing work, so that orders can be processed continuously without the need to build up a queue.

Information processing in batch mode is generally less expensive (due to the elimination of implementation and maintenance costs of inter-company IT interfaces), but much less efficient. It has the major disadvantage of lower flexibility, because rush orders cannot be processed directly, and lower information readiness due to delayed order processing. This discrepancy between the state of the data and the real state requires additional auxiliary files to be kept alongside the actual order processing. In contrast, information processing in real-time mode has the advantage of extremely short processing times and great flexibility. With this form of automatic order processing, information is available at all times.

Nowadays, SIM systems are usually equipped with so-called virtual marketplaces. These electronic marketplaces enable demand and supply to be brought together in virtual space. Due to the possible multiplicity of suppliers and demanders, both parties (from a microeconomic point of view rather) reach a fair price.[16] Either the virtual markets operating on the platforms are run by the suppliers or demanders, or by intermediaries.[17] In principle, however, they can be used without any interruption, i.e. at any time of day.

A distinction is made between horizontal and vertical marketplaces. In horizontal marketplaces, primarily C-parts are traded across industries. Vertical marketplaces, on the other hand, trade industry-specific products that are specific to a value chain. In these marketplaces, demanders can discuss products, online auctions can be held or specialist information can be provided by the intermediary. Either these vertical marketplaces process transactions based on long-term framework agreements or they enable so-called spot market purchases.[18]

[16] Yahsi, 2017, p. 153.

[17] Hausladen, 2011, p. 95.

[18] Hausladen, 2011, pp. 95–96.

Incoming goods	• Article-specific incoming goods recording - order matching - Assessment and inventory management • Invoice control • Marking of the goods (printing of labels)
Outgoing goods	• Article-specific outgoing goods recording (by data cash registers) • Inventory booking
Disposition and ordering	• Determination of MRP aids and order proposals • Order writing and order monitoring
Marketing and management information	• Creation of inventory lists • Creation of race lists, action lists, etc.

Fig. 4.4 Modular structure of merchandise management systems

Merchandise Management Systems

Closed-loop merchandise management systems are used in retail companies. They are based on a short-term, item-specific recording of all goods receipt and all goods issue data. They have a basic structure consisting of four modules, which is shown in Fig. 4.4. Merchandise management systems make it possible to represent the information flows accompanying goods and to provide the necessary information throughout the company.[19]

Merchandise management systems are called closed because the items in the system are guided through all phases of the information flow without the need for manual intervention in borderline cases. The prerequisite for the economic use of closed merchandise management systems is created by the development of modern information processing and communication technologies. *Four areas of development* can be distinguished here: *mobile data entry*, *scanning*, *data transmission* and *microelectronics*.[20]

Devices for *mobile data entry (MDE)* allow data to be acquired directly in EDP-compatible form with the aid of portable acquisition devices and transmitted to the controlling EDP systems by radio, via satellite or via an Internet/"Wireless-Local-Area-Network" connection (WLAN). Mobile data entry is primarily used in two task complexes. In the context of stocktaking, the inventory data is recorded and imported directly from the recording device into the EDP system. In the context of order processing, the order data is entered decentrally, e.g. at the shelf, into the data entry device and then transmitted to the supplier via a data line and an interface in the supplier's systems. There, the data can be directly included in the ongoing planning and further processed without media disruption.[21]

A further reduction in manual intervention and avoidance of data capture errors is possible through a special form of data capture known as *scanning*. Scanning is the automatic, contact-free recording of goods data by a reader using laser or radio technology. The prerequisite for this is the unique identification of the goods by means of special tags.

[19] Cf. Hertel, 1999, p. 3.

[20] Cf. Pfohl, 1997, pp. 27ff.; Diruf, 1998, pp. 189ff.

[21] Cf. Straube, 2004, pp. 155ff.

These tags consist either of numeric codes, barcodes (linear, stacked or 2D) or quick response codes (QR codes).[22] In the European retail trade, for example, EAN[23] (for "European Article Number") and scanner checkouts are used as the de facto standard for continuous recording of outgoing goods.[24] In addition to such simple and limited one-dimensional barcodes as EAN, for example, there are also newer two- and three-dimensional variants that have a higher storage capacity and can thus encode entire delivery notes, for example.[25]

Even more powerful is the so-called "radio frequency identification" (RFID) technology (in the form of "tags"), which enable localization and identification via radio.[26] A basic distinction is made between active and passive RFID technology. If passive RFID technology is used, the respective RFID tags do not have their own energy source—they obtain the energy they need from the radio waves they receive. The reader is therefore forced to transmit radio waves continuously. Active RFID tags have their own energy source, thus increasing the flexibility of the overall system. They are idle until an appropriate activation signal is received to save energy.[27]

Nowadays, they are used as part of a higher-level IT system (or "warehouse management system"). These IT systems usually collect the location and status information of individual goods (continuously transmitted by the RFID modules) in a central database in order to evaluate them in real time using planning software. Based on this data, the higher-level IT system controls the flow of goods. On the one hand, the reduction of waiting times can increase operational efficiency in logistics. On the other hand, real-time coordination also increases the agility of logistics processes, since the IT system can react to changes in the environment of each of the individual, connected goods based on the existing data situation.

RFID is more costly as a technology than the use of barcodes. Technological advances over the last decade have reduced these costs, but they continue to be highly volatile depending on the storage capacity chosen. RFID transponder prices currently range from €0.30–0.35.[28] For high-value and complex goods and machines in particular, RFID offers not only read access for pure identification but also the possibility of write access, e.g. to store instructions for use, repair logs and other data directly in the goods.[29] In some cases, the individual goods equipped with the necessary intelligent technology decide completely

---

[22] Hausladen, 2011, pp. 53–54.

[23] OCR: Optical Character Recognition, EAN: European Article Number, UPC: Universal Product Code.

[24] Cf. Hausladen, 2011, p. 54.

[25] Cf. Gleißner/Femerling, 2008, pp. 209ff.

[26] Cf. Hausladen, 2011, pp. 52ff.

[27] Cf. Hausladen, 2011, p. 56; Richter et al., 2015, pp. 251ff.

[28] Cf. Hausladen, 2011, p. 58.

[29] Cf. Finkenzeller, 2002, pp. 2ff.; Richter et al., 2015, pp. 251ff.

autonomously about their own movements, for example in intralogistics. However, this requires far-reaching and reliable control systems to ensure safety in the workplace.

For the fast transmission of recorded data between different branches of a company or across companies from customers or to suppliers, the *Internet* still plays a central role today, as does the connection of machines and databases that this makes possible. The worldwide access and the low costs enable an almost instantaneous and fast information transfer between any locations as well as a bidirectional communication between business partners, e.g. for the agreement of conditions, delivery dates or availability queries. In this way, frequently used and standardized goods and raw materials in particular can be automatically ordered and distributed between trading companies or between trading and industrial companies. For example, the existing platform solutions only work because a secure IT connection between the companies is made possible via the Internet. However, not only organizations in the form of responsible persons (buyers and sellers), but also machines and autonomous programs are directly connected via the Internet.[30]

Due to the high degree of penetration of the Internet in the private sphere, this communication option is now also playing an increasingly important role in communication with and sales to consumers. For order processing, the new communication options mean that manual order processing can be largely eliminated. It is therefore conceivable that future goods provision processes will be initiated by customer orders and controlled by (artificially) intelligent algorithms.

In addition to the Internet, other technologies exist for data transfer: these include, for example, "Universal Mobile Telecommunication Systems" (UMTS), "General Packet Radio Service" (GPRS), "Global System for Mobile Communication" (GSM) and satellite communication. Companies decide between these technologies with respect to two aspects: (1) performance in terms of information that can be transported as well as data security, and (2) the cost of using each technology. The individual technologies can also be used in combination in operational practice.[31]

The fourth area of development is computer-aided cash register systems. Due to developments in *microelectronics*, modern checkout systems are directly networked with the companies' merchandise management systems and automatically record the articles sold via scanners and electronic scales. In addition, these cash registers also enable payment via connected card readers and subsequent processing of the payment in the accounting systems. Conversely, the checkout systems obtain information on the recorded articles directly from the central systems, so that all checkouts always have the same article data and also information on current offers.

Such networked merchandise management systems enable retail chains to have a comprehensive overview of the movement of goods throughout the company. With the

[30]Cf. Pfohl et al., 2015, pp. 38–39.

[31]Cf. Hausladen, 2011, p. 62.

help of EDI[32] and the Internet, retail companies network with specific user groups and form large networks with several thousand companies. The automatic item-by-item recording of sales at the point of sale (POS) and the communication of this sales data creates the basis in these networks for ordering systems with high responsiveness, so-called quick response systems.

In addition, such merchandise management systems with item-specific recording in all movement processes and connection of all companies involved in distribution make it possible to provide comprehensive shipment tracking. Such shipment tracking is also referred to as tracking and tracing. For example, it is possible for a customer to track all movements of the ordered goods in a web front end after placing an order and to include this information in his own planning.

In addition to the networking of retail companies and the connection of suppliers and end consumers, the integration of banks, logistics service providers and market research institutes into the complex merchandise management information flows is the goal of a perspective oriented towards the optimization of the overall system.[33] For such integrated merchandise management systems, the linking or coupling of different information systems plays a major role.

4.4 Linking Logistical Information Systems

The previous presentation of the functions, tasks and forms of order processing shows that different logistical subsystems must be linked with each other in terms of information technology. The linking problem arises here both at the intra-organizational level, e.g. when linking the various phases of order processing or when linking order processing with the warehouse, and at the inter-organizational level, e.g. when linking the internal order processing system with the order processing system of a freight forwarder, the supplier or the customer. In other words, the problem of linking logistical information systems is about the *design of interfaces* between such systems.

This problem is very difficult to solve because the individual logistics subsystems are very different and use different formats for data processing. They are also often more closely linked to other functional areas of the company, such as cost accounting, book keeping or production, than to each other. In many cases, the individual information systems have grown historically and have developed into isolated solutions that are independent of each other. However, the goal should be to operate all information systems in a data network or with a database so that the logistics systems involved can access the same data.

[32] EDI: Electronic Data Interchange.

[33] Cf. the trend towards integrated information networks in Zentes et al., 2012, p. 590.

In addition, the particularly personnel-intensive and error-prone re-entry of data at interfaces as well as data breaks should generally be avoided. This can be achieved by using integrated applications, such as inter-company booking and reservation systems or by using standardized interfaces between two or more different systems (such a system in the form of an inter-company IT platform, the SIM, has already been described above). End-to-end, automatic and electronic data exchange can take place directly, e.g. by means of EDI between the systems of two business partners, or indirectly via the systems of a clearing house. Clearing houses take over the storage, processing, conversion and delivery of data, as well as some administrative tasks, such as freight billing. Connected companies place their messages in their mailboxes, for example, from which they are read and transmitted to the recipient in his mailbox.

Manual forms of business data exchange are becoming less and less important due to the rapid development of the Internet. The purely analogue forms use the means of communication letter and in very rare cases fax. Analogue technologies are used precisely when documents and records are to be exchanged, the transmission of which must be secure and trustworthy so that the shared documents can be used in internal financial accounting. In addition, there are other manual forms, such as the physical dispatch of digital data on CD or USB stick as well as the electronic dispatch of digital data attached to an e-mail.

Some basic *coupling possibilities of* logistical information systems are shown in Fig. 4.5. As an example, the information systems of a supplier and a customer are shown

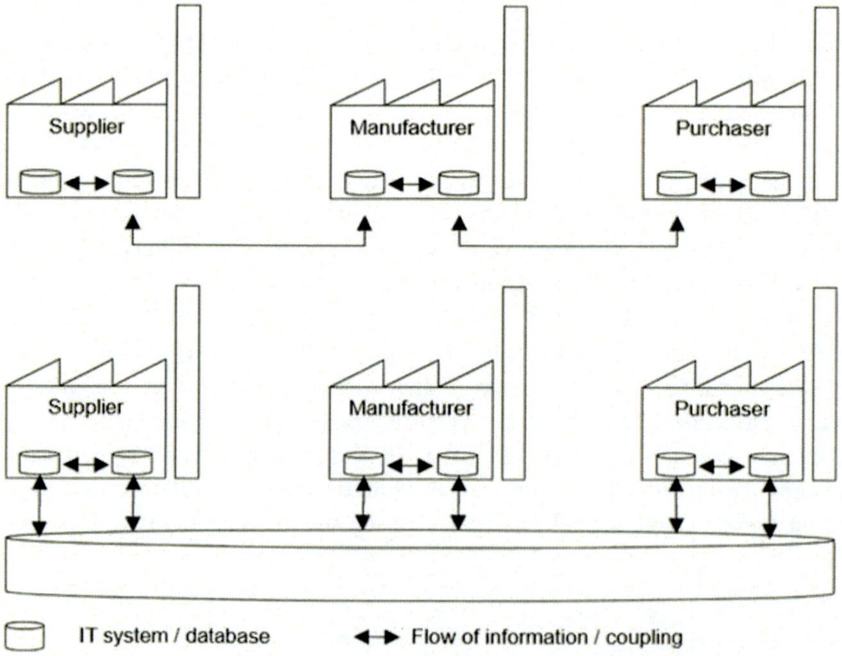

Fig. 4.5 Coupling possibilities between the information systems of different actors in a supply chain

schematically with the systems of a manufacturing company in a supply chain. However, the coupling possibilities shown apply in principle to all information systems. The companies either use interfaces between successive systems for data exchange—upper part of the figure—or a common data source, which can also be provided by an independent supplier, for example—lower part of the figure.

At the same time, the development of the Internet has increased the importance of new communication options in both the business-to-business and business-to-consumer areas. As a result, many communication processes with customers and suppliers can be automated more cost-effectively than before. However, the Internet does not automatically eliminate the problem of uniform data standards at the interfaces. Similar to EDI communication, the data of different market participants must be represented or translated in the same standard.

The basic problems that must be solved when *setting up a data network* can be summarized as follows:

- The purpose of the data link is to enable information that precedes, accompanies or follows the flow of goods to be transmitted more quickly to the appropriate recipient.
- In the data network, data and documents must be available in uniform standards. Based on the realization that many logistics subsystems require the same information, the problem of standardizing data records and documents has been tackled at various levels.[34] At the international level, the International Standardization Organization (ISO) has defined EDIFACT,[35] an industry-independent international standard for trade data. In July 1993, the German EDI Society (DEDIG) was founded in cooperation between the German government and German industry within the German DIN[36] to promote and support the EDIFACT standard. The EDIFACT Transport Message Group, a special working group within the Western European EDIFACT Board, is responsible for the forwarding and transport sector. It has created a framework for the message types booking request, booking, booking confirmation, order, order confirmation and advice. Based on this, the Federal Association of Freight Forwarders and Warehousing, together with the Central Association of Freight Forwarders, has published a guideline "Electronic Freight Forwarding Order" in order to standardize the information content along the logistics chain, which has only been done to a very limited extent so far.
- The pursuit of standardization has its limits where different information needs exist in the logistics systems. For example, information about the size of the means of transport or about the loading and unloading conditions at the recipient is only relevant in the information system for transport. Furthermore, data must be available in different aggregation forms for different management levels. This problem is addressed by

[34] Cf. Seidelmann, 1997, pp. 106ff.

[35] EDIFACT: Electronic Data Interchange for Administration, Commerce and Transport.

[36] DIN: German Institute for Standardization.

specific business intelligence systems that consolidate information and present it according to the requirements of the management level.

- In the inter-organizational linking of logistics systems, problems of data security (protection against loss, destruction and falsification), data protection (protection against unauthorized access) and liability must be solved during data exchange. Not only in the context of data transport, but also in data storage, "cyber security" is playing an increasingly important role. In this context, the need for neutrality in information processing should also be pointed out.
- The linking of information systems is all the easier to implement the earlier standards and interfaces for electronic data exchange are taken into account in the planning, programming and configuration of hardware and software.
- Beyond the requirements for data exchange, IT support is of great importance for the coordination of all activities necessary for the fulfilment of the order(s). Workflow management systems or Business performance management systems are just two examples with great potential for companies.[37]
- With regard to intra-organizational linking, integrated operational software solutions are now widespread. In Germany and Europe, SAP R/3 or the newer architecture with SAP ERP continues to represent the de facto standard. These programs attempt to map all operational processes. They contain solutions for the individual phase- and activity-specific operational subsystems and link them together so that uniform information is available in all areas. Intercompany platform solutions for order processing (such as the SIM system above) are also linked to the respective SAP applications within the companies.

One of the intra-organizational logistics systems to which the order processing system must have a close link is the inventory management system described in the following section, which is sometimes also referred to as the goods in stock management system.

References

Arnold D u.a. (Hrsg) (2008) Handbuch Logistik. 3. neu bearb. Aufl. Berlin Heidelberg

Diruf G (1998) Computergestützte Informationsund Kommunikationssysteme der Unternehmenslogistik als Komponenten innovativer Logistikstrategien. In: Isermann H (Hrsg) Logistik: Gestaltung von Logistiksystemen. 2., überarb. und erw. Aufl. Landsberg a. L., S. 181-196

Ebel G (2012) Vorgehensmodell für die Anforderungsanalyse in der Baulogistik. Dortmund

Ebert K (1986) Warenwirtschaftssysteme und Warenwirtschafts-Controlling. Frankfurt a. M./Bern/ New York

[37] Cf. Oehler, 2006, pp. 32ff.

Finkenzeller K (2002) RFID-Handbuch. Grundlagen und praktische Anwendungen induktiver Funkanlagen, Transponder und kontaktloser Chipkarten. 3., aktualisierte und erweiterte Aufl., München u.a.

Gleißner H, Femerling C (2008) Logistik. Grundlagen – Übungen – Fallbeispiele. Wiesbaden

Günthner W A, Zimmermann J (2008) Logistik in der Bauwirtschaft. Status quo, Handlungsfelder, Trends und Strategien. Nürnberg

Hausladen I. (2011) IT-gestützte Logistik. Systeme – Prozesse – Anwendungen. Wiesbaden

Hertel J (1999) Warenwirtschaftssysteme: Grundlagen und Konzepte. 3. überarb. und erw. Aufl. Heidelberg

Johnson R A, Parker D D (1961) Optimizing Customer Delivery Service with Improved Distribution. In: Business Review 21, S. 38ff

Klee J, Türks M (1970) Aufgaben und Organisation der Warenverteilung. In: Poth L G (Bearb) Praxis der betrieblichen Warenverteilung. Marketing-Logistik. Düsseldorf, S. 67-89

Klotz H (2003) Dienstleistungen rund ums Büro. In: DVZ 57 60, S. 25

Knolmayer G, Mertens P, Zeier A (2000) Supply Chain Management auf Basis von SAP-Systemen: Perspektiven der Auftragsabwicklung für Industriebetriebe. Berlin u.a.

Lienemann C (2001) Informationslogistik – Qualität im Fokus. In: Deiters W, Lienemann C (Hrsg) Report Informationslogistik – Informationen just-in-time. Düsseldorf, S. 13-34

Oehler K (2006) Corporate Performance Management. Mit Business Intelligence Werkzeugen. München u.a.

Peters S (2003) Dokumenten- und Informationslogistik für Finanzdienstleister. In: Bundesvereinigung Logistik (BVL) e.V. (Hrsg) Banklogistik – Lösungsansätze für interne und externe Dienstleister. Ergebnisse des Arbeitskreises Logistik und Finanzen. Berlin, S. 105-119

Pfohl H-Chr (1972) Marketing-Logistik. Gestaltung, Steuerung und Kontrolle des Warenflusses im modernen Markt. Mainz

Pfohl H-Chr (1997) Informationsfluß in der Logistikkette. In: Pfohl H-Chr (Hrsg) Informationsfluß in der Logistikkette: EDI – Prozeßgestaltung – Vernetzung. Berlin, S. 1-45

Pfohl H-C, Yahsi B, Kurnaz T (2015) The Impact of Industry 4.0 on the Supply Chain. In: Innovations and Strategies for Logistics and Supply Chains (Hrsg Kersten W, Blecker T, Ringle C M, S. 31-58

Prümper W (1979) Logistiksysteme im Handel. Die Organisation der Warenprozesse in Großbetrieben des Einzelhandels. Thun/Frankfurt a. M.

Richter K, Poenicke O, Kirch M, Nykolaychuk M (2015) Logistiksysteme. In: Schenk M (Hrsg) Produktion und Logistik mit Zukunft. Berlin/Heidelberg, S. 245-282

Rohweder D (1995) Informationstechnologie und Auftragsabwicklung. Potentiale zur Gestaltung und flexiblen kundenorientierten Steuerung des Auftragsflusses in und zwischen Unternehmen. Berlin

Schnedlitz P, Teller Ch (1999) Aktuelle Perspektiven der Handelslogistik. In: Pfohl H-Chr (Hrsg) Logistikforschung: Entwicklungszüge und Gestaltungsansätze. Berlin, S. 233-250

Seeck St u.a. (2014) Logistik im Handel. Struktur, Erfolgsfaktoren, Trends. Studie der Bundesvereinigung Logistik (BV2) e.V. Bremen

Seidelmann Ch (1997) Das standardisierte EDI-Interface zwischen Kunde und Operator im kombinierten Verkehr. In: Pfohl H-Chr (Hrsg) Informationsfluß in der Logistikkette: EDI – Prozeßgestaltung – Vernetzung. Berlin, S. 101-128

Specht G, Fritz W (2005) Distributionsmanagement. 4., vollst. überarb. und erw. Aufl. Stuttgart

Stock J R, Lambert D M (2001) Strategic Logistics Management. 4. Aufl. Boston u.a.

Straube F (2004) e-Logistik. Ganzheitliches Logistikmanagement. Berlin u.a.

Toporowski W (1996) Logistik im Handel: Optimale Lagerstruktur und Bestellpolitik einer Filialunternehmung. Heidelberg

Türks M (1972) Auftragsabwicklung. In: Klee J, Wendt P D (Hrsg) Physical Distribution im modernen Management. München, S. 65-85

Wildemann H (o J) Informationsflußintegration in der Auftragsabwicklung. Forschungsbericht Technische Universität München. München

Yahsi, B (2017) Financial Supply Chain Management. Erfolgsfaktoren der Gestaltung von Finanznetzwerken. Zugl.: Darmstadt, Technische Universität, Diss., 2017. Universitäts- und Landesbibliothek Darmstadt: Darmstadt

Zentes J, Swoboda B, Foscht T (2012) Handelsmanagement. 3., neu bearb. Aufl. München 2012

5 Inventory Management (Stockkeeping)

5.1 Definition and Functions of Inventory Management

Definition

Inventory management deals with all decision facts that have an influence on stocks, which is why it is also referred to as stockkeeping. Stocks are *buffers* between input and output flows of goods. These buffers arise as soon as the temporal and quantitative structure of the input flows differs from that of the output flows. Such buffers can arise as a result of the different structure of the input and output flows at the most diverse points in the logistics chain, as can be seen in Fig. 5.1.

Only through the complete synchronization of input and output flows can inventories be made superfluous. But, this will only succeed in individual cases. However, the definition of inventory as a buffer must not lead to static thinking that fundamentally does not question the existence of such buffers. Based on logistical systems thinking, inventories can also be defined as partially undesirable interruptions of the flow of goods. However, this already addresses the functions of inventory management.

Functions

The functions of inventory show why inventories are held. The functions of inventory[1] summarized below demonstrate the benefits that can be associated with inventories.

Inventories are necessary if a company wants to exploit *economies of scale in* the purchase, transport or production of goods. Procurement inventories can be created because a company wants to obtain volume discounts from the supplier or more favorable transport conditions from the carrier. Similarly, the buildup of inventory in distribution warehouses can be used to obtain more favorable transportation conditions for larger

[1] Cf. Stock/Lambert, 2001, pp. 228ff; Stölzle et al., 2004, pp. 13ff.

H.-C. Pfohl, *Logistics Systems*,
https://doi.org/10.1007/978-3-662-64349-5_5

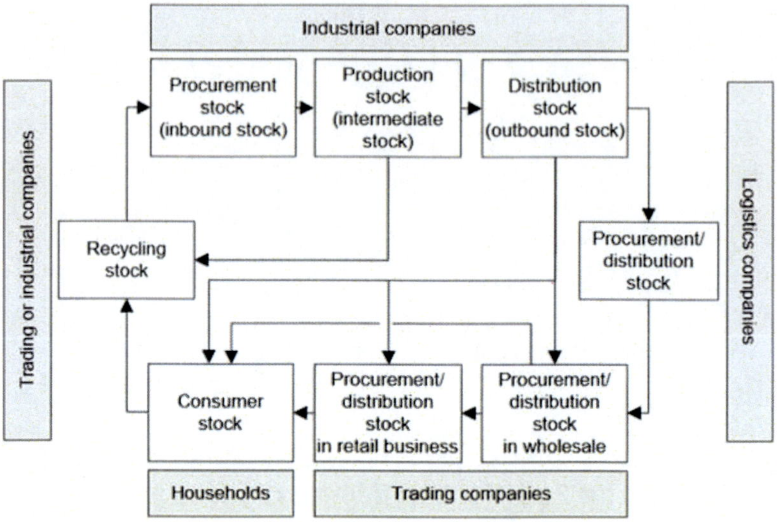

Fig. 5.1 Inventory in the flow of goods through the logistics chain

shipment quantities. Like inventory in production warehouses, inventory in distribution warehouses can also serve the purpose of reducing unit production costs through larger production lots. In this case, higher inventory costs are accepted because lower setup costs are incurred in production.[2]

Another function of inventories is to *balance the mismatch between supply and demand.* An example of this is the seasonal demand for certain consumer goods at Christmas time. In this case, the build-up of inventories in distribution as well as in production warehouses serves to continuously utilize production capacities despite seasonal demand. In the case of agricultural goods, on the other hand, there is often a seasonal supply. In order to be able to sell the goods continuously throughout the year, inventories must then be built up in distribution or production warehouses.

Stocks also generally facilitate the *specialisation of* production in different plants of a company or the division of labour in an economy or in the world economy in general. Specialisation of production, for example in different plants of a company, on certain parts reduces production costs. If it is not possible to deliver the parts to the assembly plant at the needed time, this specialisation is only possible by accepting higher stock levels.

Stocks are also used for *speculation.* Thus, inventories are built up in both procurement and distribution warehouses if an increase in the prices for these goods is expected. In this case, the procuring company still wants to supply itself with goods at the currently lower price. The supplier may speculate that the shortage of supply will drive prices even higher, so he keeps stock in his warehouse. Speculations that lead to inventories do not always

[2] See Fig. 2.5 in Part I.

relate to price. More generally, inventories arise as a result of speculation about the scarcity of goods. For example, inventories can also result from the expectation that a strike at the supplier companies will affect the supply situation.

Ultimately, one also holds inventories as a *protection against uncertainty*. If the input and output flows are different than expected, the demand for goods can only be satisfied from inventories. The need to build up such inventories exists in procurement as well as in production and distribution warehouses. They are a consequence of the fact that one cannot always reliably forecast the demand of the customers or one's own production and that the delivery by the suppliers or by the production is not always certain.

The functions mentioned are initially generally valid independently of a specific inventory type, although some functions are of greater importance for certain inventory types than for others. An attempt can also be made to work out more specifically for individual inventory types the functions that apply to them.[3]

5.2 Tasks of Inventory Management

In designing the buffer between input and output flows, four closely related questions need to be answered:

- Which goods are to be stored?
- How much of a good should be stored?
- How much should be ordered to replenish stock?
- When should orders be placed to replenish stock?

Obviously, the answer to these four questions determines the level of stocks. By answering the first question, it should first be clarified in principle whether stocks are to be held for all goods or whether they are only built up for very specific goods in the sense of selective stockholding. By answering the remaining three questions, the level of stocks for these goods is then determined by means of measures to supplement and safeguard stocks. In this context, it is useful to distinguish between the various components of the inventory that make up the stock.

Components of the Stock

As can be seen from Fig. 5.2, one component of the inventory results from the *order quantity* with which the warehouse stock is replenished. The larger the order quantity or the less frequently it is ordered, the larger the average stock available in the warehouse resulting from the order quantity, which is referred to as the *medium (median) stock.* To be distinguished from the medium stock level is the *total average stock*, which still contains a stock level for securing supplies. If the demand for replenishment could be predicted with

[3] See the functions of the production warehouse in Fig. 10.4 in Part III.

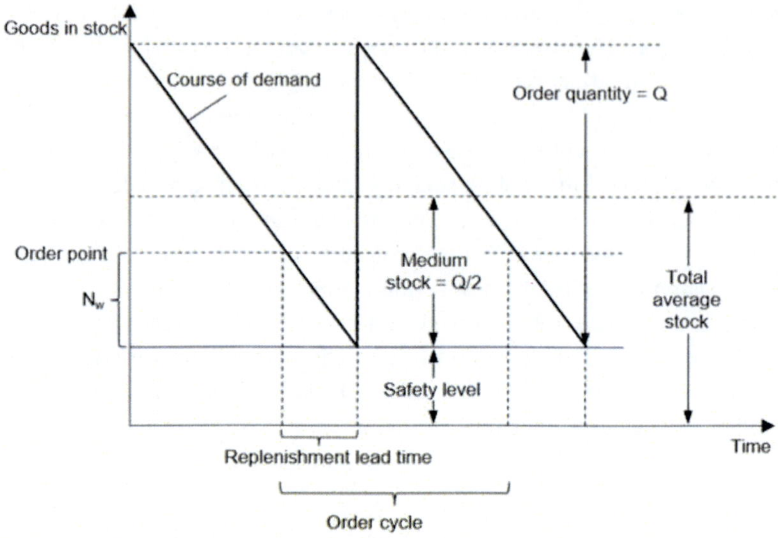

N_W: Demand (need) during ordering period. (In order to not impair the clarity of the overview, the course of demand is only portrayed for the particular case of constant demand per unit of time where the predicted value matches the actual value.)

Fig. 5.2 Components of the inventory due to stock replenishment and safeguarding

complete accuracy, the medium stock level would still be sufficient to satisfy the demand that occurs after the reorder point has been reached. The order quantity triggered at the reorder point would arrive in the warehouse at the end of the replenishment period just when the old inventory level is exactly reduced to zero. Since the forecasted course of demand (stock decrease) often does not correspond to the actual course of demand and the planned delivery of goods (stock receipt) does not correspond to the actual delivery, the *safety stock* must be kept in stock as additional stock. The medium stock level therefore results from the replenishment of stocks when the planned and actual demand as well as the planned and actual replenishment lead times match. The safety stock results from uncertainties in demand and replenishment, which lead to stock withdrawals and replenishment lead times being exceeded.[4]

If one wants to know how much capital is tied up in inventories, one should also take into account the *inventories during movement* (movement inventories, pipeline inventories, inprocess inventories) in addition to the average inventory as shown in Fig. 5.2. They arise during the transport and transshipment of the goods. For example, if it takes 2 weeks to transport and handle an item from the factory warehouse to the distribution warehouse and

[4]Cf. Grochla, 1990, pp. 101ff; Liesegang/Wohlgemuth, 1997, p. 963.

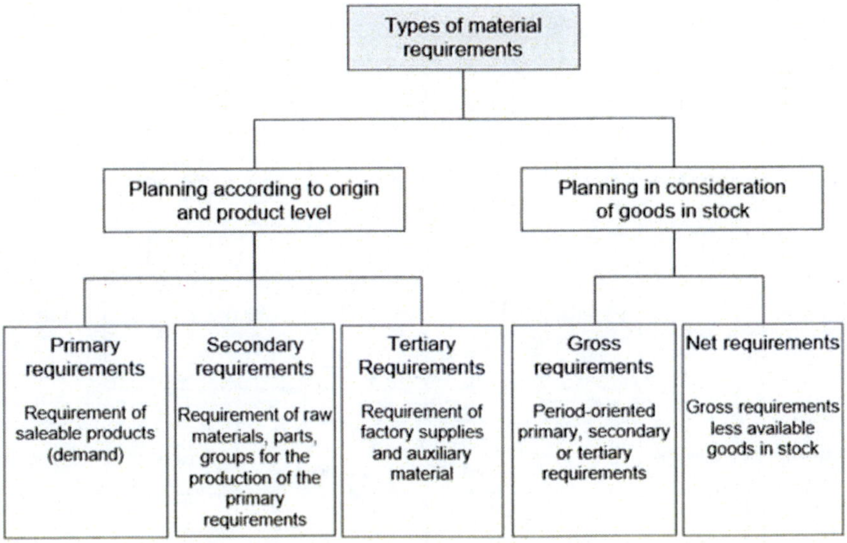

Fig. 5.3 Compilation of material requirement types (Source: Hartmann, 2002, p. 278)

100 units of the item are sold from the distribution warehouse each week, the average level of movement inventory is 200 units.[5]

Before the tasks of replenishing and safeguarding stocks as well as selective stockholding are discussed in more detail, the determination of demand—and thus the determination of the demand curve shown in Fig. 5.2—is discussed. The determination of demand is the basis for the performance of the other inventory management tasks.

Requirements Planning

The way in which the task of determining requirements can be performed depends essentially on the type of requirement.[6] In Fig. 5.3, the material requirement types to be distinguished in principle are compiled, whereby the term material in this case is identical with the term good.

Primary requirement is the market demand, i.e. the demand for saleable goods (finished products, spare parts, merchandise). In retail companies, the primary requirement is the basis for further inventory planning. In industrial companies, this only applies to inventory planning in distribution logistics. For production and procurement logistics, the primary requirement must be broken down into a *secondary requirement* of raw materials, purchased parts and assemblies, which is created for production in accordance with the primary requirement. *Tertiary requirement* is the requirement for auxiliary materials and operating supplies as well as wear and tear tools for production.

[5]Cf. Stock/Lambert, 2001, pp. 233 f.; Magee et al., 1985, p. 86 and pp. 280ff.

[6]Cf. Hartmann, 2002, pp. 275ff.

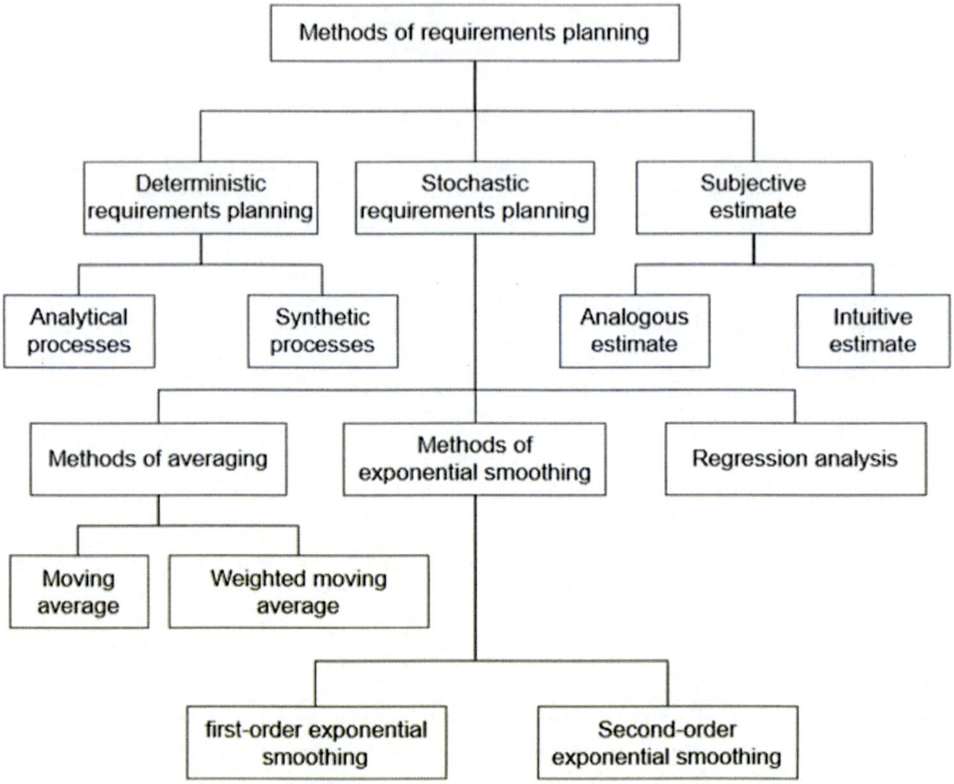

Fig. 5.4 Methods of requirement planning (Source: Hartmann, 2002, p. 284)

Gross requirements are the requirements for material occurring in a period without taking into account the material in stock. *Net requirement* is determined by subtracting available inventory from gross requirement. The methods summarized in Fig. 5.4 are available for determining the gross requirement.[7]

The methods of requirements planning differ according to the data basis on which they are based. *Deterministic or program-based* requirements determination assumes a primary requirement of marketable products for specific periods (planned product program) or for a sales order. The dependent requirements are then calculated deterministically using bills of material or where-used lists. If you use bills of material, you have an analytical requirements explosion. The products contained in the product range are broken down step by step on the basis of BOMs via various assemblies to individual parts and raw materials. While the BOM specifies which materials are required in which quantities for the production of an assembly or a saleable product, the parts where-used list, as the reverse BOM, specifies in which assemblies and saleable products a particular material occurs. If

[7] Cf. Grochla, 1990, pp. 40ff.; Hartmann, 2002, pp. 282ff.

parts where-used lists are used as an aid to deterministic requirements determination, this is a synthetic requirements explosion. In both cases, deterministic requirements determination is very computationally intensive and requires the use of electronic data processing for companies with broad and deep production programs. Because of the large amount of work involved in this method of requirements planning, it is only used for important materials. Which materials are considered important is discussed in the section on selective inventory. Limits to the use of deterministic requirements determination also lie in the uncertainty involved in determining the product range for a period. In this context, it should be noted that the term deterministic should not lead you to assume that the dependent requirements can always be determined deterministically, that is, with certainty. This is only possible in cases where the product range is completely fixed after customer orders have been placed. If, on the other hand, the forecast of the product range is subject to uncertainties, then deviations between planned and actually required dependent requirements can certainly occur.

The basis of *stochastic or consumption-based* requirements determination is the material consumption of comparable periods in the past, which has been recorded in consumption statistics. Based on this data, the demand is determined using forecast methods that are suitable for short-term forecasts. Methods of averaging, exponential smoothing and regression analysis are frequently used here. Whereas deterministic requirements determination is mainly used for high-value main product materials—that is, materials that make up a significant part of the finished product—stochastic requirements determination is used for low-value common main product materials as well as for auxiliary materials and operating supplies (tertiary requirements). It is less time-consuming and can also be used in cases where the movement of requirements only slightly follows the change in the production program, as is the case with many operating materials. The prerequisite for its use is that the historical values are sufficient and reliable and that the time stability hypothesis is valid. The latter states that the complex of causes that caused the development of the variable to be forecasted in the past will continue to have the same effect in the future.[8]

If these prerequisites for the use of stochastic demand assessment are not met, the methods of *subjective estimation* remain. Their basis is the personal opinion of one or more persons. If the opinions about the presumed demand in the future are given purely intuitively, then an intuitive estimation is present. In contrast, the analogue estimate attempts to draw on logically justifiable and therefore intersubjectively verifiable relationships. For example, to forecast the requirements of a particular material, you use the development of requirements for a comparable material.

In the context of demand determination, Anglo-Saxon literature also distinguishes between *independent and dependent (coordinated) demand systems.*[9] In independent

[8]Cf. Pfohl/Stölzle, 1997, pp. 48 f.

[9]Cf. Schary, 1984, pp. 117 f., pp. 175ff. and pp. 192ff.

demand systems, the demand originating from the receiving point is uncertain for the delivery point[10] and must be forecast by the latter. In contrast, in a dependent (coordinated) demand system, the demand for the delivery point from the receiving point is known. In this case, the problem in a multi-stage logistics system is to derive the demand (requirement) at the upstream delivery points from this demand. In the logistics subsystem Material Logistics, this is done on the basis of the production program (Master Production Schedule), which is assumed to be known, using the method of deterministic requirements determination, which is summarized in Anglo-Saxon literature under the term Material Requirements Planning (MRP).[11] In the logistics subsystem Distribution Logistics, the problem is to combine the requirements that are assumed to be known at several receiving points (e.g. local distribution warehouses) over several stages until they are required at a central delivery point (e.g. central warehouse or factory warehouse). In Anglo-Saxon literature, this is referred to as Distribution Requirements Planning (DRP).[12]

5.3 Stock Replenishment and Safety Stock

Stock Replenishment

The decision problem of inventory replenishment is to determine when and how much to order for the identified material requirements so that the sum of inventory holding costs and ordering costs is minimized. A number of ordering rules are given in the literature that specify when and how much to order.[13]

The question of when an order is placed can be answered both by a specific quantity specification and by a specific time specification. Thus, an order is placed either when the *stock level* s (the order point in Fig. 5.2) falls below a certain level or when a *certain period* t (the ordering cycle in Fig. 5.2) has elapsed. The question of how much of an order can also be answered from two points of view. The ordered quantity is either a *predefined order quantity* Q or a variable quantity that supplements the inventory up to a *certain order level* S in each case. Using these four decision variables, one usually distinguishes the ordering rules shown in Fig. 5.5.

In the (s,Q) rule, the *optimal order quantity* or *optimal lot* or *order size* is the main decision variable.[14] When determining the optimal order quantity, a *conflict of objectives* typical for logistics must be resolved. This is because the amount of capital tied up and thus the inventory costs depend on the size of the order quantity. In contrast to these order variables, there are order-fixed (lot-fixed) costs that are incurred only once with each

[10] See also Fig. 1.2 in Part I.

[11] Cf. Pfohl, 2016, pp. 138ff.

[12] Cf. Pfohl, 2016, pp. 144ff.

[13] Cf. on the following Hammann/Palupski, 1997, pp. 88ff; see also Thonemann, 2015, pp. 193ff.

[14] Cf. Grochla, 1990, pp. 69ff; Arnolds et al, 2016, p. 73.

		Order quantity	
		Q	S
Order point	s	(s,Q)-rule	(s,S)-rule
	t	(t,Q)-rule	(t,S)-rule

Fig. 5.5 Ordering rules

inventory receipt. These are the order costs that are regressive with respect to the order quantity. In extreme cases, the order quantity could be equal to the annual requirement quantity, so that only one order is placed per year. As a result of the high average stock level, very high inventory costs are incurred in this case, but only low ordering costs. In the other extreme case, the purchase order accounts for only one unit of quantity of the requirement. In this case, the minimum inventory costs are compared to the maximum ordering costs. The *classic order quantity formula* for minimizing the sum of the two opposing costs is:

$$\text{Optimum order quantity} = \sqrt{\frac{200 \cdot \text{Annual demand} \cdot \text{Ordering costs}}{\text{Cost price} \cdot \text{Inventory cost rate}}}$$

Ordering costs include the costs of all activities necessary to prepare and process an order. The inventory cost rate includes the imputed interest rate and the warehousing cost rate. The former is to calculate the imputed interest on the average capital tied up in inventory. The latter is used to capture the other costs associated with warehousing. The classical order quantity formula is based on the following *premises*: Constant demand, constant demand pattern (constant inventory retirement rate), constant cost price, constant inventory cost rate, constant fixed order quantity costs, and no inventory or financing constraints. On the basis of these partly unrealistic premises, *modifications* of the order quantity formula were developed which, for example, assume variable cost prices (consideration of price scales due to quantity discounts and lower quantity surcharges, transport cost scales and low-priced special offers), variable requirement quantities or storage space and financing restrictions.

The considerations for the optimal order quantity apply analogously to the optimal lot size in production. The order costs are then replaced by the setup costs for the one-time setup of the machines for the lots to be produced. The cost price is replaced by the cost of goods manufactured without the setup costs.

A purchase order with the *(s, Q) rule* is triggered when the warehouse stock has fallen to a reorder point or reorder level. This is therefore also referred to as the reorder *point procedure*. The reorder point is set so that the stock level can cover the demand during the replenishment lead time.

A completely different method of replenishing stocks is used in the *ordering cycle method.* This procedure is based on a constant ordering cycle, which is determined, for example, by the delivery cycle of the supplier or the production cycle. If stock decreases have taken place, the system always reorders after the ordering cycle has expired. With the *(t,S) rule*, for example, the order quantity is set so high that it corresponds to the demand during the ordering cycle and the replenishment lead time minus the remaining stock.

Safety Stock

The decision problem of securing inventories is also characterized by a *trade-off* that arises in the fixing of safety stock.[15] The larger the safety stock, the greater the inventory holding costs caused by it. In contrast, the probability of shortages occurring, and therefore the shortage costs, decrease as safety stock levels increase. The shortage costs include all costs that arise when a requirement that occurs in production (shortages in material logistics) or at the customer (shortages in distribution logistics) cannot be covered with the existing warehouse stocks. These include, for example, costs for special measures to satisfy the demand that cannot be met from inventory, costs of a production shutdown, costs of a production plant conversion, costs of lost sales and lost orders, as well as costs incurred in the longer term due to loss of image.

Basically, three possibilities can be distinguished for *determining the optimal safety stock:*

- Determination of the safety stock level with the help of the shortage costs,
- Determination of safety stock by explicitly taking into account the change in demand due to the occurrence of shortages,
- Determination of the safety stock level by specifying the service level, which limits the probability of shortages occurring.

The first method, which attempts to determine the common minimum of the shortage cost and the inventory cost, is used in many inventory models.[16] It has great importance for inventory theory, but less for practice, since in most cases it is very difficult to quantify the shortage cost.[17] The second method has found less entry into inventory theory. For its applicability in practice, however, the problem of quantifiability also arises.

The third method avoids the quantification difficulties associated with the occurrence of shortfalls. It is therefore preferred in practice. However, the application of this method is not without problems. It is not difficult to determine the safety stock level with the help of a

[15] See Fig. 2.5 in Part I.

[16] On such models see Schneeweiß, 1997, pp. 488ff.; Zwehl/Kramer, 1997a, pp. 492ff.

[17] Cf. Zwehl/Kramer, 1997b, pp. 272 f. Theoretical approaches to the quantification of shortage costs were developed by Schmid, 1977.

predefined service level. The real problem then lies in the way in which this service level is determined.

Similar to stock replenishment, where it was shown that the mean stock level depends on the four influencing factors listed in connection with ordering rules, *four influencing factors for the level of safety stock* can also be named for stock protection. These are:

- Length of the replenishment time,
- Probability of replenishment time overrun and withdrawal overrun (errors in forecasting reliability of meeting replenishment time and errors in forecasting demand),
- Service level,
- Number of warehouses.

The incorrect deliveries due to a lack of delivery accuracy, which are also partly mentioned as an influencing factor of the safety stock,[18] are reflected in the second influencing factor.

Length of Replenishment Lead Time

It is assumed that, on the basis of past values of demand development, the average demand expected during the replenishment period can be calculated and the maximum demand that appears possible can be estimated. The safety stock will then have to be high enough to cover the difference between the average demand that is actually expected and the maximum demand that still seems possible. Thus, the safety stock can be represented as a function of the replenishment lead time[19]:

$$S = N_{max} \cdot t_w - N_{\varnothing} \cdot t_w = t_w \cdot (N_{max} - N)$$

where

S = safety stock,

N_{max} = maximum possible demand/unit of time,

$N_{\varnothing}$ = expected average demand/unit of time,

t_w = replenishment time.

Thus, the shorter the replenishment time for a store, the lower the safety stock level that can satisfy the demand that appears possible. Since the safety stock has a much lower turnover rate or circulation speed than the mean stock—which is therefore also called circulating stock—a reduction in the safety stock can lead to a noticeable reduction in inventory costs. However, a *reduction in the replenishment lead time* normally also results in higher costs in the logistics system, for example through the use of faster means of communication and transport. A reduction in replenishment lead time is therefore only ever

[18] Cf. Grochla, 1990, pp. 115 f.

[19] Cf. LaLonde/Grashof, 1969, p. 55.

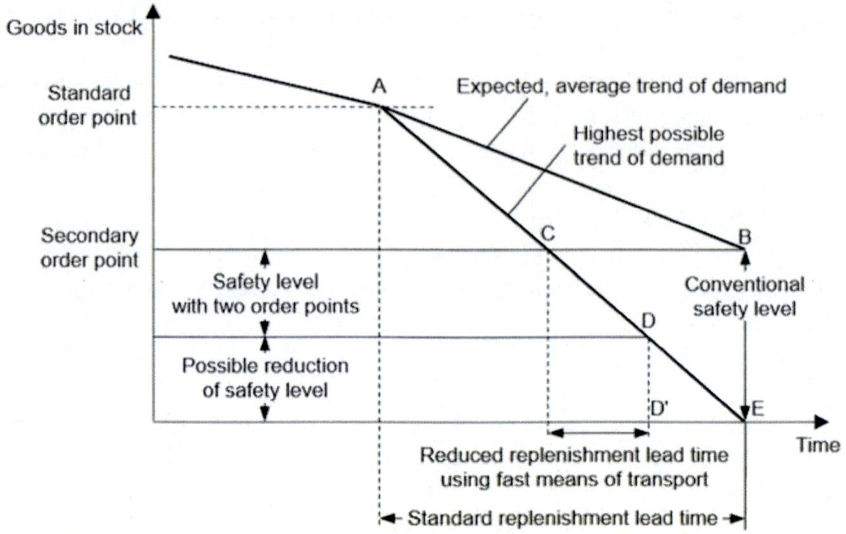

Fig. 5.6 Reduction of safety stock by using two order points (Source: Magee et al., 1985, p. 109)

beneficial if the increase in these costs is more than offset by the reduction in the inventory costs associated with the safety stock.

If it is not possible to shorten the replenishment time by transporting the entire quantity of goods to replenish the warehouse using fast means of transport because of the high transport costs involved, there is another way to try to exploit the advantage of a short replenishment time with regard to the safety stock. In this case, not only one order point is used, but *two order points, as shown* in principle in Fig. 5.6. This makes it possible to transport the majority of the quantity of goods to the warehouse with cheap slower means of transport and to use the expensive fast means of transport only when an exceptionally high demand makes this necessary.

Line AB in Fig. 5.6 shows the expected average demand pattern and line AE the maximum possible demand pattern.[20] BE is the conventional safety stock that must be maintained if only one order point is used. If the standard order point (A) is reached by the stock issues, a purchase order is triggered and the order quantity arrives at the store at the end of the—due to the use of cheaper slower means of transport—long standard replenishment time (E). If two order points are used, the secondary order point is determined in such a way that when it is reached due to stock issues as a result of the maximum possible demand pattern (C), the then triggered order quantity (DD′)—due to the shorter replenishment time as a result of the use of faster means of transport—still arrives at the store in time (D) to be able to satisfy demand by the end of the standard replenishment time (E). As can be seen from Fig. 5.6, the safety stock can be reduced considerably at two order points. The

[20] Cf. Magee et al., 1985, pp. 108 f.

safety stock is now only held for a part of the demand in excess of the expected average value and no longer for the entire demand in excess of it up to the maximum value that appears possible.

The likelihood of the warehouse being supplied by fast means of transport depends on how the secondary order point is determined. Here, too, the balance between the reduction in the storage costs of the safety stock and the increase in the transport costs will have to be taken into account. The use of two order points is particularly useful for reducing the safety stock by the percentage that has a particularly low inventory turnover. The length of the replenishment lead time also influences the safety stock indirectly via forecast errors. This is because the size of the forecast error depends on the length of the replenishment lead time.

Forecast Error

The error made in forecasting demand or requirements for a period is the difference between the predicted value and the value that actually arrived. The random fluctuations in demand cause these forecast errors, which lead to *withdrawal overruns*. Similarly, errors in forecasting the reliability of meeting the replenishment time lead to *replenishment time overruns*. In the following, using the example of demand forecast errors, it is shown that the more accurate forecasts are made, the smaller the necessary safety stocks can be. The explanations can be applied in the same way to the errors in the forecast of the reliability of keeping the replenishment lead time.

The frequency distribution of forecast errors can generally be described well enough by a normal distribution.[21] The normal distribution is characterized by the mean μ, which characterizes the position of the distribution, and by the standard deviation σ, which characterizes the spread of the distribution ($N[\mu,\sigma]$-distribution).

For any forecast method suitable for forecasting a particular event, the mean μ of the forecast error must be zero. Thus, for the distribution of the forecast error there should be an $N[0,\sigma]$-distribution, which exactly, however, is true only in the ideal case. When using forecasting methods, however, constant care must be taken to ensure that a check of the forecast errors produced with them shows whether their mean value actually does not deviate too much from zero. For this purpose, so-called warning signals can be built into the forecasts.[22]

The areas below the distribution curve in Fig. 5.7 indicate the probability with which the forecast error falls within a range to be read off the horizontal axis. For the safety stock, only the forecast errors by which the predicted average value of demand is exceeded are decisive. Thus, to determine the level of safety stock, only the probability of not exceeding a given demand is of interest. It is called one-sided statistical certainty. It can be seen that, based on the distribution curve, 50% of the time the predicted average value of demand will

[21] For the rationale, see Pfohl, 1972, p. 100 and the literature cited there.

[22] Cf. Gudehus, 2010, pp. 301 f.

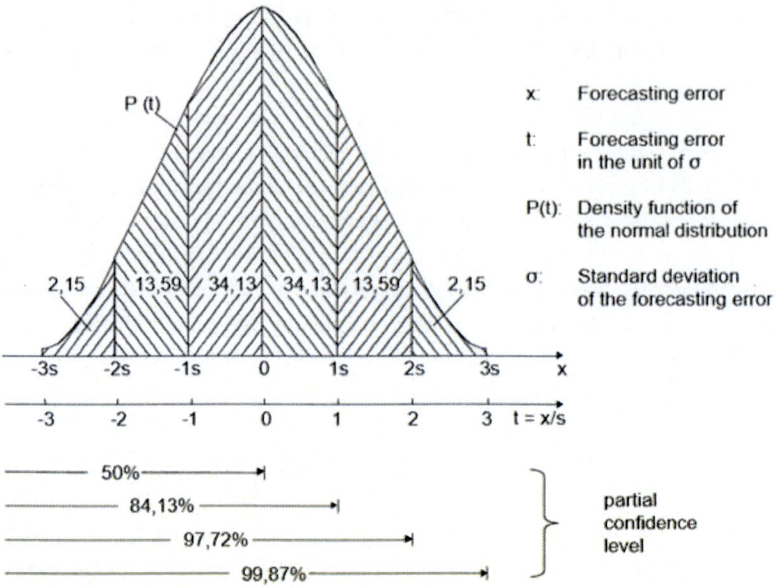

Fig. 5.7 Normal distribution of the forecast error

not be exceeded. If one is satisfied with meeting the demand arising during the replenishment period from the stock only in half of the cases, then there is no need to hold any safety stock. If safety stock is held at the level of one standard deviation, then demand can be expected to be satisfied 84.13% of the time. If one wants to have the statistical certainty that in 97.72% of the cases the demand occurring during the replenishment period can be satisfied, then a safety stock in the amount of two standard deviations must be held. The error or overrun probability, which indicates the probability that the demand cannot be satisfied, i.e. shortages occur in the stock, is only 2.28% in the last case. With an arbitrary frequency distribution, however, the statistical certainties are much less favourable and correspondingly low error probabilities can only be achieved by a much larger number of standard deviations than with the normal distribution.[23]

The level of the safety stock therefore depends on the size of the forecast errors and the probability of their occurrence. Using the standard deviation and based on the frequency distribution of the forecast errors, the safety stock can thus be calculated as follows:

$$S = k \cdot \sigma$$

where

[23]Cf. Inderfurth, 1996, pp. 1031ff.; on the distribution of demand, see also Thonemann, 2015, pp. 246ff.

S = Safety stock,
k = safety factor,
σ = standard deviation of the distribution of forecast errors.

The more accurate the demand forecast, the lower the value of the standard deviation[24] and the lower the safety stock must be in order to satisfy demand with a certain probability. The safety factor indicates how many standard deviations correspond to a certain probability. The probability then corresponds to the service level, which expresses the extent to which the actual demand is to be satisfied by the stock during the replenishment lead time.

Service Level

The service level therefore influences the level of safety stock via the safety factor. As already explained in the discussion of the service components, the service level can be measured in very different ways.[25] However, the type of definition of the service level determines the mathematical relationship between the service level and the safety stock. Two very commonly used definitions of service level are as follows:

- The service level is measured as the percentage of the number of replenishment lead times in which the inventory is sufficient to satisfy demand out of the number of all replenishment lead times. In other words, you measure the percentage of replenishment lead times in which there are no shortages. The size of the shortages is irrelevant.
- The service level is measured as the percentage of demand during the replenishment lead time that can be satisfied by inventory. This definition is therefore not aimed at the frequency of occurrence of shortages, but at the size of the shortages.

Assuming the first definition of service level, which corresponds to one-sided statistical certainty, the safety factor can be derived directly from the distribution of forecast errors in Fig. 5.7. For example, if the service level is to be 97.72%, the associated safety factor is k = 2. Thus, one can expect that no shortages will occur in 97.72% of the replenishment periods if a safety stock equal to 2 σ is maintained. The amount of shortages that occur in, say, a year depends on the frequency of orders for the replenishment of an item. If 100 orders are placed by a warehouse per year for an item, then with a 98% supply availability, shortages are likely to occur twice a year before delivery arrives to replenish stock. If, on the other hand, orders are placed only once a year, a shortage can be expected to occur in 2 out of 100 years.

One can no longer determine the safety factor directly by simple derivation from the function of the normal distribution if it is not the frequency of occurrence of shortfalls but the size of the shortfalls themselves that is of interest. In this case, the safety factor must be

[24] On the dependence of the standard deviation on the length of the replacement period and the size of the sales area, see Pfohl, 1972, p. 102 and the literature cited there.

[25] See Fig. 2.5 in Part I.

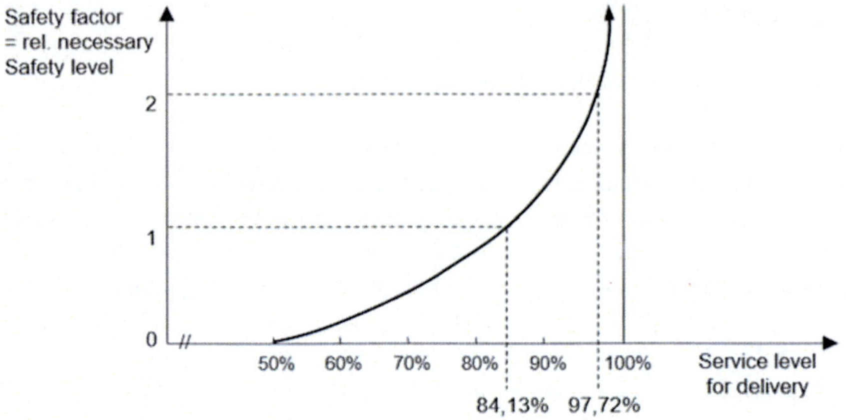

Fig. 5.8 Relationship between safety stock and service level (The figures apply to the relationship between safety factor and service level according to the first definition. In principle, however, the course of the curve also applies to the service level according to the second and other definitions)

determined via the so-called Brownian service function—which is also referred to as the supply function.[26]

If we examine the influence of different values of service level on inventory costs, we find that these increase much more for large values of service level than the service level itself. It can be seen from Fig. 5.8 that a small improvement in an already high service level is associated with a disproportionately large increase in safety stock and hence in inventory costs. This is because for large safety factors, the probability of shortages occurring decreases only slightly as the safety factors are increased. In theory, 100% service level can only be achieved with an infinitely large safety stock.

Number of Warehouses

So far, only the safety stock in one warehouse has been considered. However, the total safety stock that must be held for an item also depends on how many warehouses are used to satisfy a given demand. If the number of customers to be supplied from a sales warehouse, for example, decreases, then the effect of the balancing effect,[27] which is always present when demand is divided among several customers, decreases. If only one customer is supplied from a warehouse, the warehouse must be prepared to satisfy the peak demand of this customer. If, on the other hand, two customers are supplied, it is very unlikely that the peak demands of both customers will coincide, so that the warehouse only has to be prepared to meet a peak demand that is lower than the sum of the peak demands of

[26] Cf. Pfohl, 1972, pp. 104 f. and the literature listed there.

[27] Cf. Flaks, 1967, p. 266.

the two customers. Thus, if several smaller sales warehouses are maintained instead of one large sales warehouse to satisfy demand, the sum of the safety stocks in the small sales warehouses will be greater than the safety stock in the large sales warehouse.[28]

The change in safety stock for a given demand by a change in the number of sales warehouses can be calculated using the standard deviation of the forecast error, as can be shown in the following example.[29] If the demands to be satisfied by two warehouses x and y are statistically independent of each other, the variance characterizing the spread of forecast errors for the total demand is equal to the sum of the variances of forecast errors for the individual demands. Thus, for example, for the standard deviations of the forecast errors that determine the safety stock in each sales warehouse:

If

$$\mathrm{Var_x} = 9 \text{ und } \mathrm{Var_y} = 16$$

$$\mathrm{Var_{ges}} = \mathrm{Var_x} + \mathrm{Var_y} = 25$$

thus

$$\begin{aligned} \sigma_x &= \sqrt{9} = 3 \\ \sigma_y &= \sqrt{16} = 4 \\ \sigma_{ges} &= \sqrt{25} = 5 \end{aligned}$$

The following applies to the total safety stock of an article in the case of two warehouses (the safety factor should be: k = 1):

$$S = 1 \cdot \sigma_x + 1 \cdot \sigma_y = 3 + 4 = 7$$

In the case of one warehouse, the following applies:

$$S = 1 \cdot \sigma_{ges} = 5$$

In general, the effect of an increase in the number of warehouses on the safety stock can be estimated by the following formula[30]:

[28] Cf. Bowersox et al., 1986, pp. 286 f.

[29] Cf. King, 1967, p. 536.

[30] Cf. Bowersox et al., 1968, p. 221.

$$S_n = S_1 \cdot \sqrt{n}$$

where

S_1 = Safety stock to satisfy a specific demand by one warehouse,
n = number of warehouses, if this demand is to be satisfied by several warehouses,
Sn = total safety stock for n warehouses.

The relationship between safety stock inventory costs and the number of warehouses is an aspect that must be taken into account when deciding whether an item should be delivered centrally from one warehouse or decentrally from several warehouses. This issue leads to the problem of selective storage.

5.4 Selective Storage

Basic Idea of Selective Storage

Cost considerations lead to the question of whether or not it is necessary to treat all goods equally with regard to storage and thus to engage in selective storage. A good is *economically storable* if the shortage costs resulting from non-storage are greater than the costs associated with storage.[31] The level of shortage costs depends initially on the importance of delivery time and readiness in a market and the level of storage costs depends on the type of product.[32]

The decision on the storage of a good will also depend to a large extent on its consumption structure. Thus, in the case of regular consumption, delivery synchronous with use, in the case of fluctuating consumption, storage and in the case of irregular consumption, individual procurement as required are the best options.[33]

In addition to the consumption structure, the sort quantity/value ratio of the[34] materials in material logistics or the article quantity/sales ratio in distribution logistics are important for the decision on selective storage. In the following, the problem of classifying goods according to this criterion, which is used very frequently in practice, is discussed using the example of distribution logistics. The considerations apply analogously to material logistics.

ABC Analysis

Research in many industries has shown that the majority of total sales are achieved with only a relatively small number of the items that make up a company's product range. An

[31] Cf. Vahrenkamp, 2007, pp. 195ff.; Schieck, 2008, pp. 357ff.

[32] See the hypotheses on the importance of service and logistics costs in Part I, Sect. 3.4.

[33] On the consumption structure, cf. Grochla, 1990, pp. 31 f. and the supply principles presented in Part III, Sect. 9.1.

[34] Cf. Grochla, 1990, pp. 29 f.

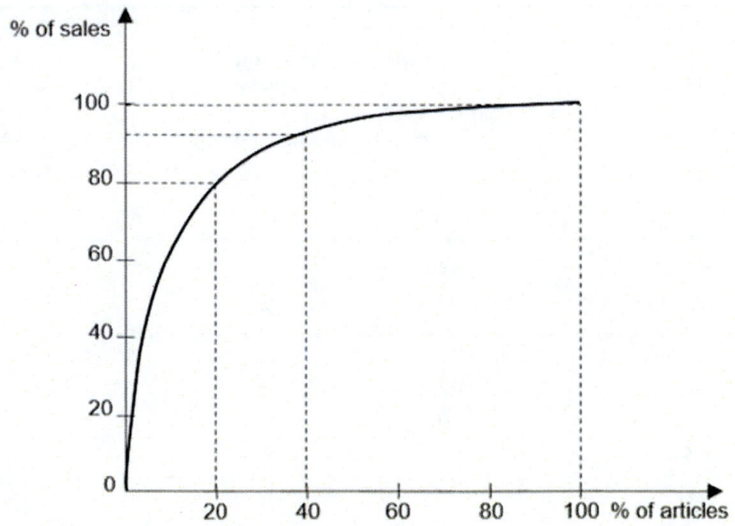

Fig. 5.9 Example of a Lorenz curve to indicate the sales concentration in a product range

often-cited rule of thumb is the 80:20 rule, which states that 80% of sales are driven by 20% of the items.[35] Making item sales statistics allows one to determine the items on which sales are concentrated. One can represent the sales concentration by a concentration curve, the so-called Lorenz curve. Figure 5.9 shows an example of such a Lorenz curve to indicate the concentration of sales in a product range.

If one compares different industries, it can be seen that the more the storage is oriented towards the end consumer, the less curved the curve is—i.e. the less sales are concentrated on a few articles. The Lorenz curve of the retail trade therefore generally has a lower curvature than that of the wholesale trade or that of the manufacturing industry.

The share of sales of each article depends on its unit price and the quantity sold. Thus, the value of the article and its sales volume determine whether it can be counted among the main contributors to sales in the product range or whether it only accounts for a smaller share of total sales or even none at all. Because of their great importance for the company, the main contributors to sales should generally be given considerably more attention in inventory management than the articles with a smaller share of sales.

It is also very instructive to classify the articles according to their *contribution to profit*. A corresponding investigation will generally reveal that even the largest part of the profit is only generated by a relatively small number of articles and that the profit is often reduced by a considerable number of articles sold at a loss.[36]

---

[35] Cf. Magee et al., 1985, p. 61.

[36] Cf. Stölzle et al., 2004, pp. 54ff; Homburg/Krohmer, 2017, pp. 1143ff.

Article	Sales (in $)	Sales (in %)	Sales category and percentage of articles	Ranking of articles according to share of sales	Critical value	Evaluation of articles according to sales and critical value (7) = (5) x (6)
(1)	(2)	(3)	(4)	(5)	(6)	(7)
I	20.000	56	A-articles	1	3	3
II	18.000		20%	2	1	2
III	8.000	34	B-articles	3	2	6
IV	6.000		40%	4	3	12
V	5.000			5	1	5
VI	4.000			6	2	12
VII	3.000	10	C-articles	7	3	21
VIII	2.500		40%	8	1	8
IX	1.000			9	2	18
X	500			10	1	10

Fig. 5.10 Classification of articles using sales analysis and critical value analysis (Source: The table is a summary of the two tables in Bowersox and others, 1968, p. 201. Critical value from 1 to 3 means from critical to non-critical)

However, a profit analysis of all articles involves a great deal of effort, so that in most companies one will generally be content with the classification of articles according to sales revenue, which is much easier to determine. If the price differences of the articles of a product program are only small, the classification of the articles can also be made according to the *sales quantities* instead of the revenues.

If sales is the evaluation criterion, the articles are divided into different classes according to their importance for the total sales. In most cases, three classes are ranked and designated A, B, and C. One speaks therefore with the sales classification also of the ABC principle, of the ABC analysis or of the ABC system of the stockkeeping. In Fig. 5.10, columns 1 to 4 show an example of class formation according to the ABC principle.

Based on the ABC sales analysis, A-items are given much more attention in selective storage than C-items due to their significant share of sales, because the inventory value of A-items is very high and the occurrence of shortages can be associated with large losses.

Now, it is possible that even an item that falls into the C category because of its low share of sales requires special attention in inventory management because it plays a particularly critical role for the customer.[37] For example, a small, inconspicuous spare

[37] Cf. Bowersox et al., 1968, pp. 197 f.; Steinbrüchel, 1971, pp. 198 f.

part for a car's ignition is much more important to the customer than a fender, because a car can still be driven with a dented fender, but not without a functioning ignition. Thus, in selective storage, the supplier must consider the critical value that an item has to the customer as another evaluation criterion, in addition to the item's share of sales. Figure 5.10 shows in column 6 an example of an analysis of the critical value or importance factor.

The critical value of an item can be quantified by its valuation. If a supplier has to stock spare parts for the machines he supplies, he may give critical value 1 to the items that must be immediately available for the machine to function. The items whose failure will not affect the functioning of the machine for a certain period of time will be given critical value 2; and the items which are not of immediate importance to the functioning of the machine will be given critical value 3. Thus, the evaluation criterion for determining these weighted critical values is the importance of the item to the functioning of the machine. There are also other evaluation criteria for determining the critical value. For example, if one relates the critical value to the phase in the life cycle of a product, one will have to pay much more attention to the stocking of a product in the growth phase than to the stocking of a product in the degeneration phase. One will weight the critical values for the products accordingly. From columns 5 and 6 in Fig. 5.10, it can be seen that the importance of items is estimated differently based on the proportion of sales and based on the critical value. In order to arrive at a ranking of the articles, the evaluation based on the two evaluation criteria must be combined into a single value. Multiplication is a good way of doing this, as the range of values is then widely spread and a clear ranking is possible. Column 7 shows the unified valuation of the items using the sales analysis and the critical value analysis. The result is a different ranking of the items than if the sales analysis were used alone. This ranking can be used to make a new classification into A, B and C items. The ABC principle can be extended in this way.

Application of Selective Storage

The application of the ABC principle means that the A, B and C articles are treated differently in the entire area of storage or in certain subareas. Thus, for each article class, the ordering process can be different, the service level can be different, and the number of delivery points can be different. For example, if a company ordered all articles in an eight-week ordering cycle before implementing selective storage, it can work with different ordering cycles after performing an ABC analysis. The A-items are ordered every 4 weeks in order to reduce the very high costs of capital commitment. For the B-items, the eight-week ordering cycle is maintained, while for the C-items, an ordering cycle of 16 weeks is introduced because the capital commitment is not very significant for this item.[38] Accordingly, the effort required for stock control can also be differentiated.

There is no single policy of selective storage for all companies; rather, the nature of selective storage may vary from company to company. However, the *basic idea* is always

[38] Cf. Constantin, 1966, pp. 331ff. and pp. 409ff.

to reduce storage costs by treating items unequally without significantly impairing the delivery service. This basic idea will be explained in more detail using the example of the question at which delivery points the articles are to be stored, since this question—which is also to be taken up in the function of warehouses to be discussed in the next section—is aimed at the *problem of centralization or decentralization of the delivery system*, which has an effect on the entire decision-making area of distribution logistics.

MAGEE et al. formulate an ABCD policy for stocking items at different delivery points.[39] They assume the general rule that cost considerations lead to storing articles with a low share of sales only at a few delivery points. It was shown in the treatment of factors influencing safety stock that the more warehouses used to satisfy a given demand, the greater the safety stock of an item. However, in addition to the cost of tying up capital, the number of warehouses also increases personnel costs, space costs, and costs resulting from fewer opportunities for rationalization. Thus, the total storage costs in a warehouse increase relatively as the turnover decreases. Since turnover per warehouse increases the fewer warehouses there are, warehouse costs argue for centralization of warehouses. Transportation costs generally argue against centralization because the lack of warehouses in individual markets means that the opportunity to bring products as close as possible to individual markets in large transportation units is lost. However, transportation costs increase relatively equally when high and low turnover items are centralized. Therefore, a comparison of the reduction in warehousing costs and the increase in transport costs suggests that high-turnover articles should be stored decentrally in several delivery points, while low-turnover articles should be stored as centrally as possible. If the articles are divided into four classes, the following *ABCD policy can be* formulated:

- A-articles have such a high turnover that it is justified to stock them in all local warehouses,
- B-articles have lower turnover and are stored in a few selected regional distribution warehouses,
- C-articles are only stored in factory warehouses because of the low turnover,
- D-articles account for such a low proportion of turnover that they are not kept in stock at all and are only produced to order.

Assuming that centralization of warehouses increases the delivery time of the goods to the customer, the delivery service for the centrally stored items decreases. However, since only the lower-turnover items are affected, the delivery service for the entire turnover is only insignificantly affected. It can even be increased if, as a result of the cost reduction made possible by the centralization of the low-turnover items, the funds freed up are used to increase the safety stocks and thus the delivery readiness of the high-turnover items.

Before introducing such an ABCD policy, the critical value of the items still needs to be analysed and the ABCD classes may need to be modified accordingly. It is also necessary

[39] Magee et al., 1985, pp. 286 ff.

to examine the extent to which high-turnover and low-turnover items are purchased by the same customers. This aspect may also force a change in the ABCD policy if the customers expect the items ordered to be delivered to them together from one warehouse. In the case of the high-turnover items, it is also necessary to consider whether the turnover arises from the demand of many customers who purchase the item in small quantities or whether it is a small number of customers who purchase the item in wagon or truckloads. If the second case applies, centralized delivery will be advantageous even for high-turnover items. It can therefore be seen that in order to answer the question of which articles should be stored at which delivery points, a whole series of influencing factors must be taken into account in each case. Last but not least, this also includes the warehouse itself, with its location, the size of the market area it serves and the facilities for storing and handling the goods.

References

Arnolds H u.a. (2016) Materialwirtschaft und Einkauf. 13., durchges. Aufl. Wiesbaden 2016

Bowersox D J, Closs D J, Helferich O K (1986) Logistical Management. A Systems Integration of Physical Distribution, Manufacturing Support, and Materials Procure-ment. 3. Aufl. New York/London

Bowersox D J, Smykay E W, LaLonde B J (1968) Physical Distribution Management. Logistics Problems in the Firm. Rev. ed. New York

Constantin J A (1966) Principles of Logistics Management. New York

Flaks M (1967) Total Cost Approach to Physical Distribution. In: Marks N E, Taylor R M (Hrsg) Marketing Logistics. Perspectives and Viewpoints. New York/London/Sydney, S. 264-269

Grochla E (1990) Grundlagen der Materialwirtschaft. Das materialwirtschaftliche Optimum im Betrieb. 3., gründl. durchges. Aufl., unveränd. Nachdruck. Wiesbaden

Gudehus T (2010) Logistik: Grundlagen, Strategien, Anwendungen. 4., aktual. Aufl. Berlin u.a.

Hammann P, Palupski R (1997) Stichwort "Bestellpolitik". In: Bloech J, Ihde G B (Hrsg) Vahlens Großes Logistik Lexikon. München, S. 88-90

Hartmann H (2002) Materialwirtschaft. Organisation, Planung, Durchführung, Kontrolle. 8., überarb. und erw. Aufl. Gernsbach

Homburg C (2017) Marketingmanagement. Strategie Instrumente Umsetzung Unternehmensführung. 6., überarb. u. erw. Aufl. Wiesbaden

Inderfurth K (1996) Stichwort "Lagerhaltungsmodelle". In: Kern W, Schröder HH, Weber J (Hrsg) Handwörterbuch der Produktionswirtschaft. 2. Aufl. Stuttgart, Sp. 1024-1037

King W R (1967) Quantitative Analysis for Marketing Management. New York 1967

LaLonde B J, Grashof J F (1969) Computer oriented information systems provide effective P. D. management. In: Handling and Shipping 10 10, S. 53-59

Liesegang D G, Wohlgemuth E (1997) Stichwort "Sicherheitsbestand". In: Bloech J, Ihde G B (Hrsg) Vahlens Großes Logistik Lexikon. München, S. 963

Magee J F, Copacino W F, Rosenfield D B (1985) Modern Logistics Management. Integrating Marketing, Manufacturing, and Physical Distribution. New York u.a.

Pfohl H-Chr (1972) Marketing-Logistik. Gestaltung, Steuerung und Kontrolle des Warenflusses im modernen Markt. Mainz

Pfohl H-Chr (2016) Logistikmanagement. Funktionen und Instrumente. 3., neu bearb. und aktual. Aufl. Berlin/Heidelberg

Pfohl H-Chr, Stölzle W (1997) Planung und Kontrolle. 2., neu bearb. Aufl. München
Schary Ph B (1984) Logistics Decisions. Text and Cases. Chicago u.a.
Schieck A (2008) Internationale Logistik Objekte, Prozesse und Infrastrukturen grenzüberschreitender Güterströme. München u.a.
Schmid O (1977) Modelle zur Quantifizierung der Fehlmengenkosten als Grundlage optimaler Lieferservicestrategien bei temporärer Lieferfähigkeit. Frankfurt a. M./Zürich
Schneeweiß Ch (1997) Stichwort „Lagerhaltungsmodelle". In: Bloech J, Ihde G B (Hrsg) Vahlens Großes Logistik Lexikon. München, S. 488-492
Steinbrüchel M (1971) Die Materialwirtschaft der Unternehmung. Bern/Stuttgart
Stock J R, Lambert D M (2001) Strategic Logistics Management. 4. Aufl. Boston u.a.
Stölzle W, Heusler K F, Karrer M (2004) Erfolgsfaktor Bestandsmanagement: Konzept, Anwendung, Perspektiven. 1. Aufl., Zürich
Thonemann U (2015) Operations Management: Konzepte, Methoden und Anwendungen. 3., aktual. Auflage. Halbergmoos
Vahrenkamp R (2007) Logistik. Management und Strategien. 6., überarbeit. und erweiterte Aufl. München
Zwehl W v, Kramer D (1997a) Stichwort "Lagerhaltungsmodelle, erweiterte". In: Bloech J, Ihde G B (Hrsg) Vahlens Großes Logistik Lexikon. München, S. 492-496
Zwehl W v, Kramer D (1997b) Stichwort "Fehlmengenkosten". In: Bloech J, Ihde G B (Hrsg) Vahlens Großes Logistik Lexikon. München, S. 272-273

Warehouse

6

6.1 Definition and Functions of the Warehouse

Definition

A warehouse is a node in the logistics network where goods are temporarily held or transferred to another route through the network. From Fig. 1.2, which summarizes the basic structures of logistics systems, it can be seen that warehouses can be points of delivery and receipt as well as points of breakup or concentration in the logistics system. Warehouse and movement processes take place in the warehouse. Which processes dominate depends on the function of a warehouse. The function thus strongly determines the location and the technology to be used in the warehouse.

Functions

In Fig. 6.1, a distinction is made between three types of warehouse according to the functions primarily to be fulfilled by a warehouse. The basic distinction between storeroom, transshipment and distribution/supply warehouses[1] does not exclude the possibility that mixed forms may occur in practice.

Storeroom warehouses are usually assigned to the production plant. Their most important function is to provide capacity for receiving goods to be used in production, but also finished goods (e.g. when receiving seasonal production). In storeroom warehouses, the storage processes dominate over the movement processes. They can be procurement warehouses as well as production or distribution warehouses.[2] *Transshipment warehouses* (transit terminals) are intended to hold goods at short notice between transshipment from

[1] For a more comprehensive distinction in terms of warehouse types and functions, see Klaus, 1996, pp. 1013ff.; Schulte, 2013, p. 246.

[2] See Fig. 5.1.

H.-C. Pfohl, *Logistics Systems*,
https://doi.org/10.1007/978-3-662-64349-5_6

Types of warehouse	Most important function	Locations	Stored goods
Storeroom warehouse	High storage capacity	Production-oriented	Material, seasonal semi-finished and finished products
Transshipment warehouse	High handling capacity	Transport-oriented	Material, semi-finished and finished products, merchandise
Distribution warehouse			
-Supply warehouse	High concentration capacity	Procurement-oriented	Material, products Merchandise
-Delivery warehouse	High breakup capacity	Sales-oriented	

Fig. 6.1 Differentiation of warehouse types according to their function

means of transport to means of transport. Such warehouses are mainly found in logistics companies. In the meantime, however, they are also playing an increasingly important role for trading companies. In so-called crossdocking, the manufacturers deliver the goods picked for the individual branches to the central warehouse of the retailer. There, the shipments from different manufacturers for the respective branches are combined and delivered together. In the central warehouse of the retail company, branch-specific picking is thus eliminated and inventories are also eliminated completely or at least to a large extent.[3] Accordingly, the movement processes dominate in the transshipment warehouse, so that it is not the warehouse capacity but the achievement of a high transshipment speed that is to be achieved in its design. In *distribution warehouses*, the flow of goods is changed in its composition. In such warehouses, both storage and movement processes are of equal importance. The most important objective is the capacity to restructure the flow of goods. The distribution function can be either a supply or a delivery function, so that supply and delivery warehouses can be distinguished.[4] *Supply warehouses* are concentration points in logistics systems. They are used to collect goods from different suppliers and distribute them to one or more production sites—in the case of a retail company, to one or more retail sites. *Delivery warehouses* are breakup points in logistics systems. They are where goods from production are collected and delivered to the customer.

A distinction is made between central, regional and local *distribution warehouses* according to the area served by them. *Central* distribution warehouses are generally

[3]On crossdocking, cf. Schulte, 2013, pp. 502ff.

[4]See Fig. 1.6 in Part I.

referred to as central warehouses, while decentralized *regional or local* distribution warehouses are referred to as supply or delivery warehouses in the narrow sense.

A warehouse function is to be described in more detail using the example of the *delivery warehouse in the narrow sense.* The warehouse occupies a key position in the logistics system. This is because the delivery warehouse splits up the flow of goods from the supplier to the customer. It receives the goods in wagon or truckloads and forwards them to the customers in much smaller units according to the orders. Both cost and delivery service considerations can speak for the establishment of a delivery warehouse.

If we compare the transport costs per unit of goods when transporting goods in full wagon or truck loads (carload transport) and when transporting smaller quantities (less-than-carload transport), the transport costs are considerably lower in the first case. Figure 6.2 shows the development of transport costs per unit of goods as a function of distance. In practice, the transport costs per unit of goods will not follow such an even course, since they depend not only on the transport distance, but also on other factors. In principle, however, the transportation cost curve that increases degressively with distance is correct.[5] When a unit of goods is delivered to a customer from a delivery warehouse, the additional costs incurred in the delivery warehouse and the transport costs from the delivery warehouse to the customer in less-than-carload transport are incurred in addition to the transport costs from the factory to the delivery warehouse in carload transport.

A comparison of these costs with the transport costs of direct delivery to the customer in less-than-carload transport shows that the area between A and B can be supplied more cheaply from the delivery warehouse. From the one-dimensional representation it is already apparent that the delivery warehouse needs by no means be located at the center of a circular sales area, as is often assumed. "The best middle location between two or more points of provision or use is usually not at their geographical center, but at one of the two main points of emergence."[6] This is referred to as the Hoover effect. The reason for this is the fact that transport costs always increase degressively depending on the distance. If the expanse of the area to be supplied is determined with the help of the specified transport cost curve, it becomes apparent that this area is not circular but rather drop-shaped. The wide, flattened part of this area, where the delivery warehouse is located, faces the factory warehouse. The narrow, pointed end of the area faces in the opposite direction furthest away from the delivery warehouse. Since costs are also incurred in the factory warehouse that are to be taken into account for direct delivery to the customer, the system does not calculate the total costs incurred in the delivery warehouse, but only the additional warehouse costs incurred in comparison to direct delivery. This correction for the additional costs incurred makes the transportation costs comparable.

---

[5] For an empirically determined transport cost trend, see Magee et al., 1985, p. 253. See also Fig. 3.7 in Part I.

[6] Klaus, 2005.

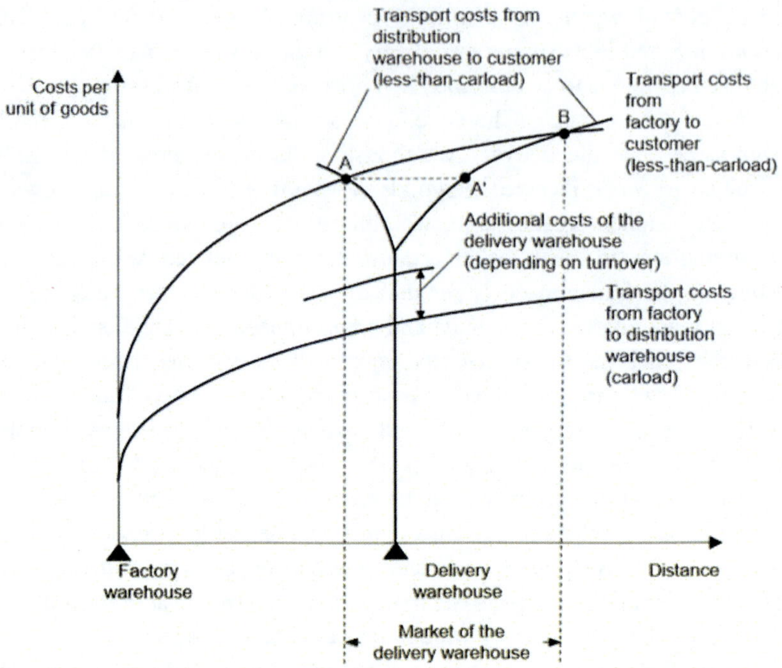

Fig. 6.2 Cost reduction in the supply of an area by a delivery warehouse (Source: Cf. Frederick, 1957, p. 81; Bowersox/Closs, 1996, p. 503)

Figure 6.2 also shows that the establishment of a delivery warehouse enables a company to compete on price even in distant submarkets which it was previously unable to supply because its prices were too high due to the transport costs incurred. For example, if customers will not accept a higher price than the one at point A and if a price reduction can only be achieved by lowering transport costs, the sales area can be enlarged by using the delivery warehouse up to point A′.

However, the cost advantages of setting up a delivery warehouse arise only if demand is sufficiently high in the submarket it is to supply. The lower the quantity of goods handled, the higher the costs per unit of goods incurred in the delivery warehouse. The establishment of a delivery warehouse is therefore only worthwhile if demand is so high that the costs incurred in the delivery warehouse do not compensate for the reduction in transport costs per unit of goods.

When deciding whether to set up a delivery warehouse, it is important to consider not only its impact on costs but also, and above all, on demand. Often the establishment of a delivery warehouse is the only way to be able to supply a distant submarket faster and thus shorten the delivery time. By increasing delivery service in this way, it is possible to offset a competitive advantage of competitors who previously had a more convenient location. The delivery warehouse is therefore an important instrument in the battle for market share.

6.2 Warehouse Tasks

Warehouse Location

Decisive for the functional performance of a warehouse are its functionally appropriate location and its functionally appropriate operation. The warehouse tasks therefore first include the *choice of location*, which must be made at two stages. In the case of *interlocal* site selection, a decision is made as to where in an economic area a warehouse is to be built. Once a decision has been made, e.g. for a municipality, it must then be determined in the *local* site selection where the warehouse is to be located.

The interlocal location of a warehouse depends on the primary function to be performed by the warehouse. Apart from cases in which warehouses are located in a production-oriented location and are generally directly linked to the production plant as factory warehouses, the interlocal choice of location of a warehouse is influenced to a far greater extent by logistical considerations than the choice of location of a production plant.[7] As the function of the warehouse has already been explained in more detail using the example of the delivery warehouse, the factors influencing the choice of location will also be explained using this type of warehouse as an example. The following six influencing factors are important for the choice of location of a delivery warehouse:

- Delivery service: The delivery warehouse should ensure that customers are supplied quickly. What delivery service do customers expect in this respect and what delivery service does the competition offer? An answer to this question already gives an idea of where a delivery warehouse should be set up in the first place and which area can be supplied by it in order to achieve the required delivery service.
- Type of sales area: Is demand concentrated at certain points in the sales area or is it evenly distributed over the entire sales area? Is the sales area divided by geographical barriers (mountain ranges, lakes or national borders) which more or less define the area to be supplied by a delivery warehouse?
- Development of demand: The choice of location must be made on the basis of a forecast of the future development of demand. It must be examined how the level of demand and its distribution will change as a result of shifts in demand in the sales area.
- Transport connections: What transport links (motorway, railway, airfield, waterway) are necessary to supply the delivery warehouse and to supply customers from the distribution delivery and where are such transport links available?
- Transport and warehouse costs: What transport and warehouse costs are incurred at different delivery warehouse locations? What are the freight rates of the individual means of transport at different locations? What are the effects on the utilization of the own vehicle fleet?
- Labor force: What are the differences in labor supply in different possible locations?

[7] Cf. Heskett et al., 1973, p. 415.

When comparing different plots of land available for the construction of a delivery warehouse, other criteria besides the price of the land must be taken into account. The distribution centre does not have to be located in an industrial zone, but can be built completely on its own outside of a built-up area. However, it is necessary to take into account the conditions imposed by the regional planning authorities and the local building inspectorate on the plots of land available for selection, as well as the condition of the subsoil, the access to the site and the energy supply. The smooth transport of goods to and from the site must be ensured by good access and exit facilities and, if necessary, by a railway siding. Care must be taken to ensure that the main traffic arteries can be easily reached from the site. Inconvenient access for the workforce will have a detrimental effect on the procurement of labor. Consideration should also be given to whether, for example, the close location of a site to a busy motor road will enable the warehouse building to be used as an advertising space. The shape of the lot must allow for easy maneuvering of trucks in front of the loading docks. It must also offer the possibility for a later extension of the delivery warehouse.

Warehouse Operations

The storage and movement processes necessary for the functional performance of a warehouse require the performance of a number of warehouse tasks, which can be subdivided according to the warehouse areas indicated in Fig. 6.3.[8]

In the warehouse area *incoming goods*, the tasks of receiving goods from the supplier and preparing the goods for storage are performed. This includes in detail: the unloading of the incoming goods, the identification of the incoming goods, the incoming goods inspection as well as the preparation of the goods for storage (e.g. reloading onto the correct loading equipment or repacking). The *movement processes* are at the forefront of task fulfillment. The time the goods spend in the incoming goods area should be as short as possible.

In contrast to incoming goods, in the *unit storage* area the *storage processes* are in the foreground. Unit storage is used exclusively to bridge the time between goods that are placed into and removed from storage in the same unit. Because of the resulting high degree of uniformity of activities, unit stores are highly automatable. In addition, since the movement processes are very much in the background, extreme space utilization can be achieved. If the units do not go directly to outgoing goods after disbursement, but to a picking store, unit stores are also referred to as reserve stores, in which the goods are stored in large quantities and units for a relatively long time.

In the *picking* warehouse area (seize store, work store), goods are stored in small quantities and units for only a short time. Primarily in this warehouse area, movement processes take place that serve to concentrate or dissolve the flow of goods.[9] The term

[8] Cf. for the following Bowersox/Closs, 1996, p. 397; Stock/Lambert, 2001, pp. 396ff.; Arnold et al., 2008, pp. 378ff.

[9] See Fig. 1.2 in Part I.

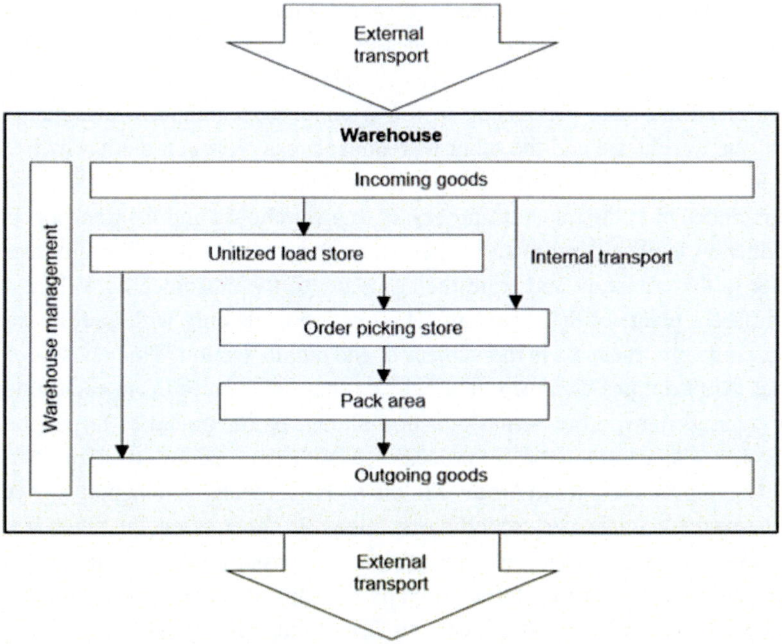

Fig. 6.3 Warehouse areas (Source: Adapted from National Council of Physical Distribution Management, 1984, p. 190; Bowersox/Closs, 1996, p. 397)

picking has become established to describe these processes. As a result of picking, the goods do not leave this warehouse area in the same condition in which they were put into storage. Even today, picking is still a manual process in many cases. In order to nevertheless keep picking times as short as possible, attention must be paid to time-saving picking processes and short transport routes when designing this warehouse area.

As a result of the different dominance of storage and movement processes in the unit and picking store, both warehouse areas must therefore be designed differently. In cases where a separation between the two warehouse areas does not appear to make sense, the goals of space utilization and short picking times should of course be pursued with equal weighting.

In the warehouse area of *packing*, the picked order is assembled into a unit ready for shipping, whereby shipping also means transport to other internal locations. The packaging tasks are dealt with in the next section in the logistics subsystem Packaging.

The *outgoing goods* warehouse area comprises the tasks of delivering goods to the recipient as well as the associated preparatory work. This includes the receipt of goods from the packing department, interim storage according to customer or shipping type until collection and the scheduling of the collecting means of transport as well as loading. As in incoming goods, the focus here is also on *movement processes*. A longer time bridging of the goods in the outgoing goods takes place only in special cases.

The warehouse area *warehouse management* comprises the tasks of controlling and coordinating the storage and movement processes in the other warehouse areas. Warehouse management forms the interface to the logistics subsystem order processing. If we transfer terms from electronic data processing to the warehouse, warehouse management is the software of the warehouse and the other warehouse areas represent the hardware.

In Fig. 6.3, a distinction is made between internal and external transport. *Intra-company transport* includes both the transport processes in a warehouse and the transport processes in the production facility. Non-plant *transport*, on the other hand, includes transport from the supplier to the company and from the company to the customer, as well as transport between different plants of the company. This section deals only with internal transportation. Extra-company transport is the subject of the eighth section. This does not mean, of course, that the two types of transport should be seen in isolation from each other. When designing internal transport, external transport should be integrated as far as possible. In order to avoid storage and movement processes, which cause additional logistics costs, the aim should be to produce directly from or to the external means of transport. However, this will only be possible under very specific conditions. In other cases, the internal transport, which has to cover much shorter distances than the external transport, must be carried out using special means of transport. Intra-company[10] means of transport are also referred to as means of *conveyance*, whereby this term can also include the means used to carry out the handling processes. When transport or conveying means or equipment are referred to in the following, all means for overcoming horizontal and vertical distances are meant.

Before giving an overview of the storage and transport equipment that can be used to rationalize the storage and movement processes, the storage bin assignment should be discussed. The assignment of storage bins to goods essentially determines the possible use of storage and transport equipment.

6.3 Storage Bin Assignment

The problem of storage bin allocation—also dealt with in the literature under the keywords warehouse organization or warehouse arrangement—can be understood as the third stage of the warehouse location problem. It concerns the determination of the storage locations for the goods to be stored in the warehouse.

Unit and Picking Warehouse

The *storage location of* the goods determines the length of the route that has to be covered when transporting these goods in the warehouse and thus the transport costs as well as the time required for the transport operations. For a delivery warehouse, for example, it is typical that the goods are received in large storage or transport units and returned to the

[10] Cf. ten Hompel et al., 2007, pp. 122ff.

customer in smaller units. The storage and transport units therefore become smaller as the goods flow through the warehouse. The more the flow of goods is split up in this way, the greater the costs associated with it. Therefore, the transport operations are of much greater importance when a large number of small units are removed from storage than when a relatively small number of large units are placed in storage. The shorter the distances to be covered during picking, the lower the transportation costs incurred in the process and the faster the order can be assembled and delivered. It is therefore desirable to accommodate all stored goods in the smallest possible storage space, which is as close as possible to the picking location and the outgoing goods area in the warehouse. Another argument against too large storage space is the big investment involved for the storage and transport equipment to be used in picking. The desire to store goods in a small space and to bring them as close as possible to outgoing goods is countered by the need to maintain larger stocks. For this reason, it often makes sense to distinguish between the picking store and unit store in the sense of a reserve store in the warehouse.[11]

The incoming goods are stored in large quantities and units in the unit store. The picking store is located between the unit store and the outgoing goods, whereby it is also conceivable that the unit store encloses the picking store on three sides.[12] From the unit store is replenished the picking store, where the goods are only stored in relatively small quantities so that the entire assortment can be stored in a relatively small space.[13] The goods are seized in this picking store in small quantities during picking.

However, the separation of the warehouse area into a unit store and a picking area also has the major disadvantage that a transfer process must be switched between the storage and disbursement processes of the goods. The more frequently the stored goods have to be moved, the higher the costs. This disadvantage is particularly significant in the case of large assortments, so that it is often no longer practical to set up a separate picking store there.[14]

The decision for or against a separation of unit and picking store generally concerns the storage area in the warehouse to be provided for the entire assortment; the exact storage location of the individual goods must then still be determined. Various influencing factors can be specified for this, the most important of which are listed below.

Factors Influencing Storage Bin Assignment

Figure 6.4 shows the basic options for storage location assignment. However, there is only a choice between them insofar as the material properties of the goods to be stored (e.g. odor or climate sensitivity) do not restrict the free choice of storage location in the warehouse. Here, the fixed storage bin arrangement and the completely free storage bin assignment

[11] See Fig. 6.3.

[12] Cf. Gudehus, 2010, pp. 632 f. and pp. 883ff.

[13] For the quantity of a good to be stored in the picking warehouse, see Heskett and others, 1973, p. 631.

[14] Cf. Gudehus, 2010, p. 724.

Term	Description	Effect	Requirement
Fixed assignment of storage space	Each article is assigned a fixed storage space	Access security in case of loss of inventory master file; separation of commodity groups	None
Transverse assignment of storage space	Several standard loads of an article are distributed among various aisles	Access security in case of failure of a rack serving unit	Warehouse management system or shelf file; organisational separation of aisles
Completely free assignment of storage space (single bin storage, chaotic storage)	Standard loads are taken into stock in random shelves	Increased utilisation of storage capacity	Warehouse management system or shelf file
Free assignment of storage space within a fixed area	Standard loads are taken into stock freely only within fixed areas	Separation of commodity groups; reduction of capacity utilisation compared to completely free assignment of storage space	Warehouse management system or shelf file

Fig. 6.4 Possibilities of storage bin allocation (Source: Eggenstein et al., 1981, p. 259)

(single bin storage, chaotic storage) are opposed to each other. The other two options are merely variations of the fixed and free storage bin assignment.

Various influencing factors can be named for the *fixed storage bin assignment*, such as route length, turnover frequency, value, weight, access frequency, volume and dimensions of the goods to be stored,[15] but also the volume of the unit in which the goods are sold to the customer. Two influencing factors, namely the access frequency and the volume of the sales unit, which can also be combined with each other, will be described below as examples[16]:

- Access frequency (withdrawal or order frequency): When storing the goods according to this influencing factor, it is assumed that the handling costs related to the order quantity of the customers are independent of the storage location. Thus, the only thing that changes with the storage location is the transportation cost, which depends on the distance to be covered and the frequency with which this distance must be covered. Therefore, the more frequently they have to be accessed the closer the goods are to the picking location.
- Volume of the sales unit of a good: When storing goods according to the volume of the unit in which the good is sold, the items with a large volume per sales unit are stored

[15] Cf. ten Hompel/Schmidt, 2008, pp. 31ff.

[16] Cf. Heskett et al., 1973, pp. 626ff.

away from the picking location and the items with a small volume per sales unit are stored close to the picking location. The aim is to store as large a percentage as possible of the sales units to be stored as close to the picking location as possible in order to shorten the distances that have to be covered in order to transport the sales units. Storage according to this factor also avoids goods with different dimensions being stored next to each other. The advantage of this is that goods stored next to each other generally place the same demands on the technical aids for their storage and disbursement and on the space required to serve the storage bins.

- Volume-per-order index: This index combines the volume of the sales unit of a good with the access frequency. It can be calculated using the following information: (1) the sales unit volume of a good, (2) the average number of sales units ordered per order, (3) the number of orders per day, (4) the number of daily deliveries (demand per day) to be stored in the picking area. From this information, it is possible to calculate the volume required in the picking area to store each good. Dividing this volume by the access frequency per day gives the volume-per-order index of an item in the form: volume required in the picking warehouse/access frequency per day.

The index therefore indicates how much space is required in the picking store for an item per order. The lower the index value for an item, i.e. the smaller the volume and the greater the access frequency per day, the closer to the picking location in the warehouse the item is stored. This ensures that, taking into account the influence of the access frequency and the space required to store an item, the transport costs for picking, which change with the storage location, are minimized.

If there is no storage bin assignment defined according to the type of goods, this is referred to as *free bin assignment*, single bin storage or chaotic storage. In this case, any good can be stored in any free space. The aim is to make optimum use of the storage space. This problem arises particularly in the case of strongly fluctuating demand. Since the goods are stored in randomly free storage locations in chaotic storage, it requires the use of electronic data processing to control and monitor storage and disbursement when there are a large number of storage locations. The electronic data processing system takes over the task of assigning a storage bin of the required size to a good to be stored. It registers which goods, in which quantities, are stored in which storage location. During picking, the picking personnel is partially automatically guided to the storage bin where the required quantity of the good can be picked. Thus, this type of storage bin assignment is often combined with a specific technique in the warehouse.

6.4 Technology in the Warehouse

Technical Storage Systems

The technical storage and transport equipment that can be used in the warehouse depends on the technical storage system to be implemented in a warehouse. Technical storage systems differ first of all according to the type of stored goods, which are broken down into piece goods, loose goods, gases and liquids[17]:

- Piece goods are items which can be treated as a unit during transport, handling and storage and which do not change their shape and form during these operations, or only slightly. Piece goods include solid bodies of the most varied dimensions (e.g. components or semi-finished products), storage units consisting of individual piece goods (e.g. loaded pallets or stacks of goods shrink-wrapped without pallets), loose goods, liquids or gases contained in containers (e.g. drums or bags) and also packaging as empties (e.g. containers or drums).
- Loose goods change their shape during the movement processes. These are loose goods in pourable form, which include the most diverse granular and dusty goods such as ores, coal, sand, cement, grain, coffee etc.
- Gases and liquids are goods for which, during transport, handling and storage, material properties must be taken into account which require special knowledge of chemistry and process engineering.

In the context of this business management-oriented book, technical storage systems for loose goods as well as liquids and gases will not be discussed. Only a brief overview of technical storage systems for piece goods is given, for which a schematic diagram can be found in Fig. 6.5.

In technical storage systems, a distinction is made between the static system for storage and the dynamic system for the movement of goods.[18] The simplest form of *static storage* is *ground storage without a storage rack. Ground storage in the narrow sense is* used when there is no stacking. This leads to poor space utilization, so that high investments per storage unit can arise despite the lack of investment in storage equipment. The disadvantage of poor space utilization is avoided by *block storage*, in which the stored goods are placed without gaps on top of, next to and behind each other. In this case, however, direct access to any of the stored goods is no longer possible. This is also no longer the case with *row (line) storage*, when the stored goods are not only placed next to each other but also on top of each other. Two to four stacks with heights of normally approx. 5 m, but in exceptional cases also up to 10 m, are usual for stack formation.

Static rack storage systems are technically more advanced, as they permit a greater storage height and allow better use of space even for storage units that cannot in principle be stacked on top of one another. In addition, in the form of rack storage, the possibility of access to any storage unit is given.

In the *dynamic storage system*, the stored goods are moved. On the one hand, movement processes take place in order to move the load units for storage (loading) and disbursement

[17] Cf. Appelt, 1997, p. 482.

[18] Cf. Jünemann/Schmidt, 2000, p. 45.

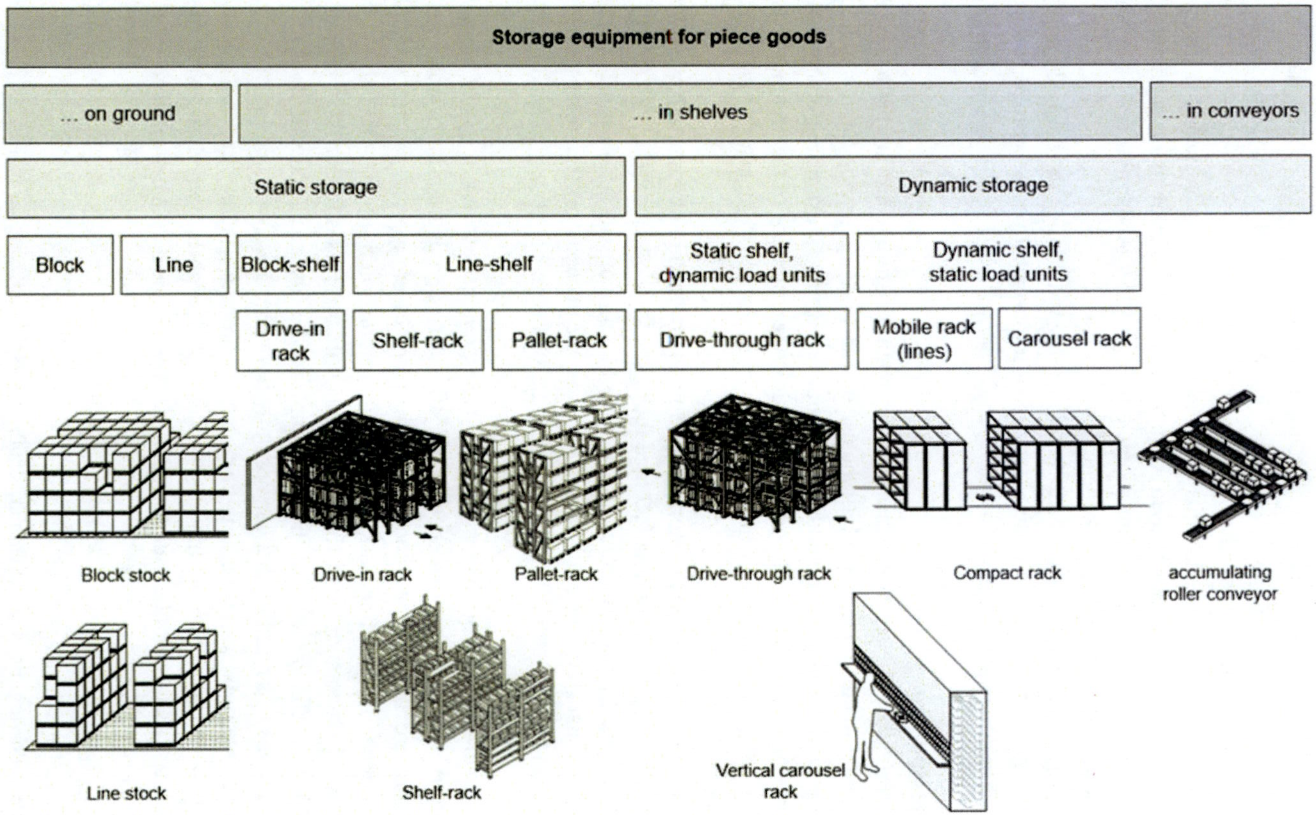

Fig. 6.5 Schematic diagram of the main technical storage systems for general cargo (Source: ten Hompel et al., 2007, p. 56ff)

(removal). In this case, storage and movement processes are separated (Cf. Fig. 6.5, dynamic rack storage). On the other hand, it is possible that storage and movement processes cannot be clearly separated. For example, movement processes take place during actual storage when the stored goods are in the storage rack or when the stored goods and the storage rack are moved together. Storage processes during the actual movement processes take place when the load units are constantly circulating on means of transport. This type of dynamic storage can be used when the upstream and downstream points in the flow of goods do not coincide in terms of cycle time and the goods involved have a high quantity volume and high turnover frequency.[19]

In order to give an impression of the technical variety with which the basically possible technical storage systems can be realized, the most important types of storage buildings, storage racks and means of transport are compared below.[20]

Warehouse Building

Outdoor warehouse: The structural measures for open-air storage are limited to securing the floor in order to create the subsoil for the use of transport equipment. Such warehouses are primarily suitable for storage goods that are not sensitive to weathering; otherwise, the storage goods must be additionally protected against the influence of the weather (e.g. by shrink-wrapping or covering with tarpaulins). In most cases, storage takes the form of ground storage.

Flat warehouse: A flat store is a storage building up to a height of approx. 7 m, in which storage is carried out with or without storage racks.

Multi-storey warehouse: A multi-level storage system consists of several floors. The transition from a flat store to a multi-storey store may be necessary if the storage area is to be increased on small plots of land. As a rule, the floors are connected to each other via elevators.

High-bay warehouse: High-bay warehouses are warehouses with a height of more than 12 m. Up to this height, some warehouses are still referred to as high-flat warehouses. In practice, there are high-bay warehouses up to a height of approx. 55 m.[21] High-bay warehouses either have a fixed structure (concrete construction) in which the storage racks are placed free-standing, or the storage racks themselves are used as a supporting structure for the walls and roof of the warehouse. In the first case, the warehouse is, for accounting purposes, a building that can normally be depreciated over a period of 50 years. In the second case, it is, for accounting purposes, a warehouse facility that can usually be depreciated over a 15-year period. In both cases, however, high-bay warehouses are single-purpose facilities that cannot be used for other purposes. This distinguishes them from flat

[19] Cf. ten Hompel et al., 2007, pp. 80ff.

[20] Cf. for the following Jünemann/Schmidt, 2000, pp. 41ff.; ten Hompel et al., 2007, pp. 55ff. Numerous illustrations can also be found there.

[21] Cf. ten Hompel et al., 2007, p. 67.

or multi-storey warehouses, which can also be used as production or office space if they are suitably constructed with the aid of minor conversion measures.

Airhall warehouse: This type of warehouse is created when a hall skin consisting of air-impermeable fabric is stretched like a balloon over a fixed base area by means of a fan. Access is via air locks. Since such warehouses can be set up and dismantled quickly, they are particularly suitable as temporary storage facilities.

Storage Racks

Shelf racks: They consist of lateral supports between which continuous shelves are fastened, whereby the shelf height depends on the stored goods. These racks are mainly used to store small parts. The racks are mainly operated manually. Typically, such racks are found in flat or multi-storey warehouses.

Pallet racks: These racks do not have shelves, but only consist of side uprights with cross or longitudinal beams to support the palletized stored goods. They are usually operated with the aid of forklift trucks, which requires appropriate aisle widths.

Drive-in racks: They are a special case of pallet racks and are used for block storage with storage rack. Since these racks only have longitudinal cross beams, the palletized goods can not only be stored above and next to each other, but also behind each other.

Drive-through racks: These are dynamic stores in which the stored goods are moved continuously or discontinuously from the loading to the unloading side. As a rule, palletized goods are moved either on inclined roller conveyors or on inclined tracks with roller pallets by gravity or with mechanical drive on non-inclined roller conveyors. With these storage racks, the FIFO principle (first in first out) is inevitably guaranteed and the degree of space utilization is relatively high. However, live storage racks require relatively large investments.

Compact racks: Compact racking includes mobile racking and carousels. Mobile racking is either compact shelving or pallet racking arranged side by side that can be pulled out into the aisle or moved parallel to it. In this way, only one or two racks are directly accessible at any one time. In carousel racking, racking units circulate horizontally or vertically. Such chain-driven circulating racks are typically used where the "goods come to the man" during order picking and not vice versa. Compact racks, in which the stored goods and racks are moved together, ensure a very high degree of space utilisation, but access is not particularly good due to the technically required high manipulation times. Therefore, these racks are only suitable for goods with low turnover frequencies. In addition, the investment corresponds to the comparatively high technical expenditure.

High-bay racks: These are shelf or pallet racks with very high shelving heights, which are loaded by means of high-tech storage and disbursement machines. Due to the large rack heights and the narrow service aisles, there is a comparatively good degree of space utilization. The access times are very short due to the generally high degree of mechanisation. High-bay racking is used primarily for large quantities of goods and wide product ranges, although it must be borne in mind that only limited adjustments can be made in the event of operational changes.

Means of Transport

Continuous conveyors: These include all means of transport with which goods are moved continuously in a horizontal, vertical or inclined direction along a fixed and usually constant path. Continuous conveyors have low energy requirements, low operating costs and high operational reliability. Continuous conveyors can be designed either overhead (load is carried by the ceiling structure of the warehouse) or on floor (load is carried by the floor structure of the warehouse). As a rule, continuous conveyors are designed as conveyor belts, drag chains, roller conveyors or underfloor drag chain conveyors. Circulating chains (circular conveyors) or swing conveyors with fixed hangers or replaceable swings (power and free), on the other hand, are often designed overhead.

Discontinuous conveyors: These include all means of transport that operate in intermittent mode. Non-continuous conveyors with predominantly vertical transport tasks are referred to as elevating conveyors. Those with predominantly horizontal transport tasks are referred to as industrial trucks, which in turn are subdivided into track-bound and trackless industrial trucks. Track-bound industrial trucks include, for example, mine hoists for underground mining. Non-railed industrial trucks include tractors and trailer trucks that do not have a lifting device as well as industrial trucks with a lifting device, which include in particular the most diverse types of forklift trucks. In contrast to tractors and trailers, vertical movement is always in the foreground in the case of lifting conveyors or hoists, although they can of course also perform horizontal movements. Elevating conveyors include the trolley, overhead cranes, gantry cranes and jib cranes, which are also known as slewing cranes if they can perform a rotary motion. In addition, elevating conveyors also include the elevator.

Storage and disbursement machines: For storage and disbursement, in particular of high-bay warehouses, special means of transport have been developed for working in narrow aisles. In principle, these are positively driven, but can be rack-independent or rack-dependent. For example, high-bay stackers and order pickers work independently of the racking and can be moved both inside and outside the racking aisle. Both have the forks generally transverse to the direction of travel to work in the narrow aisles, with the order picker truck also having a liftable operator station to provide a ride for the warehouse worker. Storage and disbursement machines that are rack-dependent and thus can no longer serve multiple rack aisles are referred to as rack conveyors. Sufficiently long rack aisles and appropriate rack heights are required to ensure that rack conveyors are sufficiently utilized. Rack conveyors are generally operated automatically. As a rule, they are connected via computer systems to the operational company software, which transmits the appropriate instructions to the rack control system.

Devices for Order Picking

Depending on the picking principle, very different devices are used for picking. According to the "*man to goods*" principle, the order picker moves to the staging areas in the order picking warehouse and picks the goods manually. Special means of transport are used here, such as horizontal and vertical order picking vehicles as well as order picking three-way stackers, shelf trucks and overhead conveyors. A number of guidelines must be observed when designing the racking so that, for example, it is ensured that even smaller employees can reach the higher storage locations without additional aids. Furthermore, storage compartments, for example, should only be so deep that the order picker can still easily reach the goods stored in the compartments further back.[22] In order to facilitate the work of the order picker, technologies such as pick-by-light (signal lamps at the storage compartments), pick-by-voice (voice communication) or pick-by-vision (data glasses) are often used for manual processes.

According to the "*goods to man*" principle, the fully automatically controlled pallet or tote stacker cranes remove the load units from the storage location and make them available to the pickers.[23] After manual picking, the storage and disbursement machines transport the load unit back to the storage location. For the pharmaceutical industry, for example, there are special automated chutes and picking robots as devices for automatic picking.[24]

Figure 6.6 provides an overview of the elements of a warehouse shown. Material flow and conveyor technology is also referred to as intralogistics in German-speaking countries.[25] "Intralogistics encompasses the organization, control, realization and optimization of internal material flows, information flows and goods handling with the aid of technical systems and services."[26] With the help of new technologies in sensor technology—technical systems for identification, location and condition monitoring—the warehouse can be developed into an *intelligent logistics space.* Two basic approaches can be defined for this: "On the one hand, the creation of intelligent infrastructures by locally equipping them with appropriate technologies and, alternatively, on the other hand, the definition of mobile, intelligent objects by equipping goods and operating resources with sensor technology."[27]

Automation in the Warehouse

The automation of processes in the warehouse can be divided into *two problem areas.* These are information processing and the handling of stored goods. Although both areas are inextricably linked and mutually influence each other, technically different problems exist.

In the *information processing* for the control and monitoring of the static and dynamic warehouse systems, there is primarily a challenge of the information technology linking of

---

[22] Cf. Pfohl, 2007, p. 795.

[23] Cf. Arnold et al., 2008, p. 683.

[24] Cf. Arnold et al., 2008, pp. 684 f. For further information on picking systems, see ten Hompel et al. 2007, pp. 251ff.

[25] Since 2003, the term intralogistics has become established in the German-speaking world, displacing the classic term material flow and conveyor technology.

[26] Arnold, 2006, p. 1.

[27] Richter et al., 2015, p. 247. On the intelligent picking workstation, see Elkmann et al. 2015.

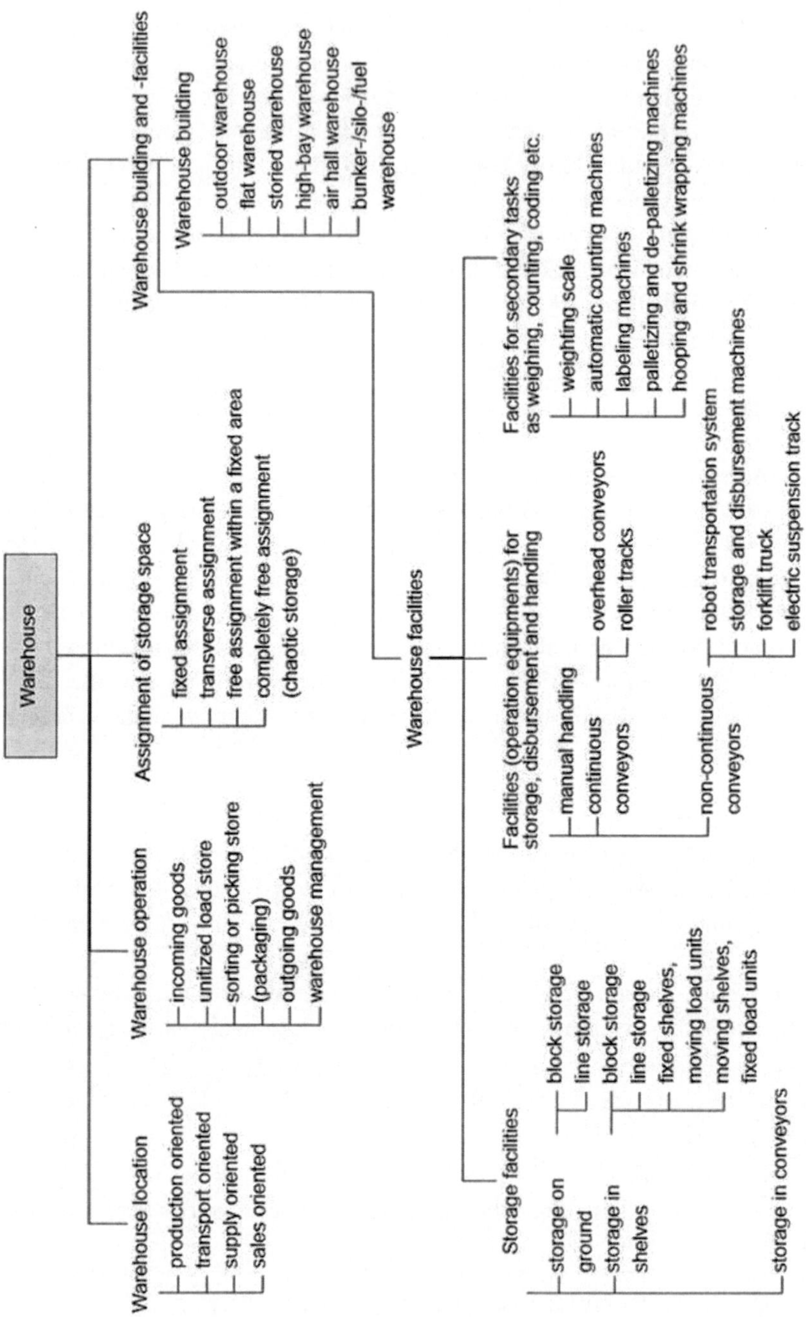

Fig. 6.6 Elements of a warehouse

the various systems and warehouse areas. Therefore, problems of physical linkage and standardization of data formats must be solved in order to realize an uninterrupted and automatic data flow.

Handling of stored goods includes all movement processes for storage and disbursement of goods as well as their transport between different warehouse areas. The possibilities for automating the storage and retrieval processes depend on the one hand on the homogeneity of the load units, e.g. goods can be stored exclusively on pallets or they can be stored in different storage units such as sacks, barrels, crates, etc.. Secondly, the possibilities for automation also depend on the picking requirements. In the simplest case, only homogeneous load units, e.g. pallets, are combined to form consignments, so that automation of the processes is comparatively easy to implement. Examples of this are modern high-bay warehouses that handle all movement processes fully automatically. In complicated cases, however, heterogeneous load units have to be combined or even large load units have to be split up and reassembled, e.g. the distribution of fully loaded pallets to partial deliveries to different stores. In these cases, it is much more difficult to automate the handling of stored goods. Often, the handling of stored goods is then only partially automated, e.g. with goods-to-man picking and pick-by-light support,[28] or even completely manual (cf. Fig. 6.7).

In contrast, there are no significant differences between the transport of large load units and smaller quantity units when it comes to the pure transport of the stored goods between the warehouse areas and their delivery to the picking location. The smaller quantity units are transported in coded collecting containers, so that, for example, the same automation requirements arise for transport on continuous conveyors as for large load units.[29]

When linking the various automation subsystems, challenges exist not only in the standardization of communication with regard to data formats, but also in the physical linking of these systems via *technical interfaces*. At such interfaces, different technical zones, such as computer control and the means of transport to be controlled, meet. It is often at these interfaces that technical malfunctions occur that can affect the entire automated warehouse operation, because without functioning interfaces the various subsystems can no longer be coordinated with each other.

The risk of malfunctions at these interfaces or the effects of malfunctions can be reduced if, when introducing automation, one company assumes responsibility for the entire plant and not just for one specific technical zone at a time. Furthermore, in addition to the computer-based control and monitoring of the operating equipment, a central, manually operated control and monitoring system can also be set up. For example, in the event of a computer failure, operations can be partially maintained.

A prerequisite for a trouble-free, automated warehouse is, last but not least, packaging of the stored goods that meets the high requirements for automatic goods handling.

---

[28] See *devices for picking in* the previous section.

[29] For the basic possibilities of automating order picking, see ten Hompel et al., 2007, pp. 119ff.

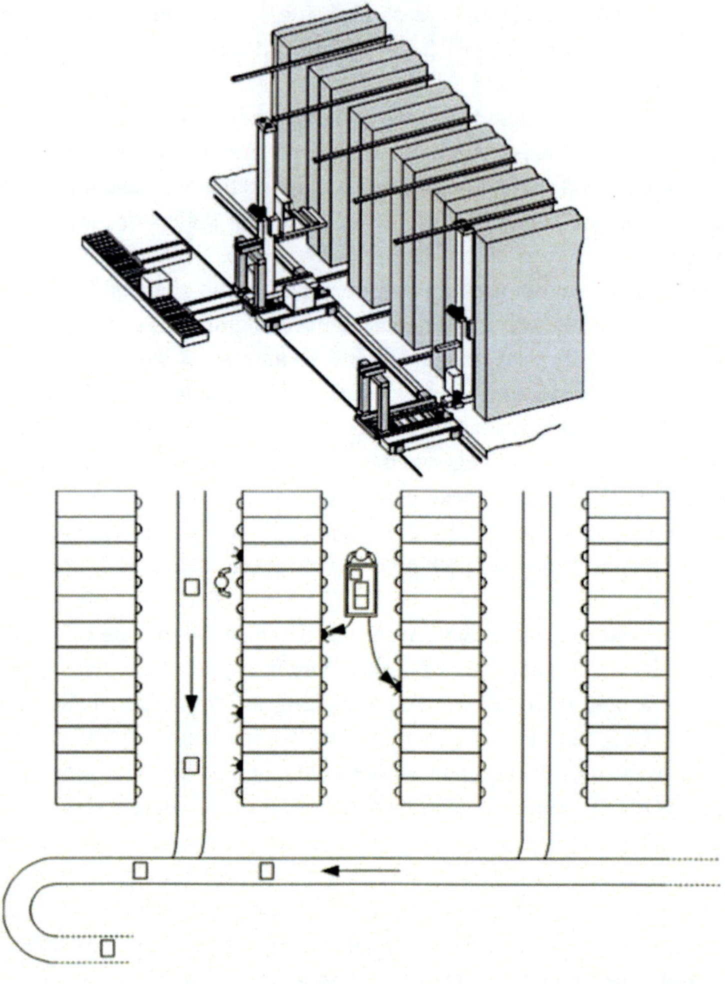

Fig. 6.7 Above is an example of a fully automated warehouse with stacker cranes and converter. Below, the process is shown for partially automated picking by pick-by-light or put-to-light. (Source: Based on ten Hompel et al., 2007, p. 194 and p. 294)

References

Appelt G (1997) Stichwort "Lagergut". In: Bloech J, Ihde G B (Hrsg) Vahlens Großes Logistik Lexikon. München, S. 481-482

Arnold D (2006) Einleitung des Herausgebers. In: Arnold D (2006) Intralogistik. Potentiale, Perspektiven, Prognosen. Berlin/Heidelberg, S. 1-4

Arnold D u.a. (Hrsg) (2008) Handbuch Logistik. 3., neu bearb. Aufl. Berlin/Heidelberg 2008

Bowersox D J, Closs D J (1996) Logistical Management. Überarb. Ausg. der 3. Aufl. New York u.a.

Eggenstein F, Herbst D, Jansen R (1981) Materialfluß und Steuerung. In: Volk P (Hrsg) Betriebsleiter Handbuch. 5., völlig neu bearb. u. erw. Aufl. Landsberg a. L., S. 159-330

Elkmann u.a. (2015) Arbeitssysteme der Zukunft. In: Schenk M (Hrsg) Produktion und Logistik mit Zukunft. Digital Engineering and Operation. Berlin/Heidelberg, S. 245-281

Frederick J H (1957) Using Public Warehouses. Philadelphia

Gudehus T (2010) Logistik: Grundlagen, Strategien, Anwendungen. 4., aktual. Aufl. Berlin u.a.

Heskett J L, Glaskowsky N A, Ivie R M (1973) Business Logistics. Physical Distribution and Materials Management. 2. Aufl. New York

Jünemann R, Schmidt T (2000) Materialflußsysteme. Systemtechnische Grundlagen. 2., Aufl. Berlin/Heidelberg/New York

Klaus P (1996) Stichwort "Lager: Funktionen und Arten". In: Kern W, Schröder H-H, Weber J (Hrsg) Handwörterbuch der Produktionswirtschaft. 2. Aufl. Stuttgart, Sp. 1019-1024

Klaus P (2005) Ringen um Investitionen. DVZ 59 (2005), 22.02.2005

Magee J F, Copacino W F, Rosenfield D B (1985) Modern Logistics Management. Integrating Marketing, Manufacturing, and Physical Distribution. New York u.a.

National Council of Physical Distribution Management (NCPDM) (1984) Measuring an Improving Productivity in Physical Distribution 1984. Oak Brook, IL

Pfohl H-Chr (2007) Logistikarbeitsplatz. In: Landau K (Hrsg) Lexikon Arbeitsgestaltung. Best Practice im Arbeitsprozess. Stuttgart, S. 795-797

Richter K u.a. (2015) Logistiksysteme. In: Schenk M (Hrsg) Produktion und Logistik mit Zukunft. Digital Engineering and Operation. Berlin/Heidelberg, S. 245-281

Schulte C (2013) Logistik. Wege zur Optimierung des Material- und Informationsflusses. 6., überarb. und erw. Aufl. München 2013

Stock J R, Lambert D M (2001) Strategic Logistics Management. 4. Aufl. Boston u.a.

ten Hompel M, Schmidt T (2008) Warehouse-Management. Organisation und Steuerung von Lager- und Kommissioniersystemen. 3. Aufl., Berlin u.a.

ten Hompel M, Schmidt T, Nagel L (2007) Materialflusssysteme. Forder- und Lagertechnik. 3., völlig neu bearb. Aufl. Berlin u.a

7 Packaging

7.1 Definition and Functions of Packaging

Definition

Packaging is understood to be the detachable, complete or partial wrapping of a good (packaged good) in order to protect it or to fulfil other functions.[1] The packaging procedure is referred to as the *packaging process.* The packaged goods, packaging and packaging process together form the *packaging system.* According to the DIN 55 405 standard, the packaging itself is a unit formed by the packaging mean and the packaging aid, which consist of different packaging materials.[2] The packaging material is the material from which packaging is made. The packaging mean is the product made of the packaging material which is intended to enclose or hold together the packaged goods. Packing aids is a collective term for aids which, together with packing means, are used to pack, seal, make ready for dispatch, etc., packaged goods. It is often useful to consider these individual packaging elements separately. This is because each of these elements has its own specific influence on the packaging system. For example, the use of a new adhesive tape may make it easier to seal the packaging material in the packaging process. The risk of accidents during handling of the packaging can be reduced by strapping the packaging with plastic straps instead of steel straps.

Functions

The properties of packaging depend on the functions it is to perform.[3] Basically, the following *four functional areas* can be distinguished:

---

[1] Cf. Isermann, 1996, pp. 2162ff.

[2] Cf. German Institute for Standardization, 2006.

[3] Cf. Isermann, 1997a, pp. 1230 f.; Jünemann/Schmidt, 2000, pp. 8 f.

H.-C. Pfohl, *Logistics Systems*,
https://doi.org/10.1007/978-3-662-64349-5_7

Production functions: Packaging enables the production input to be made available in terms of quantity and the production output to be received in terms of quantity at the production location. By selecting suitable packaging (e.g. containers), production can be carried out directly from or directly into the packaging without any further intermediate handling operations.

Marketing functions: For many products, packaging is an essential component of product policy, through which a product is made distinguishable from competing products. But packaging can also be assigned important functions in the area of communication policy as an advertising medium or in sales promotion.[4]

Use functions: This includes the reuse of the packaging by the customer or the use for other purposes. Due to the mandatory consideration of ecological effects, the environmentally friendly design of the packaging becomes an essential requirement.[5] This corresponds to the possibility of environmentally sound disposal or the reusability of packaging. However, reuse also takes into account the idea of avoidance.[6] In the organization of reusable packaging systems, a distinction is made between bilateral reusable systems (shuttle systems) and reusable pool systems. In bilateral reusable systems, the packaging cycle takes place between a sender and a recipient. In the case of reusable pool systems, the group of participants is expanded and a company specializing in this area takes over the administration of the packaging cycle as the pool carrier, the container management (ownership, exchange modalities, financing).

Logistics functions: Packaging should facilitate or enable the realization of the other logistics processes in the first place. The logistics functions of packaging are discussed below:

- Protective function: The protection of the goods by the packaging is often considered the most important logistics function of the packaging. It is part of a good service that the ordered goods arrive at the recipient in the right condition. The packaging should protect the goods against mechanical (pressure, shock) and climatic stresses (humidity, temperature) during delivery. However, packaging protection of the goods covers not only qualitative but also quantitative losses. The main aim here is to ensure that the packaging makes it as difficult as possible for the packaged goods to be stolen. In addition to protecting the packaged goods, the protective function of the packaging also includes protecting the environment. The packaging should protect the people and technical equipment used in delivering the goods, as well as the other goods to be delivered, from damage of which an unpackaged good might otherwise be the cause.
- Storage function: Packaging is required to facilitate the storage of a good. This means first of all that the packaging should be stackable. The shape and dimensions must allow

[4]Cf. Meffert et al., 2012, pp. 432ff.

[5]On the ecology orientation in packaging design, cf. Isermann, 1996, pp. 2170ff.

[6]Cf. Arnold et al., 2008, pp. 702.

the individual packagings to be placed directly on top of each other. The static friction between the stacked packages must be sufficiently great to ensure a stable stack. Stackability also requires that the packaging is strong enough to withstand the weight of the other packagings stacked on top of it, depending on the stack height. The packaging must also be able to withstand the stresses in the storage facilities, for example in a drive-through rack. The packaging must enable good use to be made of the storage space. Storage is also facilitated by matching the dimensions of the packaging to those of the storage containers. An often overlooked aspect of the storage function of packaging is the requirement for rational storage of the packaging stock (e.g. by collapsible cartons).

- Transport function: The purpose of packaging is to facilitate the transport of a product or to make the product suitable for transport in the first place. The shape and dimensions of the packaging should be as light as possible and allow optimum use of the transport space.
- Handling function: The packaging should group the goods into units that facilitate their handling during delivery. The shape and dimensions of the packaging units must also allow the use of technical aids such as forklift trucks or stacker cranes in order to rationalize handling operations. Handling operations are always interposed between the storage and transport phases of a good, so that the formation of packaging units must also always take into account the storage and transport function. If the goods are handled manually, the handling function of the packaging also necessitates packaging which is easy to grip (e.g. rough paper) or has finger holes or easy-to-grip lashing, so as not to make handling unnecessarily difficult.
- Information function: The information function of the packaging is particularly important for order picking. The packaging must be marked (e.g. by colour or inscriptions) in such a way that the order picker in the warehouse can easily identify the desired products. In addition, packaging for fragile, perishable, or similar products that require special handling during delivery should be clearly marked with pictures, signs, or explanations. Information on the packaging can lead to a reduction in the number of accompanying documents. If transport and handling processes are automated, suitable information on the packaging enables the product to be identified automatically. This is done, for example, by means of barcodes which are recognized by automatic reading devices. The performance characteristics of the reading units used in the transport chain must be taken into account here, as they are often not able to scan all sides of the packaging. Accordingly, the barcodes must be applied to certain or even several sides of the packaging. Thanks to the development of RFID technology, it is now possible to store important information in transponders that are attached to or inside the packaging. The working principle of RFID technology in logistics is discussed in Part II, Sect. 4.3.

The various logistics functions of packaging are an excellent example of the interdependencies that exist in the field of logistics. *Packaging* must therefore only ever be considered as a *component of the overall logistics system*. The correct assessment of packaging for logistics can contribute both to reducing overall logistics costs and to increasing the level of supply or delivery service. Packaging—as a very important component of service—does not only influence the condition of the delivered goods. It can also enable a more rational, faster delivery of the goods, i.e. influence the service by shortening the delivery time. Last but not least, a customer will also judge the service of his suppliers according to the extent to which they meet his needs when combining the goods into delivery units by means of packaging.

7.2 Packaging Tasks

Packaging Design

Since packaging has functions to fulfil in production, marketing, use and logistics, the entire packaging problem can only be solved by a *packaging team* in which specialists from the various areas work together. An analysis of the packaging system first involves recording all the requirements placed on the packaging in the four functional areas.[7] Then it must be examined to what extent these requirements are met by a packaging system and what costs are involved. The requirements placed on the packaging from the various functional areas can compete with each other, as shown in Fig. 7.1. Therefore, a *compromise* must be found in packaging design that takes all functional areas into account. If one functional area is to dominate for a particular reason, it must be charged with the additional costs incurred in the other functional areas of the packaging as a result.

The requirements on the packaging depend on various influencing variables, as shown in Fig. 7.2.

The primary influencing factor is always the packaged goods with its product properties, which must be assumed and to which the packaging process must also be adapted in addition to the packaging. There is a very close relationship between the packaged goods, the packaging and the packaging process. The logistical packaging problems which arise must therefore be taken into account right from the *product design.*[8] However, precisely because the weight, bulkiness, fragility, shape or other packaging properties of the goods have a major influence on the overall packaging system, even minor changes in the nature of a good can prove very beneficial to the packaging system. Often, such changes do not in any way affect the essential properties of the goods for the purchaser. Depending on the packaged good at hand, there are a plethora of legal and quasi-legal *framework factors* that must be considered in packaging design.[9] For example, there are regulations on the shape, dimensions and packaging of a product from the Weights and Measures Act and the Packaging Decree. The Railway and Motor Transport Regulations contain rules on the

[7] Cf. Isermann, 1996, pp. 2164ff.

[8] See also Part III, Sect. 11.2.

[9] Cf. Isermann, 1996, p. 2166.

Packaging functions		Requirements
Protective function		temperature-resistant tight corrosion-resistant dust-free chemically neutral holding together hardly inflammable
Protective function	Warehouse and transportation function	dimensionally stable shock-resistant shock-absorbing pressure-resistant tear-resistant
	Warehouse and transportation function	stackable slip-resistant standardized manageable automation-friendly roll-under unit forming
Marketing function	Warehouse and transportation function	space saving volume saving
Marketing function		economical
Marketing function	Identification and information function	advertise informative identifiable distinguishable
Marketing function	Usage function	easy to open reclosable
	Usage function	re-usable ecological decision-friendly hygienic

Fig. 7.1 Allocation of packaging requirements to packaging functions (Source: Jünemann/Schmidt, 2000, p. 9)

type of packaging material and the size, colour and inscription of labels for dangerous goods. In accordance with the Foodstuffs Labelling Decree, there are provisions on the textual and graphic design of packaging with the prohibition of misleading text on the packaging. The Decree on Supplements contains restrictions regarding the secondary use of packaging, i.e. with regard to its function of use.

Fig. 7.2 Influences on the design of packaging types (Source: Taken with minor changes from Frerich-Sagurna, o. J., p. 21)

Packaging Process

The packing process comprises all the steps required to pack the goods, from feeding the empty packaging and the packaged goods to the packing station, through the various stages of the actual packing process, to preparing the packaging units for transport from the packing station.[10] The tasks that must be carried out in connection with the actual packing process also include signing or labeling.

Great rationalization successes can be achieved in packaging by means of a *process-organizational analysis.* The task of such an analysis is to provide an overview of all the working parts involved in packaging as the smallest units in the packaging process and to identify potential for rationalization by means of a suitable *synthesis of* these working parts into operations which can be assigned to work carriers (man or machine). For example, it is possible to achieve a faster loading of a packaging machine by a different combination of work parts in the packaging process, which leads to a better capacity utilization of this machine and accelerates the entire packaging process. Of course, it still has to be checked whether the tensile strength of the packaging material allows the packaging machine to run faster at all.

Even this small example shows the close relationship that exists between the individual components of the packaging system. The use of packaging machines for setting up or forming the packages, for carrying out the filling and sealing process and even for automated packaging depends to a large extent on the nature of the packaged goods and the quantity of goods to be packaged. On the other hand, packaging machines also require precise matching to the shape of the packaging and place completely different demands on the dimensional tolerances of the packaging than manual packaging. For the packaging

[10]Cf. Isermann, 1997b, pp. 1238ff.

machine, it is essential to avoid even relatively minor deviations in the shape and dimensions of the packaging.

Packaging Decree

The Packaging Decree distinguishes between different types of packaging. *Transport packaging* primarily has a protective function and is intended to protect the goods from damage on their way from the manufacturer to the distributor. *Sales packaging* is used for transport from the retailer to the end consumer and, if necessary, also for storage until final consumption. *Outer packaging* consists of films, cardboard boxes or similar wrappings which serve to enable the goods to be dispensed by way of self-service, to reduce or prevent the possibility of theft or to enable advertising.[11]

The different types of packaging can place different demands on the properties of *packaging means and packaging materials*. The tear strength of the packaging material, for example, influences the possibility of using packaging machines in the packaging process, and the stackability of the packaging depends in part on the frictional behavior of the packaging material. Uniform shape and dimensions of the packaging material facilitate both the packaging process and the transport and storage of the packaged goods. The advantage of uniform shape and dimensions of a packaging material is illustrated by the example of the pallet, which is one of the best-known ways of forming logistical units.

7.3 Logistical Units

Concept and Task of Logistical Units

The term "formation of logistical units" is used to describe the process that is aptly and succinctly called *unitization* in Anglo-Saxon usage. It is therefore also referred to as the unit load concept. This refers to the grouping of goods to be delivered into larger units. Although not every package is a logistical unit, such grouping is one of the logistical functions of packaging. The basic idea is that, for a given volume, the fewer components there are in the flow of goods from the supplier to the customer, the smoother it can be. This is because the result is that fewer handling, measuring and counting operations are required.

In German literature, the grouping of goods into specific units is referred to as storage, transport, packaging, loading, ordering units, etc., and ideally requires *these units to be*

---

[11] The Packaging Decree essentially contains regulations on the take-back obligations for the respective types of packaging and is the basis for possible deposit regulations. It was the basis for the establishment of Duales System Deutschland Gesellschaft Für Abfallvermeidung und Sekundärrohstoffgewinnung mbH (green dot). See the laws and decrees in Large, 2012, pp. 245 f. See also the activity-specific subsystem of disposal logistics in Part III, Sect. 13.2.

identical.[12] The different terms show that the formation of units is important for all phases of the flow of goods from the point of delivery to the point of receipt. Lack of systems analytical thinking in the delivery of goods often prevents this from being seen. Once the interrelationships between the individual areas in the logistics system have been recognized, the logical conclusion is that the units for these areas must be as similar as possible. In practice, it will often not be possible to achieve this consistently, but it should always be strived for as a goal. This is because the logistical unit is the link between the individual phases of the flow of goods. Decisions in all areas of logistics ultimately affect this unit, which—depending on which phase of the flow of goods is being considered—has to fulfill the functions of a storage unit, a transport unit or a loading unit, for example. The term logistical unit is intended to emphasize the central importance of this unit for the entire logistical system. The grouping of goods into specific units to be moved through the logistical network is a logistical problem and not just a transport or handling problem.

The company's logistics system is not a self-contained system; rather, it has a variety of relationships with customers and with the logistics companies whose services are used. The grouping of goods into larger units therefore always requires an agreement with the customers and logistics companies. With a large number of customers and logistics companies, this means that the units to be formed must be standardized in terms of shape and dimensions. Thus, the concept of a logistical unit can be specified as follows:

Logistical units are created by combining goods into units that are standardized in terms of shape and dimensions with the aim of simplifying the flow of goods and reducing the costs incurred in the process.

The flow of goods through the logistics network is created by stringing together storage, packaging, handling, control and transport operations. The stringing together is also often referred to as a *transport chain.*[13] The formation of logistical units is a prerequisite for a rational transport chain.[14] This means that the following *important principles* must be observed:

- Grouping of goods into larger units,
- Standardization of units in shape and dimensions,
- Facilitate the use of mechanical means in the manipulation operations,
- Stackability of units,
- Choice of unit that enables a largely uninterrupted transport chain from supplier to customer.[15]

[12] Cf. Arnold et al., 2008, pp. 702ff.

[13] See Sect. 8.2.

[14] Cf. Jünemann/Schmidt, 2000, pp. 20 f.; Large, 2012, pp.43ff.

[15] Cf. Jünemann/Schmidt, 2000, pp. 21 f.; Isermann, 1998, p. 245.

Possibilities for the Formation of Logistical Units

In principle, any packaging can serve to form a logistical unit. In a borderline case, a packaged good wrapped only with a plastic film, which has a sufficiently large weight and volume and whose shape and dimensions permit the use of mechanical means in the handling operations and the formation of stacks, itself constitutes a logistical unit. In the case of stacks of bags, it is possible to form logistical units by shrink-wrapping them with plastic film, thus creating shrink-wrapped packs which are as easy to handle as palletised stacks of bags. It is also possible, for example, to form logistical units of building materials by packing them into units standardized in shape and dimensions with the aid of strings or straps. Such a packaged unit can be provided with bases (feet, bracing members) to form a type of pallet. The various types of pallets, folding boxes, small containers and large containers give an idea of the variety of possibilities for forming logistical units.

Figure 7.3 shows the basic possibilities for the formation of logistical units. It would be too much of a stretch at this point to describe all the possibilities for forming logistical units in detail and to compare them with each other. For this reason, only the pallet and the container will be dealt with in more detail by way of example. However, a key factor in the choice of logistical unit will always be which unit allows a largely uninterrupted transport chain.

Palette[16]

According to DIN 15 145 and DIN 55 405, the *pallet* is defined as a "portable platform, with or without bodywork, designed to group goods together to form a load unit for transport, storage and stacking by means of industrial trucks or other mechanical equipment. It is provided with devices for being driven under by the insertion organs of industrial trucks (forklift trucks, pallet trucks, etc.) the underride height is usually about 100 mm."[17] This definition by the German Standards Committee identifies the pallet as a typical logistical unit. In it, direct reference is made to the *functions of the pallet* as a *transport, storage, loading or manipulation unit.* These functions of the pallet are widely known and it is already used from these points of view in many areas of industry and commerce to rationalize the delivery of goods.

The *pallet* is also widely used *as a packaging unit*. There are three main advantages here, relating to the packaging material, the packaging time and the packaging machine. The packing material costs can be reduced compared to the costs of the otherwise common packing method, where the pallet is not used as a packaging unit. If one palletizes semi-automatically and dissolves in the same way, it is possible to reduce the packaging time. If one uses a carton packaging unit and the pallet only as a transport unit, for example, then a system for carton packaging and palletizing is required. If the pallet is also used as a packaging unit, the costs incurred by the necessary machines and equipment for palletizing

---

[16] Cf. Jünemann/Schmidt, 2000, pp. 22ff.

[17] German Institute for Standardization, 1987.

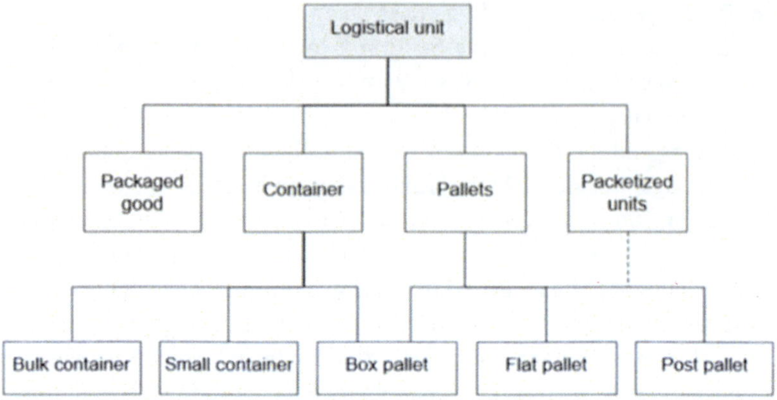

Fig. 7.3 Basic options for forming logistical units

and depalletizing are only about half the costs of a system for carton packaging and palletizing.

If one discusses the *pallet as a logistical unit*, then above all its function as a delivery unit (shipping unit, sales unit) must also be seen. Therefore, the choice of logistical unit can only be made in close contact with the customer. The customer must agree to the quantity of goods grouped together by the pallet. He must have the facilities for manipulating the pallets and for storing them (pallet racks). These facilities also require agreement on the dimensions of the pallet.

Pallets can be divided into different *pallet types* according to certain distinguishing features.[18] In Fig. 7.3, a distinction is made between flat pallets, post pallets, rolling pallets and box pallets on the basis of the construction of the pallet with or without body. While the flat pallet is merely a portable platform, the post pallet has posts at the corners onto which another pallet can be placed to form a stack. Roller pallets are flat pallets with roller bases, often used in drive-in and drive-through racks. The box pallet has vertical walls on at least three sides and may have a lid. A pallet with a folding collapsible frame has similar characteristics to a box pallet. The folding collapsible frame is advantageous, as it takes up considerably less space when transported empty.[19]

The variety of pallet types on offer reflects the fact that the choice of pallet as a logistical unit can be very much adapted to the specific requirements of delivering a good. At the same time, however, the need to standardize the dimensions of pallets also becomes clear. Only in this way can the pallet fulfil the functions of a logistical unit during the delivery of the goods to the supplier, to the customer and to the logistics companies. *Standardization* attempts to harmonize the dimensions of pallets with the dimensions of the technical equipment necessary for their storage, handling and transport, and to reduce the number

[18] Cf. Jünemann/Schmidt, 2000, pp. 22ff.

[19] Cf. ten Hompel et al., 2007, pp. 25ff.

of types. In order to make full use of the advantages of palletization through the possibility of inter-company exchange of pallets, standards have already specified not only the dimensions but also precise details of materials and construction. Such standardization should be pursued to an even greater extent than hitherto in order to further increase the freedom of pallet use. If uniform pallets are used by suppliers, customers and logistics companies, the pallets can be exchanged between them step by step, i.e. a full pallet for an empty one. Such an exchange of pallets can be realized within a *pallet pool* to which suppliers, customers and logistics companies belong. This creates a uniform pallet system in which a pallet remains as a logistical unit within the entire distribution channel for as long as possible. The logistical unit is transported from one stage in the distribution channel to the next stage.

In 1961, a European pallet pool was established. The European exchange pallet or pool pallet (Euro pallet) has the dimensions 800 mm × 1200 mm. In the chemical industry and in the hollow glass industry, the pallet dimension 1000 mm × 1200 mm and in the beverage industry the pallet dimension 800 mm × 1000 mm are also used.

Large Container

Large containers include the following *container types* (cf. Fig. 7.4):

- ISO containers for international transport,
- Inland containers for transport throughout Europe,
- Swap bodies or swaps,
- ULD (Unit Load Device) for aviation.

The *ISO container* is available in the main designs of 20 feet (6055 mm) and 40 feet (12,190 mm) in length (also in 10 and 30 feet),[20] a width of 8 feet (2435 mm) and a height of 8 to 9 feet (2435–2745 mm). It can be stacked six high and is usually loaded and unloaded through a double-leaf door on one end. The ISO container permits optimum use of space only if loaded with the ISO pallets commonly used in the USA, for example. For the pool pallet commonly used in Europe, there is a loss of space utilization in the ISO container.

The inland container was developed for German inland traffic and European traffic. It has a width of 2500 mm, a height of 2600 mm, a length of 20 feet (6055 mm) or 40 feet (12,190 mm) and has a permissible gross weight of 20 t or 30 t respectively. For loading and unloading, the container has a side door in addition to a rear door on the left side. It has an internal width of 2440 mm, so that a transverse stowage of two pallets of 1200 mm or three pallets of 800 mm is possible. However, since the length of the container has been based on the ISO containers, maximum space utilization is not possible with the European interchangeable pallet.

[20] Cf. German Institute for Standardization, 1981.

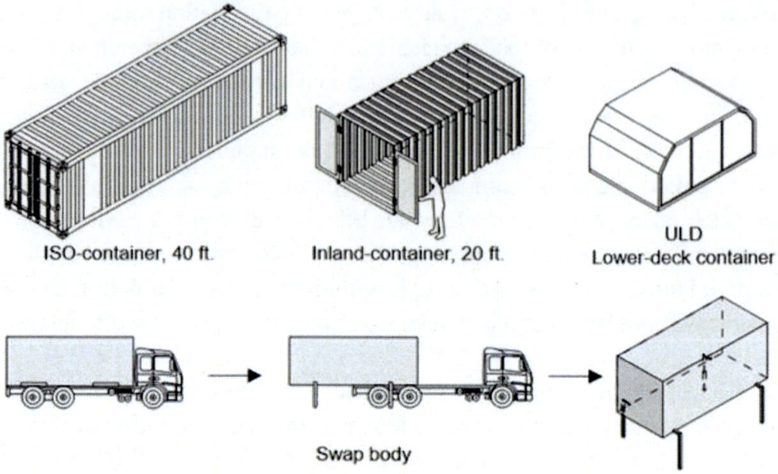

Fig. 7.4 Various large containers (Source: ten Hompel et al., 2007, p. 30ff)

Truck swap bodies have been developed for road and rail transport, which are used in German domestic traffic—and in some cases also in European traffic. The swap bodies have a width of 2500 mm, a height of 2600 mm and a length of 6250 mm or 7150 mm. As the swap bodies have an internal width of 2440 mm and an internal length of 6100 or 7000 mm, the loading area can be used to the maximum with the pallets commonly used in Europe.

Since all three types of large containers have the same connection dimensions, they can be transported with the same chassis.

In the aviation sector, *ULDs* are used as pallets and containers to carry loads. Due to the different designs of aircraft fuselages, they have a variety of different shapes. The main classes are main-deck and lower-deck containers, which are of lightweight construction matched to the pallets and are made of aluminum or aluminum and plastic.

7.4 Modular Packaging[21]

In defining logistical units, the need for standardized dimensions was pointed out. In distinguishing between ISO containers and inland containers, it was shown that the different internal dimensions of containers lead to different space utilization by pallets. This already addressed the *problem of coordinating the dimensions of different packages.* The need for a coordinated, modular packing system arises from the fact that different companies normally cooperate in a transport chain and thus there are inter-organizational logistics systems and that there may be different packing levels according to the contents of

[21] Cf. Jünemann/Schmidt, 2000, pp. 30ff.

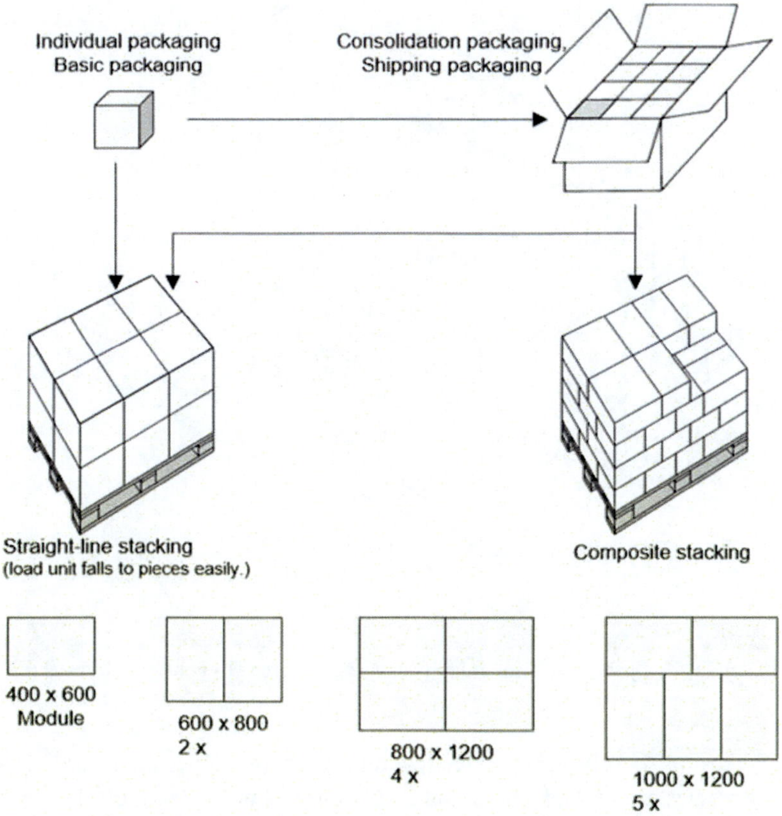

Fig. 7.5 Modular structure of packaging (Source: Taken with minor modifications from Rockstroh, 1978, p. 203)

the packaging. The internal dimensions and load-bearing capacities of the larger packages (e.g. containers) must match the external dimensions and load-bearing capacities of the smaller packages (e.g. pallets) to be placed inside them.

With a view to a modular structure of individual, collective and dispatch packages adapted to the pallet, a packaging module (footprint module) 400 mm × 600 mm has been developed by the packaging industry (see Fig. 7.5). This packaging module is adapted to both the 800 mm × 1200 mm pallet and the 1000 mm × 1200 mm pallet. Multiplying or dividing the basic module dimensions of 400 mm and 600 mm results in an overall system of modular packaging, an example of which is given in Fig. 7.6.

In order to make optimum use not only of the surface area but also of the space, the package height must be taken into account. Uniform loading heights for Euro pallets in the food industry are also recommended for cooperation between industry and trade. Two standards for loading height (load height plus pallet height) are available: Dimension I:

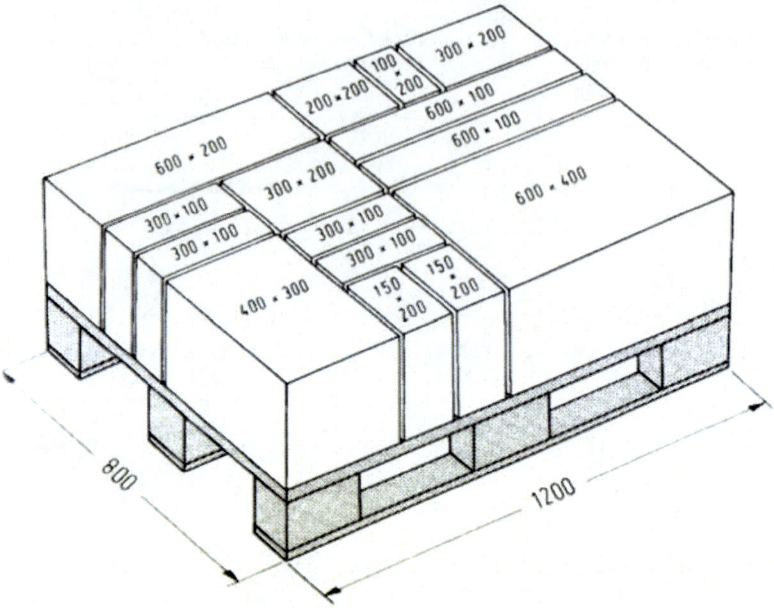

Fig. 7.6 Example of a pallet with modular packaging (Source: Eggenstein/Herbst/Jansen, 1981, p. 172)

105 cm; Dimension II: 160–195 cm. The manufacturer should decide according to which standard his articles are packed. Important *criteria* to be considered are[22]:

- the door openings or loading space heights of vehicles and containers within the transport chain,
- shelf heights in warehouse buildings,
- room heights,
- lifting heights of industrial trucks,
- door dimensions of the elevators.

One modular packaging system, for example, is the *LogistikBox*. This transport packaging is compatible with the existing technology of combined transport with swap bodies and is intended to facilitate the transition between internal transport processes[23] and local and long-distance transports through simple bundling.

The external dimensions and technology of the containers in this system are based on both the pallet packaging unit and the standard large containers. There are two modules: the

[22] Cf. Eggenstein et al., 1981, pp. 174 f.

[23] See Sect. 6.2.

4-pallet box and the 6-pallet box. The *4-pallet box* (footprint 1700 mm × 2500 mm with a height of 2470 mm) allows easy integration into the internal material flow due to its small footprint. The square footprint of the *6-pallet box* (footprint 2500 × 2500 mm at a height of 2470 mm), on the other hand, enables flexible arrangement of the packaging units on the various means of transport. For these modules, which can be loaded and unloaded with forklift trucks, there are support frames with which several units can be assembled into larger units and moved like swap bodies.

As has already been pointed out several times and is made clear by the example of the LogistikBox, the existence of modular packaging facilitates the construction of transport chains, which is the core of transport tasks.

References

Arnold D u.a. (Hrsg) (2008) Handbuch Logistik. 3. neu bearb. Aufl. Berlin/Heidelberg

Deutsches Institut für Normung e. V. (DIN) (Hrsg) (1981) Deutsche Norm, DIN ISO 668, ISO-Container der Reihe 1: Klassifizierung, Außenmaße, Gesamtgewichte. Berlin/Köln

Deutsches Institut für Normung e. V. (DIN) (Hrsg) (1987) Deutsche Norm, DIN 15 145. Paletten – Systematik und Begriffe für Paletten mit Einfahröffnungen. Berlin/Köln

Deutsches Institut für Normung e. V. (DIN) (Hrsg) (2006) Deutsche Norm, DIN 55 405, Verpackung-Terminologie-Begriffe. Berlin/Köln

Eggenstein F, Herbst D, Jansen R (1981) Materialfluß und Steuerung. In: Volk P (Hrsg) Betriebsleiter Handbuch. 5., völlig neu bearb. u. erw. Aufl. Landsberg a. L., S. 159-330

Frerich-Sagurna R (o J) Verpackung und Logistik. In. EWI – Gesellschaft für Europäische Wirtschaftsinformation mbH. (Mehr)wege aus der Verpackungskrise. Stammberg, S. 19-38

Isermann H (1996) Stichwort "Verpackung". In: Kern W, Schröder H-H, Weber J (Hrsg) Handwörterbuch der Produktionswirtschaft. 2., völlig neu gestaltete Aufl. Stuttgart, Sp.2162-2182

Isermann H (1997a) Stichwort „Verpackung". In: Bloech J, Ihde G B (Hrsg) Vahlens Großes Logistik Lexikon. München, S. 1227-1234

Isermann H (1997b) Stichwort „Verpackungsprozeß". In: Bloech J, Ihde G B (Hrsg) Vahlens Großes Logistik Lexikon. München, S. 1238-1240

Isermann H (1998) Stauraumplanung. In: Isermann H (Hrsg) Logistik: Gestaltung von Logistiksystemen. 2., überarb. und erw. Aufl. Landsberg a. L., S. 245-286

Jünemann R, Schmidt T (2000) Materialflußsysteme. Systemtechnische Grundlagen. 2. Aufl. Berlin/ Heidelberg/New York

Large R (2012) Logistikfunktionen. Betriebswirtschaftlicher Logistik. Band 1. München

Meffert H, Burmann Chr, Kirchgeorg M (2012) Marketing: Grundlagen marktorientierter Unternehmensführung: Konzepte – Instrumente Praxisbeispiele. 11., überarb. und erw. Aufl. Wiesbaden

Rockstroh O (1978) Die Abstimmung der Verpackung mit Palette und Container. In: Rationalisierung 29 9, S. 203-206

ten Hompel M, Schmidt T, Nagel L (2007) Materialflusssysteme. Forder- und Lagertechnik. 3., völlig neu bearb. Aufl. Berlin u.a.

8 Transport

8.1 Definition and Functions of the Transport

Definition

Transport is the bridging of space or change of location of transport goods with the aid of means of transport. Each *transport system* consists of the goods to be transported, the means of transport and the transport process. If a shipment is to be transported from delivery point A to destination point B and the required means of transport is not available at delivery point A, a transportation process called an empty shipment may be necessary. Empty transports are transport processes without transported goods, but they are necessary prerequisites for the subsequent transport process with transported goods.

In-house transportation is transportation in a plant from one production location to another or transportation in one area or between different areas of a warehouse. *External transportation*, on the other hand, is transportation from the supplier to the customer, transportation between different plants or between different warehouses of a company, and transportation between its plants and its warehouses. Since internal transportation was dealt with in the section on the warehouse, the following explanations refer to external transportation.

Functions

A distinction is made between the primary and secondary functions of transport.[1] The *primary functions* include the transport function and the inseparable handling function. The *secondary functions primarily* include the securing of routes, i.e. the creation and maintenance of routes.[2] The liability function is frequently mentioned as a further secondary function.

[1] For a discussion of different functions, see Aberle, 2009, pp. 1ff.

[2] Cf. Buchholz et al., 1998, p. 2.

H.-C. Pfohl, *Logistics Systems*,
https://doi.org/10.1007/978-3-662-64349-5_8

The liability function is not dealt with in the context of this book, which is oriented towards business management. Likewise, the function of securing the route will not be dealt with in the following, as this is a task for the national economy. The following remarks therefore refer to the possibilities of fulfilling the transport function and the handling function necessarily associated with it.

8.2 Transport Tasks

Transport Problem

The transport problem in a logistics network is characterized by the goods being transported, the structure and nature of the delivery area, the locations of the delivery and receiving points, and the nature of the supply and demand from these points. For a given transport problem, there are basically *two questions that* need to be answered:

- Which is the most favorable means of transport?
- Which is the most favorable transport process?

The question of the most favorable *means* of transport concerns the hardware of the transport. It must be decided with which means of transport the goods are to be transported. In contrast, the question of the most favorable *transport process* concerns the software of the transport. The software relates to the procedural rules for controlling the transport process.

In the *operations research* literature, the so-called transportation problem is usually understood to be a very specific organizational problem. If a good is available in certain quantities at the various delivery points, the transportation problem consists of determining the delivery points and the quantities of goods to be delivered from them in such a way that the total transportation costs of supplying the receiving points with the quantities of goods they demand are minimized. Other transportation problems in the transportation process that are also addressed in the operations research literature include the optimal loading of a means of transportation, the determination of the shortest route between a delivery point and a receiving point, or the determination of the optimal overall route for delivering multiple receiving points from one delivery point. In the case of a large fleet of vehicles, the problem of the optimal deployment plan for the means of transport to achieve maximum transport performance also arises.

Transport Chain

The solution to the transport problem is ultimately to set up a transport chain. According to DIN 30781, Part 1, the term transport chain is defined as a "sequence of technically and organizationally interlinked operations in which persons or goods are moved from a source

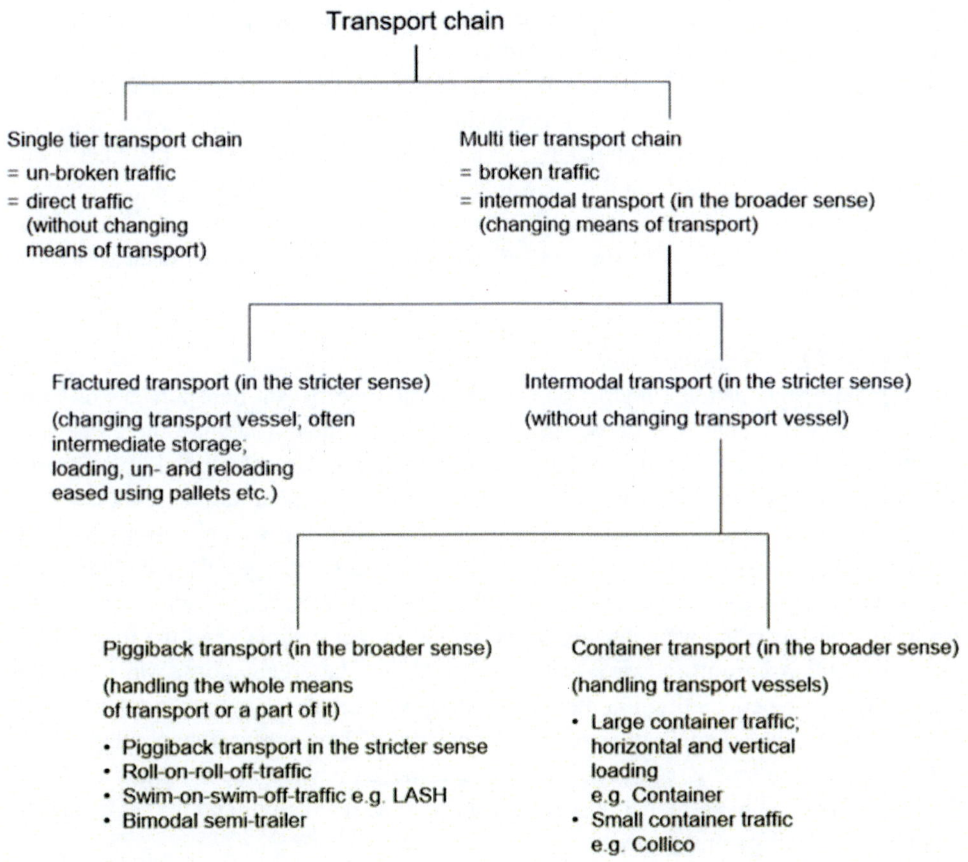

Fig. 8.1 Possibilities for the design of a transport chain (Source: Taken with slight modifications from Seidenfus, 1972, p. 79)

to a destination."[3] As shown in Fig. 8.1, transport chains can be single-tier or multi-tier. In a *single-tier transport chain*, the point of delivery and the point of receipt (source and destination) are directly connected in unbroken traffic or direct traffic without changing the means of transport. In a *multi-tier transportation chain*, on the other hand, there is a change of means of transport when the delivery and receiving points are connected. This is referred to as broken transport or combined transport in the broader sense. However, combined transport in the narrower sense is only referred to when there is no change of transport vessel. This is the sense in which the term will be used in the following.

The establishment of a transport chain for the flow of goods must be accompanied by the establishment of a corresponding *documentation chain* for the flow of information.[4] This is

[3] German Institute for Standardization, 1983.

[4] Cf. Pfohl et al., 1993.

particularly important for a smooth flow of goods in multi-tier transport chains, in which a distinction is made between *three typical phases*.

The first phase is the pre-carriage from the delivery points to a consolidation point (concentration point), which can be a forwarding warehouse, for example. This phase of the transport chain can be characterized as area transport. The main leg, on the other hand, is a drop shipment. It goes from the consolidation point to the break-bulk point (breakup point), for example, in the form of another forwarding warehouse. From the break-bulk point to the receiving points, the onward leg takes place, which can again be characterized as area transport.

Regulation of the Transport Task

Although numerous national regulations that governed the freight transport market have been repealed in the course of European integration and standardisation of the framework conditions, it is still a partially regulated market. State regulation is based on standards under public and private law, which can be found in a large number of individual laws and decrees. Special regulations under *public law* can be found, for example, in the Railway Regulation Act, the Road Haulage Act and the Air Transport Act. Special regulations under *private law* can be found, for example, in the Commercial Code, which regulates the forwarding business, the warehousing business, the freight business and the transport of goods on public transport railways and maritime trade. State regulations, which had the effect of restricting competition in the freight transport markets more than in other sectors of the economy, were largely removed during the 1990s. However, safety regulations are still subject to continuous adaptation to technical developments and scientific findings. These are expressed in particular in special working time regulations and in regulations on the transport of dangerous goods.[5]

There are also still differences in the regulation of weights and measures in Europe. Harmonisation is also required in this area in order to avoid distortions of competition.[6] So far, European integration has only led to partial harmonisation in the technical field. For example, the permissible length of trucks and tractors with semi-trailers in the EU countries, with the exception of Denmark, is 18.35 m and 16.5 m respectively. The situation is different for the permissible gross weights, which range from 38 t (Great Britain) to 50 t (Netherlands) (Federal Republic of Germany: 44 t). As far as the

[5] Safety regulations for the realization of transport tasks arise primarily from the many and varied statutory regulations on working hours in the transport industry and from the law on the transport of dangerous goods. The regulations governing the transport of dangerous goods attempt to limit risks by passing on information about the nature of the goods, the hazards they pose, the type of transport equipment and containers required, and the measures to be taken in the event of accidents in the transport chain.

[6] On deregulation activities in the transport markets see Aberle, 2009, pp. 173ff.; Kummer, 2006, pp. 226ff.

permissible length is concerned, a pilot test is under way with a long truck, the "Gigaliner", which has a length of 25.25 m.[7]

With regard to the regulations that influence competition, a distinction must be made between *works transport* (transport for own account by industrial, commercial and service companies) and *commercial transport* (transport by logistics companies for third parties). In essence, works transport is only subject to the obligation to register, whereas commercial transport is subject to authorisation or licensing. The permit for commercial goods transport is primarily linked to personal reliability, professional suitability and financial capacity and can, in principle, also be applied for by companies with works transport after a transport company has been spun off.

The liberalisation of freight transport has had a considerable impact on competition, particularly in the road haulage sector, through the abolition of quotas in cabotage[8] transport and inland freight transport, the abolition of the distinction between local and long-distance freight transport and the facilitation of the provision of commercial transport services by works transport operators.[9] These changes, which have consistently led to an increase in the capacity available on the market, are exacerbated by the increasing presence of Eastern European transport service providers on the EU market, who have considerable competitive advantages as a result of wage and social differentials.[10]

In maritime and air transport, competition is regulated by historically developed *cartels.* Liner shipping conferences and the cartel association of liner air transport, the so-called International Air Transport Association (IATA), determine market shares and/or prices.

8.3 Means of Transport

Freight Transport System

Figure 8.2 provides an overview of the freight transport system. Here, a distinction is first made between land, air and water transport in terms of the media on or in which the transport function is performed. This is followed by a breakdown according to the means of transport used. These means of transport are then subdivided according to various criteria, such as organizational criteria when distinguishing between commercial road haulage and works transport, technical criteria when distinguishing between motorised and tugboats, or goods criteria such as when distinguishing between crude oil and product pipelines. The assessments of the performance capabilities of the individual means of transport are quite

[7] See Heidmann, 2013.

[8] Cabotage is understood to mean the realization of transport operations on the territory of another country.

[9] Cf. Federal Office for Goods Transport, 2017, pp. 3ff.

[10] Cf. Federal Office for Goods Transport, 2017, pp. 1ff.

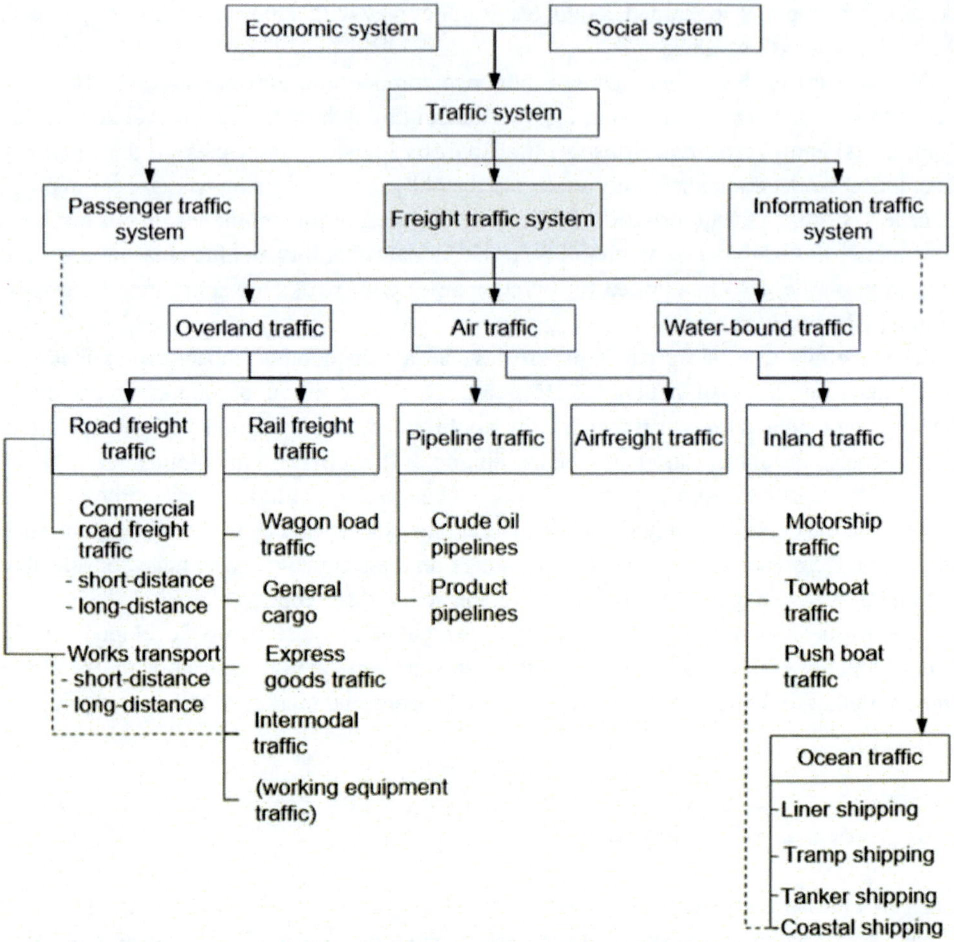

Fig. 8.2 Freight transport system (Source: Claussen, 1979, p. 15)

different, as Fig. 8.3 shows. The most important means of transport are characterized below.[11]

Seagoing Vessel and Inland Waterway Vessel

In addition to own-account *ocean shipping*, a distinction can be made between regular liner shipping and irregular on-demand shipping. The latter is also known as tramp shipping when transporting bulk goods in closed shiploads. Irregular on-demand transport is generally cheaper, but also slower than liner shipping.

[11] Cf. Aberle, 2009, pp. 18ff; Buchholz et al., 1998, pp. 105ff.

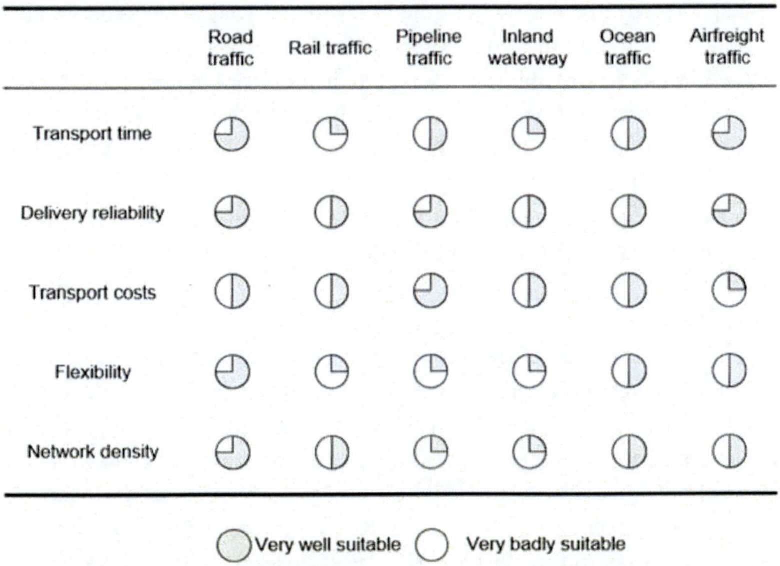

Fig. 8.3 Comparison of freight transport (Germany) (Source: based on Pfohl/Schäfer, 1998, p. 84)

Ocean freight rates are calculated in *liner shipping* according to volume or weight within value classes into which cargo is classified according to its value.[12] They are set as combined volume-weight freight rates by ship choice by the shipping lines. This means that the shipping companies always choose volume or weight as the basis for calculating the transport costs in such a way that the higher freight rate results for them. Special surcharges are charged, for example, for unusually heavy or bulky items. In *on-demand transport*, the transport costs are negotiated as part of the charter contract. Handling costs in the ports can vary and must be taken into account, especially for larger shipments.

Seagoing vessels *compete with aircraft* in offering their transport services. They have major disadvantages compared with aircraft in terms of transport time, so that they are generally not considered as a means of transport for goods that have to be delivered quickly. The comparatively high transport costs of air freight transport often do not carry as much weight in favour of sea freight because of the usually faster and also much cheaper pre- and post-transport. The disadvantage of pre- and post-carriage in sea freight transport stems from the fact that seaports are naturally located on the coast, while airports can be scattered all over the country.

Inland navigation is characterised by its price advantage in the transport of large and uniform quantities of goods over sufficiently long distances. However, this advantage is diminishing as a result of rising costs in the use of artificial waterways. Apart from the dependence on water levels and ice formation, the major disadvantage of inland waterway

[12] For the determination of sea freight rates see Korf, 1990, pp. 486ff.

transport is that it is tied to a narrowly limited network of waterways. The vastness of this network of waterways means that approximately half of the transport operations on inland waterways are carried out by broken transport in the strict sense, i.e. other means of transport have to be used to transport the goods from the delivery point to the reception point. This increases both the costs and the transport time, which is already relatively long due to the low transport speed of the ships.

Rail is the *major competitor* of inland navigation in bulk transport. The low transport costs are the main advantage of inland navigation over rail competition.

Railways and Trucks[13]

In contrast to the other means of transport, the *railways do not have any clearly defined areas of suitability* on account of their technical basis. Due to its performance characteristics and its transport prices, it can be involved in almost all types of transport, but in doing so it encounters competition from the other specialised means of transport with their strongly defined quality priorities.[14] The inland waterway vessel, for example, has the advantage of mass efficiency in bulk transport. The airplane has the advantage of the greatest speed in express traffic. In general cargo transport, the truck has time and cost advantages in local and area transport, as well as the advantage of being highly adaptable to individual transport needs.

The relatively high profitability threshold of the railways, whose use is actually only worthwhile from transport volumes of 30–35 truck units and a main running distance of 300 km,[15] further restricts their attractiveness for demanders of transport services. Because of its competitive advantages, the *truck* has displaced the railway from local and regional transport. In *short-distance transport, the* downtimes of the railways account for a very high proportion of the transport time and the time superiority of the truck is particularly great here. The greater density of the road network compared with the rail network enables the truck to form a better network than the railway. As a result, the truck is generally faster than the railway in *area transport.* The greater network-forming capacity of the truck means that transport by rail can often only be carried out in conjunction with truck transport. Additional transhipment costs are incurred in this case. Nevertheless, it can be advantageous to have the goods transport carried out by rail and truck rather than by truck alone.

In *long-distance transport*, the average transport speeds of the railways are greater. However, this transport speed advantage can be lost again due to additional times for transitions from train to train and for delivery of the goods by truck. However, as soon as

[13]For a detailed overview of the vehicles used in road and freight transport, see Buchholz et al., 1998, pp. 105ff.

[14]The quality priorities of individual modes of transport are captured in the concept of transport valences. Cf. Kummer, 2006, pp. 101ff. See also Fig. 8.3.

[15]Cf. Buscher/Hayens, 1998, p. 18.

several transport companies have to be involved in long-distance transport by truck, transport by rail often remains faster and more reliable.

Packaging often has to be more elaborate for rail transport than for truck transport. This means higher packaging costs, higher transport weight and possibly return costs for used packaging. In siding transport, additional securing of the transported goods against shunting impacts is generally necessary for rail transport.

The competitive relationship between rail and truck applies in a similar way as for general cargo traffic to wagonload traffic, where the transport volume involves a whole wagon or truckload at a time. However, it no longer applies to bulk transport, which involves several wagonloads or even whole freight trains.

Airplane

The *outstanding performance characteristics* of the airplane are transport speed, transport security and transport frequency. Further performance features are the simplicity of expeditionary handling, the manageability of the transport route and the possibility of an extraordinarily short-term shipment planning. However, the quality of the airplane's transport performance is offset by the relatively high costs of air freight. However, air freight is a good example of the need to consider the *total cost principle*. This is because the relatively high cost of air freight should not be viewed in isolation, but only in the context of the impact of the transport performance of the aircraft on other logistics costs. For this reason, the airlines also always emphasise when giving freight advice that one should not compare the pure freight rates of the airplane with those of the other means of transport, but must consider the relationship to the total costs incurred in delivering a good.

Due to the transport speed, transport safety and transport frequency when using the airplane as a means of transport, the delivery service level can be increased. However, the speed of air transport has a positive effect on shortening the delivery time only from sufficiently *long transport distances*. For short distances, especially if the place of delivery or receipt is not close to the airport, the time advantage of the airplane is lost again due to the necessary transport to and from the airport. Therefore, the airplane is not so much important as a means of transport for supplying the German market, but rather for supplying the European market and the overseas market. The recognition of the international airwaybill by more than a hundred airlines with uniform transport regulations guarantees smooth transport of the goods to be transported, even where several companies are involved in the transport.

If the airplane is used as a means of transport, the costs of capital tied up in inventories during transport are reduced because of the short transport time. The safety stocks in the delivery warehouses can be reduced because peaks in demand can be satisfied with the help of air freight from the central or factory warehouse. The number of delivery warehouses can be reduced because the various submarkets can be supplied by fewer delivery warehouses in the necessary time with the help of the airplane. This leads to a centralization tendency in the delivery network. It is also possible to achieve a high level of delivery service while reducing logistics costs through a combined application of centralized delivery of certain goods by air freight and decentralized delivery of other goods via delivery warehouses.[16]

The use of airplane as a means of transport can also reduce packaging costs, as the risk of damage is far lower than with other means of transport. Because of the comparatively low risk of damage from moisture, corrosion, breakage and theft when transported by air, the goods can be packed with less protection or even be shipped unpacked. This also reduces the transport weight of the consignments. This is significant because freight is generally calculated according to weight rates. Only when the ratio of volume to weight is greater than seven is the so-called volume weight (volume weight = volume: 7) used as the basis for freight calculation. Insurance costs are also reduced. In the case of goods which are largely exposed to the risk of obsolescence and spoilage, the associated costs can be reduced by using the airplane as a means of transport.

A number of *influencing factors* favor the use of aircraft as a means of transport in a delivery network. Before deciding for or against air freight, these influencing factors have to be analysed. Important factors favoring air freight are a large spatial extension of the market to be supplied, the difficulty of accurately predicting demand, the low turnover rate of a product, the high value of a product and the high density of a product.[17]

Air transport is also playing an increasingly important role in the transport of particularly bulky or heavy goods, e.g. industrial equipment, which cannot be transported by road or can only be transported with great difficulty and therefore usually require a rather complex transport chain of aircraft, inland waterway or seagoing vessel and truck. Helicopters, for example, are able to drop off such goods with pinpoint accuracy, but have the disadvantage of limited load capacities and ranges.

8.4 Intermodal Transport

Forms of Intermodal Transport

Intermodal (combined) transport in the narrower sense differs from broken transport in the narrower sense in that it does not necessarily involve the use of means of transport from different modes, but attempts are made to combine the advantages of different means of transport in a meaningful way and to simplify the loading of the goods necessary for transport. *Intermodal transport* is a system in which the means of transport integrated in this system are coordinated in such a way that the transfer of the goods to be transported from one means of transport to another causes as few handling operations as possible.

[16]For the model for determining the optimal combination, see Herron, 1968.

[17]For a discussion of the influence of these factors, see Herron, 1968. See also the discussion of selective stockkeeping in Sect. 5.4.

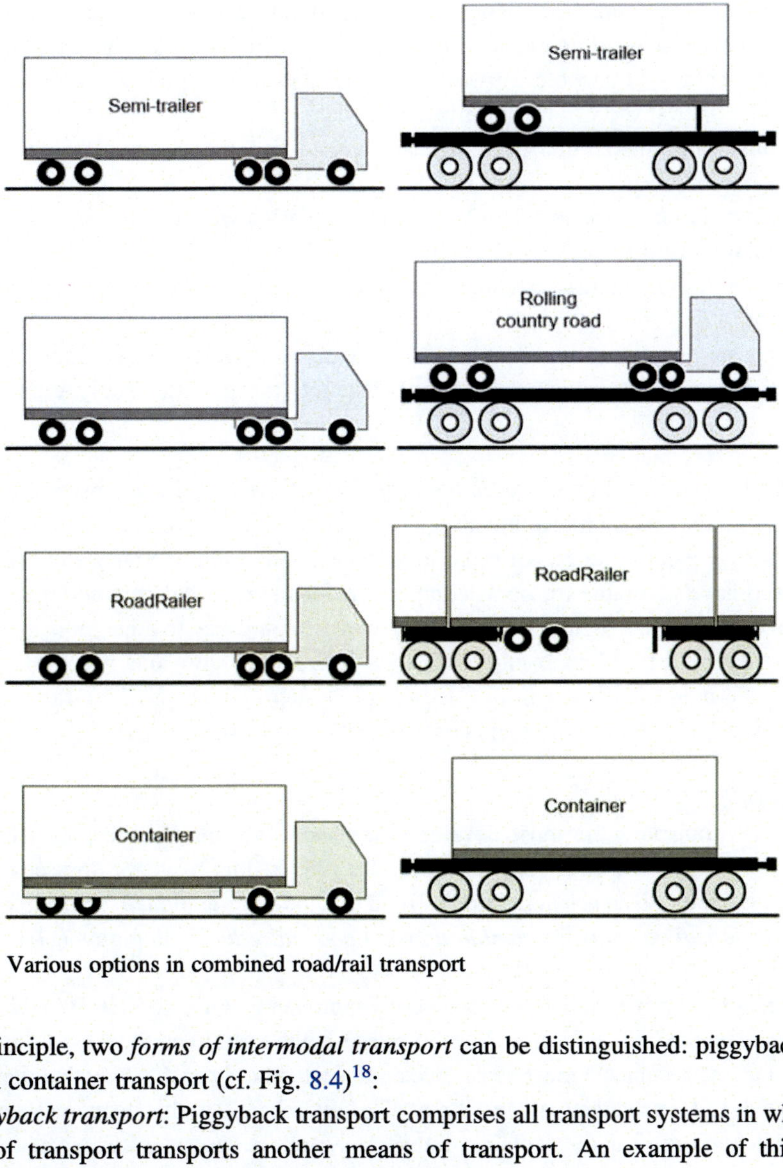

Fig. 8.4 Various options in combined road/rail transport

In principle, two *forms of intermodal transport* can be distinguished: piggyback transport and container transport (cf. Fig. 8.4)[18]:

Piggyback transport: Piggyback transport comprises all transport systems in which one means of transport transports another means of transport. An example of this is the transport of road vehicles by rail over long distances.

Container transport: Container transport comprises all transport systems in which containers are used as rationalising transport aids for the transport of goods, which cover the transport route independently on means of transport. The container is to be seen in

[18]For the forms, framework conditions and players in combined transport, see Boldt, 2009, pp. 46ff. For container transport, see also Sect. 7.3.

connection with the formation of logistical units. In addition to transport systems using large containers and small containers, transport systems using pallets are also sometimes included in container transport. However, only container transport is usually subsumed under intermodal transport in the narrower sense.

All forms of intermodal transport are characterised by three *features:*

- An attempt is being made to replace labour intensity with capital intensity through intermodal transport,
- Attempts are being made to mechanize the transfer of goods from one means of transport to another,
- Attempts are being made to combine the advantages of short-distance transport (generally trucks) with the advantages of long-distance transport (generally trains, planes, ships).

This leads to the development of *hub systems*, which are intended to ensure competitive transport times. Nodes are transhipment points (terminals) that have the necessary technical equipment (e.g. gantry cranes) for rapid transfer of goods from one means of transport to another and have favorable transport connections for the pre- and onward legs. They are located in centers with a high volume of goods and are connected to other hubs for the main leg in fast direct traffic. For example, DEUTSCHE BAHN AG runs special fast freight trains for intermodal transport in the InterCargo system, which provides fast night-time direct connections between German economic centers for urgent goods.

Piggybacking

In piggyback transport, the most developed transport systems are those in which road vehicles are transported over long distances by rail. In this way, the low costs of rail transport over long distances compared to road transport are exploited. In principle, there are *three technical variants* for *combining road and rail vehicles* in piggyback transport:

- Only the swap body (flatbed or closed body) approved by DEUTSCHE BAHN AG may be carried on the railway wagon. For this purpose, the tractor and the swap body serving as the loading receptacle must be connected to each other in such a way that they can be separated and exchanged at any time. Closely related to this technical variant of piggyback transport is the container transport system in container transport.
- The semi-trailer of an articulated truck is transported by rail. For this purpose, either the semi-trailers are shunted from the tractor unit onto the wagons or they are handled by crane.
- Complete trucks are transported by rail (rolling road). A low-floor wagon only 40 cm high was developed for this purpose. In 20 min, a 400 m long train consisting of such low-floor wagons can be joined by 16–18 trucks.

A number of time- and cost-saving methods are used in particular for the handling of semi-trailers and swap bodies, which can be divided into horizontal and vertical handling according to the type of handling. *Horizontal handling methods* are, for *example*, roll-off container transport systems (ACTS) or bimodal transport systems (Road Railer). The ACTS system has been in use in Switzerland, Austria, France and the Netherlands since the early 1990s. In this system, special roll containers can be transferred to rail without the need for additional handling equipment and with the aid of change equipment attached to the truck. Bimodal systems use either a rail running drive, which is permanently carried on the semi-trailer and is used on the rails by means of a hydraulic tightening of the road axle unit, or a separate rail running drive, which can be disposed of separately from the semi-trailer and remains permanently on the rails. This variant also includes the *trailer train* system, which represents an optimal adaptation of road vehicle technology and railway technology (cf. Fig. 8.4). The system consists of three components: Bogie, semi-trailer with coupling device and adapter as connecting element between bogie and semi-trailer. The semi-trailer of the truck is technically prepared for this purpose in such a way that, in conjunction with the bogie itself, it can take over the function of a railway wagon and thus a special railway underframe can be dispensed with.

In the case of *vertical handling* between rail and road, handling usually takes place with the aid of a crane. However, the transfer of high-bay warehouse technology to intermodal transport is also being discussed. In this case, trucks or freight trains should be able to drive directly into an automatic high-bay warehouse designed to accommodate containers, swap bodies and semi-trailers. A telescopic device will be used to pick up the containers and store them in and retrieve them from the high-bay warehouse.

Piggyback transport by rail can only be made *economically viable* over longer transport distances. As a *rule of thumb*, it is unprofitable below 200 km and generally only becomes profitable above 500 km. In competition with long-distance road transport, piggyback transport has to compete with the direct door-to-door transport of long-distance road transport.

Combining rail and road is not the only option for piggyback transport. Sea and inland waterway vessels can also be combined in the LASH (Lighter Aboard Ship) system. The LASH system consists of a mother or carrier ship, also known as a piggyback freighter, which has an on-board handling facility with which it picks up or sets down the lighters. These lighters are assembled into pushed convoys before or after sea transport and driven to the loading and unloading point on the inland waterways.

Institutionally, piggyback transport by rail is managed by KOMBIVERKEHR DEUTSCHE GESELLSCHAFT FÜR KOMBINIERTEN GÜTERVERKEHR MBH & CO. KG. (hereinafter referred to as KOMBIVERKEHR for short). The main shareholders are associations of the freight transport industry as well as road hauliers and transport companies. In addition to the shareholders, other transport companies and road haulage companies can also participate in piggyback transport, provided that the company agrees. DEUTSCHE BAHN AG holds a majority share in the company.[19] KOMBIVERKEHR has been licensed as a public rail transport company since November 1999 and provides wagons, IT and consultancy services.

Although piggyback transport is being promoted for ecological reasons and other transport service providers are developing intermodal transport services alongside the (formerly) state-owned railway companies,[20] it is not currently expected that this will lead to a significant shift of transport services from road to rail. Even if technical developments simplify and accelerate the change of transport mode when bridging the road-rail interface, current forecasts are quite contradictory.[21] Another obstacle to intermodal transport is different national standards for rail transport within Europe.[22] Only by overcoming geographical obstacles that are difficult or impossible for trucks to cross (e.g. waterways or mountains) and as a result of political measures based on ecological considerations can combined transport still gain market share in the future. This is the case, for example, with transit traffic through Austria and Switzerland.

Container Transport

Container transport can be carried out by combining all means of transport. The main leg can be carried out by rail, water or air. The preliminary leg generally goes by road. When container traffic is referred to, this primarily means the *combination of road/rail* and the *combination of rail or road with overseas ship.*

Container transport in which the main leg is carried out by ship is generally operated by special container shipping companies. If the main run is carried out by rail, special companies must be called in to carry it out. TFG TRANSFRACHT GMBH, a 100% subsidiary of DEUTSCHE BAHN AG, transports containers from German seaports to the final consignee throughout Europe in the hinterland and from the hinterland to German seaports. Another important company for container transport is INTERCONTAINER-INTERFRIGO SA, which is a company of the European railways for international container transports by rail. For transports by rail, TFG TRANSFRACHT or INTERCONTAINER are carriers. On request, it also organizes the necessary pre-carriage and onward carriage by road. It also offers additional logistics services.

The outline of the tasks of KOMBIVERKEHR and TFG TRANSFRACHT and INTERCONTAINER has already addressed institutional sub-aspects of logistics systems, which will be dealt with in detail in Part III following the discussion of the phase-specific subsystems of logistics.

[19] For the structure of DEUTSCHE BAHN AG, see Part IV, Sect. 16.5.

[20] Cf. European Commission, 2001, pp. 41ff.

[21] Cf. Aberle, 2009, p. 84 and literature cited there.

[22] In Europe, different track gauges, traction current systems and loading gauge profiles are currently still in use, which impede international rail and thus combined transport.

References

Aberle G (2009) Transportwirtschaft. Einzelwirtschaftliche und gesamtwirtschaftliche Grundlagen. 5., überarb. und ergänzte Aufl. München

Boldt O (2009) Unternehmensübergreifendes Qualitätsmanagement für korridorbezogene kombinierte Güterverkehre Schiene/Straße. Lohmar/Köln

Buchholz J, Clausen U, Vastag A (1998) Handbuch der Verkehrslogistik. Berlin u.a.

Bundesamt für Güterverkehr (2017) Marktbeobachtung Güterverkehr: Jahresbericht 2016. Köln

Buscher R, Hayens O (1998) KV-Verkehr wirtschaftlich? In: Logistik Heute 20 7/8, S. 18-21

Claussen Th (1979) Grundlagen der Güterverkehrsökonomie. Hamburg

Deutsches Institut für Normung e. V. (DIN) (Hrsg) (1983) Deutsche Norm, DIN 30 781, Transportkette: Grundbegriffe. Berlin/Köln

European Commission (2001) White Paper. European Transport Policy for 2010: Time to Decide. Luxembourg

Heidmann M (2013) Der Gigaliner in Deutschland und Europa. Entwicklung, aktuelle Diskussion und Blick in die Zukunft. Hamburg

Herron D P (1968) Buying Time and Saving Money with Air Freight. In: TDM 8 12, S. 25ff

Korf W (1990) Leitfaden für die Berufsausbildung des Spediteurs. Teil 1. 13 Aufl. Hamburg

Kummer S (2006) Internationales Transport- und Logistikmanagement. Stuttgart

Pfohl H-Chr, Krings M, Toben M-H (1993) Computergestützte Sendungsverfolgung im Kombinierten Verkehr. In: Management und Computer 1 4, S. 245253

Pfohl H-Chr, Schäfer Chr (1998) Analyse des Beschaffungsverhaltens von Industrie- und Handelsunternehmen zur Aufdeckung von Zeitpuffern im Beschaffungsentscheidungsprozeß – Ergebnisse einer Unternehmensbefragung. Arbeitspapiere zur Unternehmensführung und Logistik Nr. 24. Fachgebiet Unternehmensführung, Institut für Betriebswirtschaftslehre, Technische Universität Darmstadt. Darmstadt

Seidenfus H St (1972) Der Stellenwert des kombinierten Verkehrs in der Verkehrswirtschaft. In: Rationeller Transport 21 2, S. 79-82

Part III
Phase-Specific Subsystems of Logistics

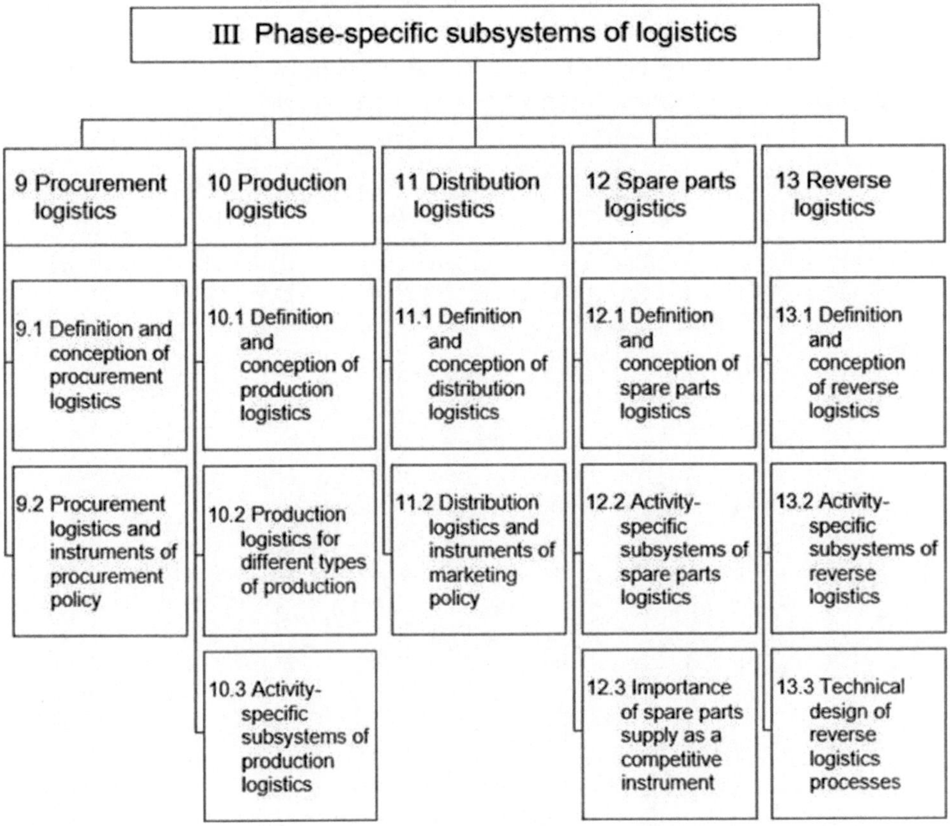

In order to further specify the decision-making problems characteristic of logistics, this chapter breaks down the overall logistics system into the following *phase-specific subsystems*, in accordance with the functional delineation made in Part I:

- Procurement logistics,
- Production logistics,
- Distribution logistics,
- Spare parts logistics,
- Disposal logistics.

Starting from the definition and characterization of the conception of these phase-specific subsystems, the problems arising from the various logistics objects and the interfaces of logistics in the various phases to the business functions of procurement, production, sales, customer service or maintenance and disposal are dealt with in particular. The activity-specific logistics subsystems are not dealt with in the phases of procurement and distribution, since the problems in this area are largely covered by the general description of activity-specific subsystems in the previous chapter.

9 Procurement Logistics

9.1 Definition and Conception of Procurement Logistics

Definition
Procurement logistics is a *market-linked logistics system*. It represents the connection between the distribution logistics of the suppliers and the production logistics of a company. The *objects of* procurement logistics are goods (raw materials, consumables and supplies, purchased parts and merchandise) that are to be made available (provided) to the company in line with requirements. The demand source is the procurement warehouse or—in the case of direct delivery—the first production stage in the company.

In order to be able to fulfil the *overall task of supplying* the company with input goods that are not produced in-house, legal availability, i.e. for example the acquisition of ownership of the goods, must be ensured in addition to physical availability. The activities associated with this fall within the scope of procurement (purchasing). The tasks to fulfill the overall task of supply are also summarized under the term materials management.[1] The demarcation between the procurement and the procurement logistics of an enterprise can be clarified by the following view. Accordingly it is the task of the procurement to provide supply capacities, to maintain supply capacities and to develop future supply capacities. Concerning the task fulfilment the marketing thought is transferred from the sales to the procurement area. This is obvious, since the organization of exchange relations between market partners, in this case on the procurement market, is to be seen as substantial characteristic of the marketing thought.[2]

[1] For the definition of materials management, see Grochla, 1990, p. 18. For a detailed presentation of the delimitation of the terms procurement, materials management and logistics, see Arnold, 1997, p. 1ff.

[2] Cf. Koppelmann, 2004, p. 77ff.

H.-C. Pfohl, *Logistics Systems*,
https://doi.org/10.1007/978-3-662-64349-5_9

The term *procurement marketing* continues to assert itself both in theory and in practice; analogously to the marketing instruments in the sales area a marketing instruments exists in the procurement area.[3] The procurement marketing fulfills the tasks set to it, by recognizing with the help of the instruments of the market study on present markets supply capacities and influencing the suppliers with the help of the procurement-political instruments (which are defined in reference to the instruments of the marketing policy in the sales area) in such a way that they supply the inquiring enterprise the desired products. The supply capacity procured in this way is to be constantly maintained by the employment of the appropriate instruments of the procurement policy, in order to receive the suppliers once won as regular suppliers. In addition the procurement range of the enterprise has to concern itself with it, which products on future procurement markets can contribute to the problem solution in the enterprise. It has to take care thus for the development of future supply capacities, whereby here regarding future supplier products and their meaning for the organization of the own products a close relationship with the research and development range becomes clear.

Procurement logistics use the existing supply capacities by generating the flows of goods and information for the provision of the input goods. Due to the market-linked, cross-company character of procurement logistics, not all decision-making processes that have an influence on the external provision of input goods are typically located within the company boundaries. Consequently, concepts and regulations are necessary that enable an overarching coordination of the flow of goods and a closer coupling between suppliers and buyers. Such regulations refer in particular to temporal and organizational aspects of the deliveries and often contain specifications about the time of the transfer of risks and costs.

This also determines to a considerable extent the *extent of the logistical span of control* in procurement logistics. A wide span of control on the part of the buyer means that he is largely responsible for the provision himself and must also take the measures to carry it out. This corresponds to a manifestation of the fetch principle, which, along with the bring principle,[4] forms one of the two basic types of staging. If, on the other hand, the task of external material staging is essentially assigned to the supplier, this is a manifestation of the bring principle with a correspondingly small span of control on the part of the buyer. An important example of regulations that directly influence the logistical span of control are the so-called Incoterms,[5] which are of course also important for distribution logistics.

Conception

Systems thinking in market-linked procurement logistics demands that, when making decisions, not only the relationships between the activity-specific logistics subsystems within a company or the relationships with other functions in the company be considered,

[3]Cf. Koppelmann, 2004, p. 271ff. On procurement market research, cf. also Large, 2013, p. 94ff.

[4]For the distinction between the bring and fetch principle in internal material provision, see Sect. 10.2.

[5]For information on the terms and conditions policy, see Sect. 9.2.

but also, in particular, the interdependencies with the logistics systems of the suppliers. For example, it must be taken into account whether measures to reduce inventories in the procurement warehouse of one's own company lead to inventories having to be increased in the sales warehouse of the supplier. In the long term, the supplier will either be forced to include the costs arising from an inventory increase in the pricing of its products or to restructure its own logistics system, which requires corresponding logistics competencies. The recognition and consideration of these interrelationships is expressed in particular in the concept of the logistics chain (from the supplier via logistics service providers to the customer) or the supply chain.[6] The associated view demands that inventories, which usually exist at several points in the logistics chain, should be grouped together, if possible, in just one stocking level, for which the most favorable location in the chain should then be determined, for example, also in a warehouse operated jointly by the customer and the supplier.

Total cost thinking can also be demonstrated using the example of inventory reduction. Lower inventories lead to smaller order lots and a higher order frequency, which can result in higher transport costs, if the demand remains unchanged compared to the previous inventory. This development depends very much on whether more frequent transport operations with lower utilization of the means of transport or smaller means of transport become necessary or whether it is possible—through suitable concepts of the transport organization—to continue to utilize the means of transport well, for example, by combining smaller order lots from different demand units into one larger transport lot. The higher order frequency leads to a higher number of order transactions, so that the costs of order processing and the corresponding costs in the area of purchasing, which are incurred when initiating and concluding a purchase contract for the material to be delivered (costs of coordination between the companies involved),[7] can tend to increase. The extent to which costs actually increase depends, for example, on whether information integration within the company, that is, between the purchasing area and procurement logistics and between the companies involved, succeeds in keeping the costs of information flow and coordination low. The use of information technology offers cost reduction potentials here, but in addition to the necessary investments, it also generates costs for the adaptation of the information systems at the supplier and the buyer. This also makes it clear that, analogous to the need described above to apply systems thinking across companies, total cost thinking must also be transferred from the logistics system of the individual company to the entire logistics chain. In addition, the supply level costs must be taken into account, which arise when a too low level of *supply service* causes disruptions in production or negative effects on the delivery service level in the company's distribution.

---

[6] See Part IV, Sect. 17.3.

[7] These costs are often referred to as transaction costs.

The *principles of external material provision* used have a significant influence on the costs incurred in procurement logistics. Three principles can be distinguished,[8] each of which places fundamentally different demands on procurement logistics.

The simplest principle of material provisioning is *individual procurement in case of need,* according to which the required material is procured only after the need arises. The great advantage for procurement logistics is that hardly any or only very little storage is required. This means that capital commitment and storage costs are very low, but production may be delayed until the material arrives. As a result, delivery times can become too long and production facilities cannot be optimally utilized. However, the resulting costs can easily exceed the savings due to the low capital commitment. The application of this principle is therefore generally only used for goods that can be procured immediately on the market and for unpredictable demand that cannot be planned. The latter case occurs, for example, in make-to-order production when certain materials are required that can only be used for a single production order.

The second supply principle is *procurement with stockpiling*, according to which the materials are[9] kept on call within the company for internal material supply and are available within a short time when the need arises. In this way, the company decouples itself from external provision and becomes less sensitive to fluctuations or unreliability in supply from suppliers. In addition, stockholding is usually associated with the purchase of larger quantities, so that economies of scale and corresponding cost savings can be achieved, for example by taking advantage of tiered pricing (volume discounts and the like). However, this comes at the price of higher capital commitment and higher storage costs. Under pure cost considerations, the problem of determining the optimal order quantity arises here.

The third principle of material staging is *production- or operation-synchronous delivery.*[10] It attempts to combine the advantages of the other two principles and avoid their disadvantages. According to this principle, the supplier must deliver the material on the dates determined by the production schedule of the procuring company. The current daily demand generally goes directly from the means of transport to the production sites, in this way short lead times of the material are achieved. Stocks are only held in the form of small safety stocks, so that capital commitment and storage costs are low. Production-synchronous delivery requires very reliable suppliers. It also requires close cooperation between supplier and customer on the basis of a more intensive exchange of information, so that any changes in the customer's production process can be passed on to the supplier immediately and he can adjust his own production accordingly. Only in this way is it

[8] Cf. Grochla, 1990, p. 23ff.; Large, 2013, p. 193ff. The term external means that the goods are provided from outside the company. Because of the possible close connection to production logistics, see also the principles of internal material provision in Sect. 10.2.

[9] The term internal means that the goods are provided by organizational units within the company.

[10] Today, one usually speaks of *just-in-time procurement or delivery.*

possible that warehousing is not merely passed on from the customer to the supplier, but that warehousing is largely avoided for both.

The prerequisites for the required close informational linkage are the existence of a cross-company planning and control system (delivery call-off system) and the extensive IT integration between supplier and buyer to enable electronic data exchange.[11] A delivery schedule system usually comprises three levels: The outline agreement, in which basic delivery conditions are agreed, the blanket order, in which the weekly or daily requirements of the recipient are specified for a period of a few months in each case, and the calling up, which, as the actual purchase order, specifies the quantities, dates, and delivery points of the material.

9.2 Procurement Logistics and Instruments of Procurement Policy

With the fulfilment of the total task of the supply of the enterprise the interdependencies between the procurement logistics and the procurement policy are to be considered. Similarly as with the instruments of the marketing policy in the sales range[12] four groups are to be distinguished with the instruments of the procurement policy: the product policy, the condition policy, the communication policy and the purchase policy.

Product Policy

Components of the product policy are the product design and the procurement program, in which it is determined which types of goods are procured in which quantities and in which temporal distribution within a planning period. Of importance for procurement logistics are initially decisions on an expansion of the *procurement program*, which are usually associated with increases in logistics costs, especially in the activity-specific subsystem of inventory management.[13] Such an expansion can result from needs for specific input goods from research and development or production, which can be justified e.g. by value analysis. This is contrasted by the efforts of procurement logistics to keep the scope of the procurement program as small as possible. This endeavor of the procurement logistics is supported by a standardization of the input goods with the help of standardization and typing as well as the use of assembled preliminary products (building groups) instead of individual parts. In the context of *product design*, attention must also be paid to the relationship between procurement logistics and packaging design or the necessary input goods (packaging materials). From a logistics point of view, packaging materials should be

[11] On the importance of electronic data exchange in procurement logistics and the possibilities of implementation, see Large, 2013, p. 195ff.

[12] See Sect. 11.2.

[13] For a detailed description of the relationship between product policy and distribution logistics, see Sect. 11.2.

used that are easy to transport and store and do not require a large amount of storage space. This is the case, for example, with foldable packaging material.

In close connection with the determination of the procurement program are the decisions on *in-house production or external procurement* (also referred to as make-or-buy) of the input goods for production. In recent years, there has been a trend in many industries to reduce the vertical range of manufacture in the company in favor of an increase in external procurement.[14] The increase in outsourcing means an expansion of the procurement program, which results in the disadvantages described from a logistical point of view. Furthermore it leads tendency to an increase of the complexity of the procurement logistics, since the number and the intensity of the supplier buyer relations increase. A procurement concept which is suitable to dissolve this logistic complexity represents the so-called Modular Sourcing. This concept provides for the procurement of entire modules, i.e. functional groups (assemblies or components) with a complex structure, from a module supplier (system supplier) instead of individual parts.

Finally, there is a close connection between the procurement program and procurement logistics with regard to *order quantities and order times*. From the short-term primary requirement of finished products, the secondary requirement of raw materials, individual parts and assemblies must be derived within the framework of requirements planning. From this, the period-specific net requirements can be calculated in further steps, from which procurement and production quantities per period are then determined.[15] As special problems of procurement logistics, the decisions about the amount of individual order quantities (order quantity problem) and the order times follow, which must be made in close coordination with the corresponding decisions in production and in production logistics.

Conditions Policy

Procurement logistics is directly related to *pricing policy* when calculating the cost price of the goods. This is because the cost price includes the costs of procurement logistics. This means that when negotiating the price of the goods and the scope of services, it is necessary to examine whether it is more favorable to procure a delivery ex works and handle the transport yourself (or hire a service provider) or to agree on free delivery, in which case the supplier provides the logistics service and includes it in the price. Purchasing goods ex works is particularly useful if positive economies of scope can be achieved by coordinating procurement and distribution logistics. For example, distribution and procurement transports can be coordinated in such a way that empty runs can be largely avoided and a high utilization of vehicle capacity can be achieved.

[14]This trend is also one of the reasons for the increased attention paid to procurement and procurement logistics in recent years. On the design of vertical integration, see also Large, 2013, p. 90ff.

[15]See the explanations on determining requirements in Part II, Sect. 5.2 and on replenishing and securing stocks in Sect. 5.3.

A survey of 560 German companies in 1997 showed that 66% of the goods procured are delivered free domicile or free unloading point. For 34% of the goods the procuring enterprises themselves take over the co-ordination of the transport. However, only 1.6% of all enterprises procure their procurement goods exclusively free domicile and only 2.1% take over the transport for all procurement goods. It can be seen that companies generally decide individually for each product who will take over the transport of goods with regard to procurement logistics.[16]

In international trade in goods, such specifications are made through the use of the corresponding *Incoterms.*[17] These are the international rules for the interpretation of customary contractual formulas. They are used by companies in the interstate movement of goods and define the terms of delivery or the obligations and rights (in particular the division of tasks, costs and risks) of the seller and buyer. The Incoterms, which consist of a total of 13 clauses, are divided into four main groups and differ according to the places where costs and risks are transferred.[18] Known clauses are:

- Ex Works (EXW), i.e. ex works sales: Here the principal (the buyer) determines all logistical services, such as the commissioning of a forwarding agent or a carrier. He also has to bear all the logistics costs incurred for this. The services of the consignor (seller, manufacturer) end with the loading of the means of transport. This is therefore a manifestation of the fetch principle.
- Free On Board (FOB): The shipper (seller) is responsible for all logistical services up to takeover of the ship. Transfer of risk and costs takes place at the ship's rail in the port of loading. The subsequent logistics activities are the responsibility of the buyer. It is proposed that this clause should only be used for the shipment of general and bulk cargoes by conventional ocean-going vessels and that the FCA (free carrier) clause should be used instead for the carriage of containers by sea, the carrier in this case being the shipping company or terminal operator carrying out the loading. Similarly, in air transport, the clause FCA should be used instead of FOB, as the seller cannot actually deliver on board the aircraft, but a handover to the air carrier already takes place beforehand.[19]
- Cost, Insurance, Freight (CIF): All logistical services up to the specified port of destination are determined and borne by the shipper (seller). The seller shall also pay the premiums for the transport insurance. The corresponding costs are to be included in the sales price by the seller. The buyer is responsible for the subsequent logistics services from the port of destination.

[16] Cf. Pfohl/Schäfer, 1998, p. 58.

[17] Incoterms: International Commercial Terms.

[18] Cf. Becker, 1992.

[19] Cf. Becker, 1992, p. 340 f.

- Carriage Paid To (CPT): The consignor (seller) determines all logistics services up to the destination station of the consignee specified in the purchase contract.
- Delivered Duty Paid (DDP): All logistical services up to the consignee are determined by the consignor (seller), who is also responsible for customs clearance. With this clause as well as with CPT, the shipper (seller) takes over the (entire) transport planning and the preparation of the shipping documents.[20] The operational activities are carried out either in the name and for the account of the shipper by contracted carriers or by own vehicles. The costs of providing the logistics services are to be included in the selling price. The application of these clauses means a realization of the bring principle.

Special consideration is given to procurement logistics considerations in price negotiations aimed at negotiating volume discounts. All too often it still happens that the buyer achieves a lower cost price by purchasing large quantities without taking into account that these quantities then tie up capital over a long period of time and cause large storage costs.

If purchase agreements are concluded in accordance with *purchase agreement forms* with special performance agreements (framework agreements, call-off agreements, successive delivery agreements, consignment reference, etc.), logistical facts in particular are often the subject of this special agreement. For example, successive delivery contracts specify the quantities to be purchased on certain dates. In the case of consignment purchasing, the vendor keeps a stock close to the consumer from which the consumer can take material as required. A typical example is also the delivery schedule system already described in production-synchronous procurement, in which the delivery schedules often contain very narrow delivery time windows.

Relationships between conditions policy and procurement logistics can also occur in *offsetting transactions* where the off-taking company's consideration is not in the form of money but in the form of goods.

Communication Policy

The communication policy is to clarify the ideas and intentions of the procurement policy to the supplier. Their goal consists of winning, of maintaining and of increasing the efficiency, the willingness to supply and the contract loyalty of the suppliers. Relations between communication politics and procurement logistics exist only in very general way, for example the close co-operation between supplier and customer with the production-synchronous supply presupposes straight efficient, delivery-willing and contract-loyal suppliers. Communication policy can thus help to win over suppliers who meet the requirements of procurement logistics. A typical instrument of communication policy is the implementation of so-called *supplier days*, to which the procuring company invites

[20]The main difference between the CPT and DDP clauses (apart from customs clearance) is that under CPT the time or place of transfer of costs can be specified in the contract of sale, whereas under DDP the transfer of costs always takes place at the named place of destination in the importing country.

current and potential suppliers and holds problem-solving discussions with them, which may also concern procurement logistics problems, or discusses possibilities for future cooperation.

Purchase Policy

As between the sales channels and the distribution logistics, so also between procurement channels and the procurement logistics particularly close relations exist. An essential decision of the procurement policy concerns the definition of the procurement *channel* (direct purchase, procurement via procurement aids or the wholesale trade), on which the input goods are to be procured, whereby this decision is strongly determined by the sales channel policy of the suppliers. Due to the analogy between the appropriate decisions in the distribution and procurement policy the remarks to the distribution policy can be transferred here.[21] Only two points should be emphasized, namely the problem of the number of suppliers and the problem of their geographical arrangement or dispersion.

A large *number of suppliers* reduces the risk of dependency and thus protects, for example, against delivery failures due to production disruptions at individual suppliers. However, it places a very heavy burden on the procurement logistics system and leads to a high level of complexity. Reducing the number of suppliers offers opportunities to reduce coordination and logistics costs. The material and information flows from the supplier to the customer can be made more efficient—especially through organizational measures and better mutual coordination through cross-company planning and control. If particularly close relationships with the supplier have to be established, as is necessary, for example, in the case of production-synchronous delivery of input goods, a reduction in the number of suppliers is inevitable. This is reflected in the trend towards single or double sourcing,[22] the restriction to one or two suppliers for the total requirements of a particular type of input goods, which is strongly linked to the concepts of just-in-time production and procurement.

Similar considerations can be made with regard to the question of the *geographical arrangement* or *spread of suppliers*. For example, international procurement—a corresponding procurement concept is often referred to as global sourcing—offers opportunities to exploit price and exchange rate advantages as well as to secure supply capacities in times of market shortage. Furthermore, the procurement risk, e.g. due to strikes, can be reduced by purchasing from suppliers in geographically different procurement markets. This is offset by longer transport distances and the associated higher transport costs, longer delivery times, larger inventories in transit and lower delivery reliability.[23] As a rule, the need to use different modes of transport and to involve an

---

[21] See Sect. 11.2.

[22] Double sourcing has the advantage over single sourcing of maintaining a certain degree of competition between suppliers and reducing dependency on them somewhat. Quotations are often made in such a way that 70% of the purchasing volume is allocated to the first supplier and 30% to the second supplier. Cf. Large, 2013, p. 135.

[23] Cf. Pfohl/Large, 1991.

increasing number of transaction partners, such as specialized logistics companies, increases with growing transport distances. Accordingly, the complexity of information flows in the form of order, transport and trade documents also increases, so that the costs of coordination also tend to rise.

Therefore, from the perspective of procurement logistics, it is generally advantageous to concentrate procurement on suppliers that are located as close as possible to the company. In the case of suppliers located further away, the establishment of an external procurement warehouse in the vicinity of the consumer, which is often operated by a logistics company, is an option in this respect.[24] The external procurement warehouse is particularly advantageous in cases where the respective delivery quantities called off from distant suppliers do not result in an economical transport lot size. The recipient is then supplied from the procurement warehouse, while the vendor transports several individual requirements from his location to the procurement warehouse in one transportation lot. If there is a large geographical spread of vendors (although at least several vendors should be located in a certain geographical proximity to each other), the *area freight forwarder concept*[25] can also be used. The prerequisite for this is that delivery ex works is agreed and the buyer thus has transport responsibility. The concept involves dividing the procurement area into regions and assigning the suppliers located in each region to a logistics service provider—the area freight forwarder. The latter organizes groupage tours, consolidates the individual consignments of various suppliers at a concentration point and transports the deliveries from there in the main leg as a single-target full load or block train to the customer. Transportation can also be organized in the form of a *milk run*.[26] The deliveries are picked up one after the other from defined suppliers and transported to the recipient in bundles. In this way, full loads can be made possible without creating additional handling at transshipment points.

References

Arnold U (1997) Beschaffungsmanagement. 2., überarb. und erw. Aufl. Stuttgart

Becker K G (1992) Incoterms und Beförderungsfrankaturen. Bedeutung und Interdependenzen aus Verladersicht. In: Internationales Verkehrswesen 44 9, S. 340-344

Grochla E (1990) Grundlagen der Materialwirtschaft. Das materialwirtschaftliche Optimum im Betrieb. 3., gründl. durchges. Aufl., unveränd. Nachdruck. Wiesbaden

Koppelmann U (2004) Beschaffungsmarketing. 4., neu bearb. Aufl. Berlin u.a.

Large R (2013) Strategisches Beschaffungsmanagement. Eine praxisorientierte Einführung. Mit Fallstudien. 5., vollst. überarb. Aufl. Wiesbaden

[24] See the classification of the procurement warehouse in the flow of goods in Part I, Fig. 1.6.

[25] Cf. Wildemann, 1995, p. 103ff.

[26] Cf. Meißner, 2013, p. 302 f.

Meißner S (2013) Schlanke Materialversorgungsprozesse am Beispiel eines Nutzfahrzeugherstellers. In: Günthner W A, Boppert J (Hrsg) Lean Logistics. Methodisches Vorgehen und praktische Anwendung in der Automobilindustrie. Berlin/Heidelberg, S. 293-304

Pfohl H-Chr, Large R (1991) Internationale Beschaffung. Einflußfaktor Logistik. In: Beschaffung aktuell o. Jg. 6, S. 22-30

Pfohl H-Chr, Schäfer Chr (1998) Analyse des Beschaffungsverhaltens von Industrie- und Handelsunternehmen zur Aufdeckung von Zeitpuffern im Beschaffungsentscheidungsprozeß – Ergebnisse einer Unternehmensbefragung. Arbeitspapiere zur Unternehmensführung und Logistik Nr. 24. Fachgebiet Unternehmensführung, Institut für Betriebswirtschaftslehre, Technische Universität Darmstadt. Darmstadt

Wildemann H (1995) Das Just-in-Time-Konzept. Produktion und Zulieferung auf Abruf. 4. Aufl. München

Production Logistics 10

10.1 Definition and Conception of Production Logistics

Definition
In accordance with the structure of corporate logistics according to the phases of the flow of goods, production logistics is arranged between procurement and distribution logistics and links them together. Production logistics comprises all activities related to the supply of input goods (raw and auxiliary materials, factory supplies as well as semi-finished products and purchased parts) to the production process and the delivery of semi-finished and finished products to the sales warehouse.

The *objects* of production logistics are characterized by the fact that they are subject to constant change within production due to machining and processing and thus place different demands on logistics in the course of the flow of goods. Production processes and logistics activities are closely linked, and in some cases even inseparable.[1] This is the case when a good also undergoes qualitative changes during storage and/or transport. Such overlaps are frequently encountered in the chemical industry. This close link is also emphasized when the (internal) material flow is mentioned as a task area of production logistics. This is because, according to the definition of the term material flow, not only the processes of transporting, storing and handling, but also the processing of the products are considered to belong to the material flow.[2]

A demarcation of production and logistics is possible if the provision of production capacities in the required capacity (quantitative and qualitative) and flexibility is defined as the *task of production*. Part of this task is also to maintain existing production capacities

[1] Cf. Günther/Tempelmeier, 2012, p. 9.

[2] Cf. Jünemann/Schmidt, 2000, pp. 5 f. In addition to the processes mentioned, testing, packaging and conveying are also part of the material flow.

H.-C. Pfohl, *Logistics Systems*,
https://doi.org/10.1007/978-3-662-64349-5_10

and to develop future ones. To this end, production processes must be designed in such a way that the production technology (e.g. various levels of automation) is matched to the performance capability and willingness of the employees. The care of existing production capacities has to be carried out by appropriate maintenance and servicing measures. Finally, production must address how future production capacities can be developed both by exploiting technical progress and by taking account of social change.

The task of logistics according to this view is to use the production capacities. To do this, production must first produce the goods required by distribution logistics. The relationships between production and distribution logistics will strongly depend on whether production is for the anonymous market (market production) or for specific customers (production to customer or order). In both cases, production logistics provides production with the material required for the production processes as part of internal material staging. External material staging concerns the material flow from the supplier to the company and is the task of procurement logistics. The interface to procurement logistics at the level of the flow of goods is formed by the receipt of goods with or without an incoming warehouse (procurement warehouse) or by the provision of the input goods at the first production stage directly by the supplier.[3] The corresponding interface to distribution logistics is provided by the transfer of finished goods to the sales warehouse or shipping.

Depending on the size of the company, it may be useful to *subdivide production logistics* according to spatial characteristics into intra-apparatus (within individual productive units), intra-plant, inter-plant, and inter-location logistics.[4] The latter two forms of production logistics, which occur when production is divided among several plants at the same location or when production takes place at several locations, have similarities to procurement and distribution logistics in terms of the design of the flows of goods and information.

Conception

In view of the aforementioned close linkage of logistical processes with production processes, *systems thinking* is an essential prerequisite for achieving optimal overall solutions. In the past, this was often neglected. Thus, although the successive production steps were optimally designed in themselves, mutual coordination was lacking, making it necessary to shield the individual production steps from each other by means of buffers.[5] Beyond the production logistics connections, the interfaces to other logistics subsystems must also be considered. One example is the delivery of production input goods by the

[3]Cf. Ihde, 2001, pp. 278 f.

[4]Cf. Endlicher, 1981, p. 26.

[5]For an approach to factory planning based on systems thinking with the dimensions of planning phases of the factory life cycle, planning objects and planning instruments, see Schenk and Wirth, 2004, p.27.

supplier directly to the first production stage. In this case, decisions in the area of production have a direct influence on the interface to procurement logistics.

The application of *overall or total cost thinking* in production logistics, i.e. the recording of all costs relevant to a production logistics decision, is particularly difficult precisely because of the close link between production and logistics processes. A typical example is the measures taken to reduce inventories in production. As a rule, they require a reduction in batch sizes in production, more frequent changes in the products produced and correspondingly more frequent set-up processes. Lower inventory costs are then offset by higher lot costs, which are determined by the fixed production run setup costs. Furthermore, the smaller production lots can lead to an increase in the number of internal transport operations and thus to higher transport costs. It is also necessary to take into account service level costs, which are caused by a too low supply service level and the resulting failures (interruptions, need to change the order sequence) of the production process. Furthermore, if inventories are reduced to a large extent and production and distribution logistics are closely interlinked, these failures can also have a negative impact on the delivery service level and lead to shortage costs.

Service thinking in production logistics means a transfer of thinking in supplier-customer relationships from a view primarily directed outwards, to cross-company flows of goods, to the flows of goods running within a company to supply production. The demand units or customers of production logistics are the individual productive units[6] of the production system. Distribution logistics can be seen as the customer at the end of the production process. Suppliers are either the respective upstream stages within the production process or procurement logistics. The requirements regarding the level of service of production logistics are determined by the goals of the named customers. In this context, the requirements of distribution logistics reflect the delivery service requirements placed on them by the company's external customers. A typical example of this is so-called customer-oriented production. It demands high flexibility and short lead times in the production areas, whereby production logistics measures to reduce lead times can primarily be applied to reducing the transition times, consisting of wait times, transport times and inspection times, between the processing stages.[7] If we consider the service requirements of the individual productive units rather than the targets derived from the requirements of distribution logistics, other supply service targets may well be derived. If, for example, the goal of maximum capacity utilization is being pursued for a particular productive unit, the supply service elements of time and flexibility will be less important from this customer's point of view. Which goals are pursued is influenced to a considerable extent

[6]Cf. Küpper/Helber, 2004, p. 6; Günther/Tempelmeier, 2012, p. 7. Productive units (or work systems) are the smallest independent organizational units in a production system.

[7]On lead times in production and the possibilities of influencing them, see Vahrenkamp, 2008, pp. 181ff.; Thonemann, 2015, pp. 161 f.

by production planning and control, which forms an important interface between logistics and production.[8]

10.2 Production Logistics for Different Types of Production

Influence of Production on Production Logistics

Both the design of the production system and the design of the production planning and control system (PPC system) have a direct influence on production logistics, and there are interdependencies between the two design areas.[9] In order to describe the different real manifestations of production systems, it is useful to differentiate between *program-related and process-related production types,*[10] since each production type is associated with different influences on the flows of goods in production logistics. The program-related production types are oriented towards the output side of the production system. The product characteristics, the number of products and the run size serve as characteristics for type formation. The differentiation of process-related production types is based on the organizational arrangement of the work systems (organizational types of production) and the structure of the production processes (form and continuity of the material flow as well as local binding of the products).

One result of the decisions about the production program and process is the volume and structure of the flow of materials and goods. From this, the *transport intensities* that are important from the point of view of production logistics can be determined. The transport intensity m_{ij} indicates the quantity of goods related to a specific transport good unit (for example, standard pallet) to be transported from a productive unit i to a productive unit j (with i, j = 1 ... m, where m equals the number of productive units) per unit of time. The representation of the volume and structure of the material flow and the corresponding transport intensities can be done by material flow matrices (also called from-to matrices in the context of factory planning[11]) and material flow graphs—with productive units as nodes and the material flows connecting them as edges of the graph. The importance of the transport intensities is to be seen in the fact that they—in connection with the corresponding transport distances—determine the transport performance to be achieved and have an influence on the level of the material flow costs. Furthermore, their knowledge

---

[8] On the question of assigning the tasks of production planning and control to the areas of production or logistics, cf. Hahn, 1989, p. 41. For an overview of different systems of production planning and control, cf. Thonemann, 2015, pp. 279ff.; Pfohl, 2016, pp. 138ff.

[9] Cf. Ihde, 2001, pp. 278 f. On the relationship between production types and production planning and control procedures, cf. the detailed presentation in Vahrenkamp, 2008, pp. 110ff.

[10] On the formation of production types, cf. Günther/Tempelmeier, 2012, pp. 10ff.

[11] Cf. Grundig, 2008, pp121ff. There, a detailed description of further representation techniques for transport relations is also given.

is the basis of layout planning (in-house location planning) of the production and logistics system.

The determination of transport intensities must be supplemented by a consideration of the relationship between the structure of the production process and the *number and arrangement of stocks* in the flow of materials and goods. Of importance here are the staging and vergence types—smooth, converging, diverging and regrouping[12]—of production and the distinction between continuous and discontinuous production. In many cases, the establishment of production stocks (intermediate stocks) is necessary due to technical conditions (process-related idle times) or makes economic sense.

After the presentation of the basic relationships between the production types and the production logistics material and goods flows, individual production types are discussed in more detail below.

Organizational Types of Production

The organizational types of production can be distinguished as shop, flow and center production.[13]

The characteristic feature of *shop production* is the arrangement of the productive units according to the activity principle. The productive units that perform similar machining tasks are combined spatially and organizationally into a single unit, the shop. Each production order must be transported to the individual shops according to the sequence of the machining operations to be performed on the workpieces. It can happen that an order has to be transported several times to the same shop. This leads to a large number of transport operations, as Fig. 10.1 illustrates. Typical here is the discontinuous transport of the material in batches of different run lengths to the next productive unit, resulting in the need for intermediate storage. Another reason for the occurrence of intermediate storage stocks is the difficulty of precisely coordinating work and transport processes, so that orders are either waiting to be processed before a productive unit or waiting to be transported onwards after processing has been completed. In addition, if there is little possibility of process standardization—different product types and a high degree of individuality in the production orders place different demands on capacities—there is the problem of optimally coordinating capacity requirements and available capacity while taking delivery dates into account. As a rule, there is a competing goal relationship here between the process-organizational goals of minimizing the (average) lead time of the orders (and thus also of the material) and maximizing the capacity utilization of the production system. This situation is referred to as the *dilemma of process planning or control.*

The characteristic feature of *flow production*, as Fig. 10.2 illustrates, is the arrangement of the productive units according to the object principle, i.e. according to the routings of the

[12] Cf. Küpper and Helber, 2004, p.9.

[13] On the organizational types of production, see Günther/Tempelmeier, 2012, pp. 13ff.

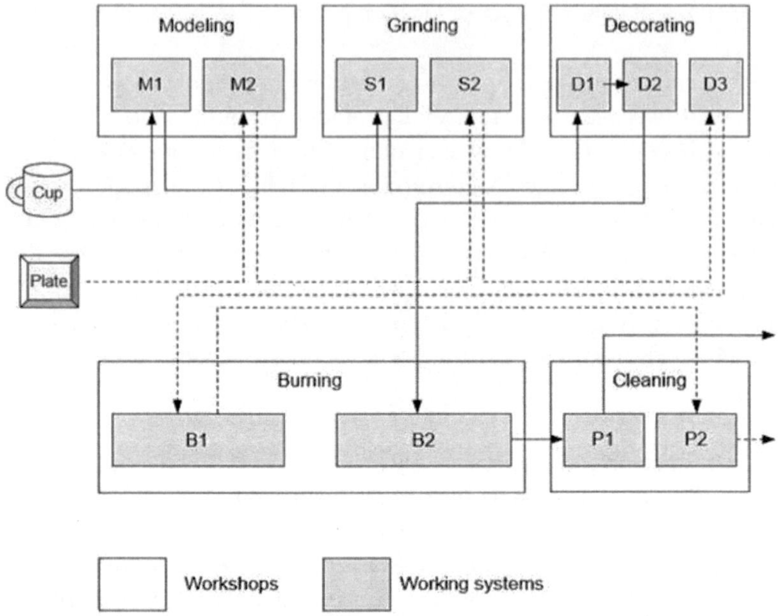

Fig. 10.1 Production of porcelain as an example of material flow in shop production

products to be processed. If the sequence of machining operations is the same for each workpiece and the productive units are arranged according to this sequence, the flow principle is realized. One form of the flow principle is sequence production, in which the work progresses without a direct time constraint on the operations. If there is a time constraint (cycle), a distinction is made between flow production in the narrow sense, in which the productive units are linked by independent conveying equipment, and the transfer line, in which the processing stations are linked to form an automated overall system. A typical feature of these two types of organization is the continuous transport of the material. Since, in contrast to shop floor production, shortages affect the entire production process due to the interlinking of the productive units, material supply is primarily subject to the requirement of permanent availability of the input materials. This is constantly threatened by equipment downtime, tool and personnel failure, etc., which is why buffer stocks are also set up. Since the production process is carried out under time constraints due to the cycle time, great demands are also placed on mechanized transport and handling.

Center production (group production) stands between shop and flow production. In this method, production units with different functions are spatially grouped together in order to enable the most complete possible processing of a specific group (parts or product family) of similar products or products that are related to each other in terms of production

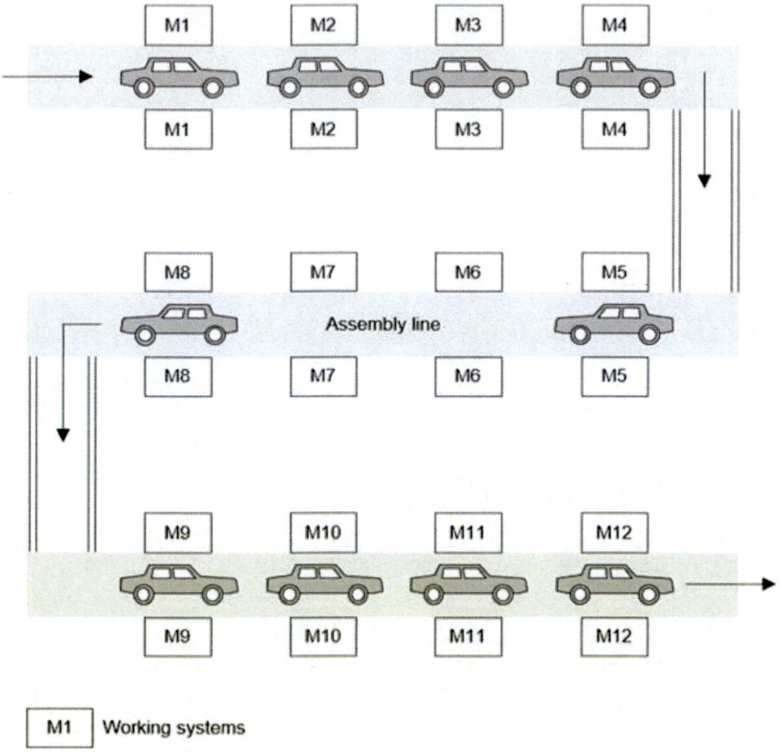

Fig. 10.2 Assembly of passenger cars as an example of material flow in flow production

technology. This procedure is also known as *production segmentation.*[14] A distinction is made between flexible manufacturing systems (FMS), in which numerically controlled machines are linked by an automated material flow system, and manufacturing islands, which are characterized by a lower degree of automation. Spatial aggregation leads to the fact that transport distances can be considerably shortened and the transport performance required to create a product (or process an order) can be reduced accordingly, as Fig. 10.3 shows. The spatial proximity enables simplifications in the execution of material transports and also contributes to an increased clarity of the production process and thus to a simplification of production planning and control. Potential advantages of center production are the reduction of waiting times for orders and the associated stocks in intermediate storages as well as the reduction of throughput times. A typical feature of manufacturing islands is function integration (complete machining). In terms of production logistics tasks, this means that the employees working in a production island are responsible not only for

[14]The separation according to product similarity (product types, production volumes, production processes, sales structure) is also referred to as vertical segmentation. Horizontal segmentation is the harmonization of capacities along the material flow. Cf. Vahrenkamp, 2008, pp. 55ff. and 274ff.

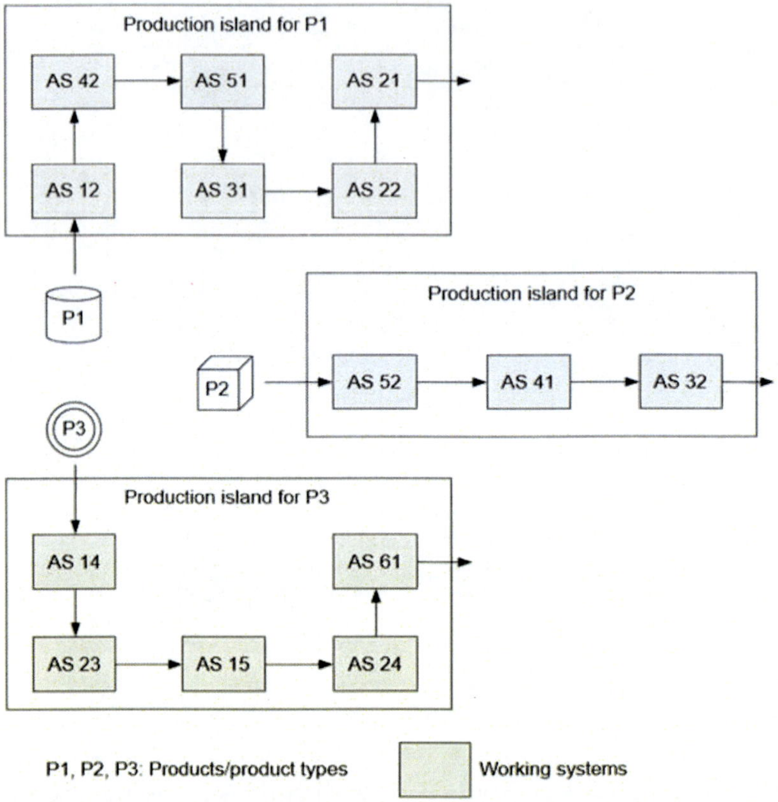

Fig. 10.3 Material flow in centre production (Source: Wäscher, 1993, p. 260)

their production tasks, but also for transport within the production island, handling and storage of the material. The material flow between the production islands is based on the flow principle.[15]

Production Program-Related Production Types

According to the degree of conformity of the products or the run size, a distinction is made between mass production, batch production, series production and make-to-order production. In *mass production*, production logistics has the task of supplying the productive units with the same input goods over long periods of time. In this case, logistics systems are required that can continuously provide the same service with as little disruption as possible

[15]For an example of material flows before and after manufacturing segmentation, see Vahrenkamp, 2008, p. 276.

while utilizing a high degree of mechanization. In *batch production*,[16] a special case of mass production, the production process must be interrupted each time a batch is changed and the production plant must be switched to a new batch. Logistics-relevant problems here are the determination of the variety sequence and the production lot sizes, which have an influence on the level of the resulting inventories. In *serial production*,[17] the problem of converting production facilities arises even more frequently. Therefore, the planning of production lot sizes is an important task. Logistics systems must be more flexible for this type of production than for mass production. Finally, the need for flexibility in production logistics is greatest in *make-to-order production*, which is almost always based on an individual customer order. The production logistics system must be able to supply the production units with constantly changing input goods in terms of type and quantity.

Principles of Material Staging

Another important factor influencing the design of the production logistics system, which is also indirectly related to the production types described, is the principles underlying the control of the material flow. A distinction must be made between the principles of physical material provision (bring and fetch principle) and the principles of control (demand and consumption-controlled material provision). If material staging is implemented using the *bring principle*, the productive units are supplied from the material store by employees specially assigned to this task. If, on the other hand, the *fetch principle* is used, the employee assigned to a productive unit must provide the unit with material from the material store himself.

The distinction between requirements-based and consumption-based material staging is linked to who triggers the material flow for material staging.[18] In *requirements-based material staging*, the material requirements for the production of the scheduled orders are determined on the basis of the production plan. Material withdrawal slips (also known as storage orders) are then created, which trigger material removal from storage in the store. The information can be transferred manually (in document form) or by means of EDP. The material is picked in the store according to the order quantity and forwarded to the productive units. Ideally, only the material that is necessary for processing the current production orders is in the productive unit.

In *consumption-based material staging*, the material flow is always triggered by consumption in the productive unit. The material flow is therefore "sucked in" by the consuming unit. This generally occurs in an order-neutral manner, i.e. not related to a specific order. The aim of this system is to ensure supply safety in the productive unit by

[16] In batch production, different versions of a product type are produced one after the other on the same machines. Cf. Bloech et al., 2007, p. 258.

[17] In serial production, different but similar types of products are produced in fixed quantities of similar products. Cf. Bloech et al., 2007, p. 258.

[18] See Pfohl, 2016, p. 138ff. and the methods of needs assessment in Part II, Sect. 5.2.

means of sufficiently large, but not oversized, material stocks. For this purpose, a reorder point is defined for each material item. If the stock level falls below these defined material stock levels, a material delivery is triggered and the stocks in the production unit are replenished. This control system is often linked to container fill quantities, so that empty containers are sent to the material store from the consuming units and these return filled. This system is also known as KANBAN, where KANBAN is the Japanese term for card. This card is often associated with the container and contains information about the material to be delivered, the quantity, the delivery time, and the point of consumption. The replenishment time from the store and the consumption in the productive unit are the most important factors in determining the reorder points.

The above-mentioned indirect connection between staging principles and production types can be seen in the fact that there are certain combinations that fit better with each other in terms of their respective characteristics. As an example, material staging in manufacturing islands typically follows the fetch principle, as this harmonizes with the above-mentioned idea of function integration. There is also a close connection between the principles of material staging and production planning and control. Thus, the KANBAN system—as a realization of consumption-controlled material staging—represents an important concept of decentralized production control.

Layout of the Production System

The layout of the production system is also closely related to the production types. *The object* of layout planning (in-house location planning) for production systems is the determination of the spatial arrangement of location-bound production subsystems on a usually specified area of a production site.[19] Different *levels* of planning can be distinguished. On the upper level, the internal locations for production segments (subsystems of the production area),[20] i.e. for example the arrangement of different departments—for example shops—within the production area, are to be determined. At the lower level, it is a question of determining the locations of the productive units within the production segments, i.e., for example, the arrangement of the machines together with the corresponding workplaces within a production department. In addition to the productive units, the stores must also be arranged. For this reason, productive units and stores together form the set of arrangement objects.

The *objective* of the arrangement is to minimize the material flow costs. However, since the recording of the relevant costs is usually difficult and subject to considerable uncertainties due to the length of the planning reference period, the transport performance, determined as the product of transport intensity and distance, is often used as an assessment criterion. Further criteria for assessing alternative layouts are the clarity and uniformity of

---

[19] Layout planning is often seen as an essential part of factory planning. Cf. Schenk and Wirth, 2004, pp. 273ff.

[20] Cf. Günther/Tempelmeier, 2012, pp. 82 f.

the flow direction of the material flows, the flexibility with regard to changing production requirements and the degree of space utilization.[21]

In *shop production*, the upper level of layout planning, i.e. the arrangement of the shops within the plant, is of particular interest. This is because it is particularly the transport services between the shops that need to be recorded. Within the shops, there are normally no material flows between similar productive units, so that the question of the arrangement of the productive units is of less importance here.[22] However, the location of intermediate storage (buffers) in the shops remains to be determined.

In *flow production*, the arrangement of the productive units is determined by the processing sequence on the workpieces. There is no typical layout problem in this respect. However, buffer stores play an important role in flow production because of their backup function, so that the determination of the number, locations and capacity of the buffer stores is of considerable importance in the context of layout planning.[23]

In *center production*, a distinction can be made between determining the locations of the centers and determining the arrangement of the productive units within the centers (center-internal layout planning) in accordance with the levels of layout planning. Ideally, due to the complete processing of part families, there are no flows of goods between the centers, so that only center-internal planning is necessary. The relative position of the productive units to each other is typically oriented to the material flow relationships described by the transport intensities. The type of transport system used, which is automated in the case of the Flexible Manufacturing System, for example, has a considerable influence on the choice of a particular basic layout structure. The locations are then arranged along the routing of the transport system.[24] If there are also flows of goods between the centers, layout planning outside of the center is necessary, which is analogous to determining the shops locations in shop floor production.

10.3 Activity-Specific Subsystems of Production Logistics

Transport

Factors influencing the *design* of a plant's *internal transport system* are the goods to be transported, the transport intensity, the transport route, legal regulations and, in particular, the type of production organization.[25] In flow production, for example, completely different means of transport and technical aids are used than in shop production. While forklifts, electric carts, hand carts, elevators and cranes are usually used there, flow production

[21] Cf. Wäscher, 1998, p. 324.

[22] Cf. Wäscher, 1998, p. 331.

[23] Cf. Thonemann, 2015, p. 337.

[24] Cf. Wäscher, 1998, p. 333.

[25] For inter-plant transport, see Part II, Sect. 8.1.

makes it possible to use continuous conveyors such as conveyor belts, moving tables, roller conveyors and chutes. The decision between the use of continuous and discontinuous conveyors represents a fundamental problem in the selection of means of transport.[26] In addition to the technical evaluation criteria (transport performance, load capacity, dimensions of the means of transport, etc.), the costs play a special role, as they can be considerable in the case of automated, capital-intensive transport systems. A high degree of mechanization and automation of the transport systems is favorable for goods flows that have a high transport intensity and temporal stability. If, on the other hand, a high degree of flexibility is required—both in terms of capacity and in terms of the course of the flow of goods—the extent to which a flexible form of automation, such as a driverless transport system, can be used or whether a simpler, manual solution is more suitable must be investigated. Finally, the flexibility of the transport system also has a direct influence on the resulting relocation costs when the layout of the production system is changed.

The following options are available for the information integration of the transport system, i.e. ensuring the *flow of information* to accompany the transport process: Stationary means of transport can be connected to a control center via data networks, while non-stationary means of transport (e.g. forklifts) receive the information by means of data transmission via radio or infrared.

Handling processes should be reduced and simplified as far as possible, as they account for a significant proportion of logistics costs in production logistics. To achieve this, it is necessary to combine measures in the various functional logistics subsystems. Thus, handling processes between two transport processes can be significantly simplified through the appropriate formation of logistical units, if this enables the use of technical aids, for example.

The *transport processes depend* on where the stores for supplying the productive units (work systems) are located—this will be discussed below in the warehouse subsystem—and how the supply takes place. The supply can take place in worker self-supply, in direct transport or as a "milk run".[27] The transport intensity is greatest in the case of worker self-supply, because the production worker has to move with his means of transport from the productive unit to the store and back again. In direct transportation, the demand is reported by the productive unit to logistics, which then delivers the requested quantity directly. In the "milk run", the transport intensity is the lowest. The supply is organized in such a way that a certain number of productive units are approached along a fixed route at predefined times and the required quantities are delivered. If necessary, empties are collected and transported back to the material store.

[26] For the classification of means of transport, see Part II, Sect. 6.4.

[27] Cf. Brungs, 2012, p.15.

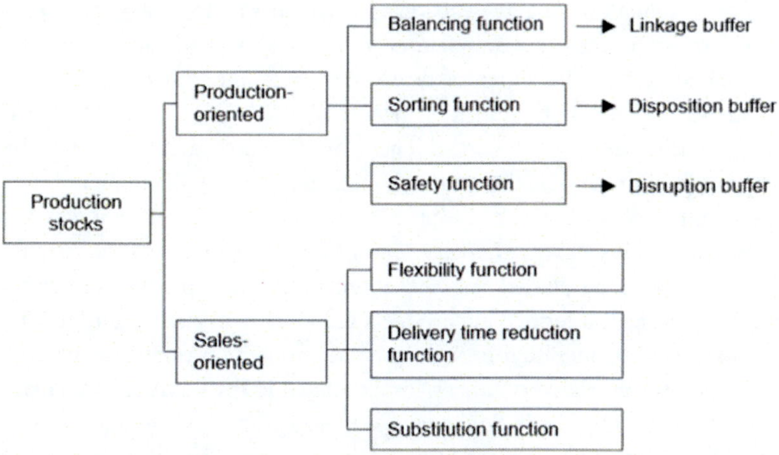

Fig. 10.4 Functions of production stocks (intermediate store)

Inventory Management

The functions of the inventory management subsystem within production logistics are compiled in Fig. 10.4.[28] In the case of *production stocks*, it makes sense to first distinguish between production-oriented and sales-oriented functions. Production-oriented production stocks have their causes in the production area. Sales-oriented production stocks have their causes in the sales market.[29]

The *balancing function* of production stocks is to link two processing stations with different inputs and outputs, which is why the resulting inventories are also referred to as linking buffers. Disposition buffers, on the other hand, serve the sorting function and disruption buffers serve the safety function. The *sorting function* is fulfilled by the possibility of changing the sequence of incoming and outgoing storage objects and by grouping or splitting lots. A typical example of disposition buffers are the intermediate storage areas upstream and downstream of a paint shop: "In order to change colors as rarely as possible, the parts run through the paint shop grouped by color. For this purpose, the batches are divided up, combined with each other and brought into a new sequence. After painting, they are sorted again in order to be delivered ready for assembly."[30] The *safety function* of disruption buffers means that any malfunctions that occur are localized.

Sales-oriented production stocks contain finished parts or assemblies. In contrast to production-oriented production stocks, sales-oriented production stocks do not arise primarily due to a lack of synchronization between successive machining processes, but rather

[28] See Part II, Sect. 5.1, for information on storage functions that are independent of the phase-specific subsystems.

[29] For a breakdown of these functions, see Salzer, 1981.

[30] Salzer, 1981, p. 7.

as a result of the production strategy decision as to whether and to what extent certain parts for products that are ultimately manufactured to order should already be completed and temporarily stored before the order is received due to internal orders. Sales-oriented production stocks thus mark the transition from make-to-stock to make-to-order production.[31] The *flexibility function* of sales-oriented production warehouses is to be able to manufacture a large number of end products from a relatively small number of intermediate products or from just a single body product in order to meet individual customer requirements. Inventories have a *delivery time reduction function* if the first processing stage does not have to be started and the entire production process run through after incoming orders. Instead, intermediate products are already available and only have to be assembled, completed or modified in features. The production times for the intermediate products are saved for the delivery time. Closely related to the flexibility and delivery time reduction function is the *substitution function* of sales-oriented production stocks. The storage of finished products is replaced by the storage of semi-finished products. Due to the smaller variety of types of objects to be stored, less capital is tied up. In addition, storage space is saved because semi-finished products can generally be stored more easily and with less space than finished products.

The view that inventories have certain functions that are important for the progress of production is opposed by another view that sees inventories as the "root of all evil",[32] since they *conceal weaknesses in the production and logistics processes*. According to this view, the aim should be production on demand, also known as just-in-time production. However, the elimination or extensive reduction of inventories may make it necessary to replace their delivery time reduction function with an increase in production capacity (deliberate creation of excess capacity) and/or an acceleration of logistics processes in order to be able to respond quickly to customer demands. Similarly, in the event of (seasonal) fluctuations in demand, it may be necessary to compensate for the loss of the inventory function of balancing supply and demand by aligning production capacity with peak demand. Thus, while cost-cutting advantages can be achieved by reducing inventories, they are to some extent offset by disadvantages due to the need for sufficiently high capacity. Overall, the reduction of inventories, in conjunction with the reduction of average lead times, is a suitable objective for production logistics. However, it must always be examined at which points in the flow of goods inventories are necessary for the provision of the required supply service and at which points inventories are only built up due to insufficient coordination between different areas. The planning of lot sizes in particular serves to reduce production inventories and improve coordination.[33]

---

[31] In this context, cf. the principle of postponement—here in the sense of postponing the decision to manufacture the final product—in Pfohl, 1994, p. 145; Pfohl/Pfohl, 2000, pp. 40ff.; Pfohl, 2016, pp. 110ff.

[32] Zäpfel, 1991, p. 217. Zäpfel explains here in great detail characteristics and also weaknesses of this material flow oriented view.

[33] On lot-size planning in general, see Part II, Sect. 5.3. On production lot sizes, see Vahrenkamp, 2008, p. 153ff.; Herrmann, 2009, p. 207ff.; Günther/Tempelmeier, 2012, pp. 195ff.

Warehouse/Store

In production logistics, the warehouse subsystem has to fulfill the functions of both providing storage capacity (*supply store*) and providing handling services (*handling store*). This handling service is necessary if the store is primarily used in production to hold goods for a short time between two transportation processes. The stocks located there will typically fulfill a buffer function, so that the focus is not on the storage process but rather on the movement processes.

The *in-house location decision* must be made as part of the layout planning. For this purpose, the number of necessary storage locations and the required storage area per store must be determined. The storage areas in the production area should be kept as small as possible in order to be able to use the areas for production on the one hand and to prevent larger quantities of goods from being stored there on the other. This is to prevent unnecessary stocks from being built up or orders from being "left lying around", which leads to long lead times. The decision on the degree of centralization of inventories and the arrangement of stores in production will depend, among other things, on the type of organization of production and production control, specifically also on the principles of material staging. For example, shop floor production often has centralized semi-finished product stores through which the issuing points are disposed of and the receiving points are supplied. In the organizational type of center-based production, each production center typically has a decentralized store, and there may also be a central store.

The *supermarket store* also serves to reduce the intensity of transport in supplying the productive units.[34] A supermarket is a store close to the productive unit from which its material requirements are picked. In a single-level supermarket, this is supplied with production material directly from the external procurement warehouse. In a two-stage concept, the supermarket is replenished from an upstream internal reserve store.[35] Single-stage systems are suitable when it is possible to limit inventory. "This primarily concerns components with low demand fluctuation and without spatially distributed multiple use, as well as with a low number of containers per delivery from the supplier with high delivery quality."[36]

Packaging

The requirements placed on the packaging by the production department are primarily directed towards the storage, transport and handling functions, whereby the latter is of particular importance. On the one hand, *mechanized* or *automated handling* of the entire packaging unit must be possible, and on the other hand, *accessibility of the individual goods* (workpieces) must be ensured. The problem of accessibility must be considered above all if the workpieces are to be accessed by a manipulator, for example an

[34] Cf. Meissner, 2013, pp. 294ff.

[35] See also the distinction between unit and picking warehouses in Part II, Sect. 6.3.

[36] Meissner, 2013, p. 298.

appropriately equipped industrial robot. In this case, a suitable arrangement and orientation of the workpieces is advantageous in order to facilitate the positioning of the manipulator. However, also in the case of manual access by employees, the accessibility and removal of the goods can be substantially facilitated by the suitable design of the packaging. For example, access to workpieces located at the bottom of a large container is comparatively laborious if the employee must bend over a side wall of the container to access. A partially foldable side wall can facilitate access here. The aspect of accessibility of the goods is closely related to the requirement to design the packaging in such a way that the goods can be fed directly from the packaging to the production processes. An example of packaging that meets this requirement is the small load carrier system (VDA-KLT-system), which consists of plastic containers (boxes) of different sizes. These boxes are used in the German automotive industry to transport piece goods such as washers, springs, screws, rubber and plastic parts.

Order Processing

The function of the order processing subsystem in production logistics is to ensure the flow of information associated with the flow of materials and goods.[37] This definition implies considerable overlap or a close link with operational production planning as well as production management and control.[38] The subject of order processing are *internal orders*, which can be divided into production orders and logistics-related transport and storage orders.

Production orders are either the result of lot-size planning or, in the case of customer-specific order production (customer production), they arise directly from a sales order. In scheduling, the production orders are given dates for processing, whereby these dates must be based on the agreed delivery dates, especially in customer production. This is followed by the actual production control with the determination of the order sequence, the provision of the resources (including the input goods) and the initiation of the processing of the orders. The realization of the information flow, which accompanies the physical provision of the input goods, is the actual task of order processing. The requirement for input materials at a particular time is specified in an internal (staging) order for production logistics. From this, transport and storage orders are generated, which trigger the execution of the corresponding logistics processes.

---

[37] The information flow-oriented definition of order processing used here represents a differentiation from a definition originating from production management. This definition refers to order processing as the service creation process induced by the customer, which includes all planning, management, execution and control activities that are directly related to the service to be provided. See also the definition of order processing in Part II, Sect. 4.1.

[38] For this reason, or because of the aforementioned difficulty of assigning production planning and control to production or to logistics, examples of companies can be found in practice in which the tasks of production planning and control are organizationally assigned to logistics, while only the day-to-day production control is reserved for the production area. Cf. Hahn, 1989, p. 44.

Transport orders control the transport system, i.e. they specify which means of transport has to pick up a certain quantity of a good at which location and to which destination this quantity is to be delivered. *Storage orders* define which quantity of a good is to be put away or removed from storage. The order information can be transmitted directly to the corresponding technical system (transport or storage system) or it can be contained in an information carrier (machine-readable sticker, microchip, transponder, and so on) assigned to the loading equipment or the workpiece. Transport orders can be transmitted, for example, by a corresponding control computer to a driverless transport system or a forklift truck equipped with a corresponding receiving device with a display for displaying the order information. The information flows downstream of the flow of goods typically comprise *feedback messages* given after the internal orders have been executed. They contain data on the time of completion of transport, handling or storage operations, as well as information on any errors in the execution of the order, and perform an important function in the control of the logistics and production system.

In order to achieve precise coordination of the various logistics processes with each other and with the production process, it is important to *integrate* the information flows associated with them. On the one hand, integration efforts are directed towards the creation of a common database (parts, means of transport, container, handling and storage data) that can be accessed by the control of logistics processes and production. On the other hand, they are directed towards the possibility of one function of a control system automatically triggering another function. An example of this is the triggering of a transport process by a productive unit reporting the completion of a production process and the provision of the workpiece for further transport.

The need for integration is particularly evident in the example of the information flows in the subsystems inventory management—the corresponding information system is referred to as the inventory control system—and warehouse (here specifically the store control system). This is because an internal order to remove a certain quantity of goods from stock requires both a change in the stock in the store and a physical removal process, which is initiated and monitored by the store control system. It is therefore advantageous here if there is a direct flow of information between the two systems. However, a close link should not only exist within production logistics, but also with the information systems of procurement and distribution logistics.

Integration through digital networks is a key feature of *Industry 4.0.*[39] The concept has not yet been conclusively defined and its implementation is based on the use of various technologies, such as machine-to-machine communication, robotization or embedded systems. The basic idea is the integration of information and goods flows through lean technologies. Such extensive integration is most likely to be possible within a factory. It is much more difficult to implement digital networks in external logistics systems, such as distribution logistics.

[39] Cf. in detail Pfohl et al., 2015; on digitalization as a logistics driver, see Pfohl, 2016, pp. 15ff.

References

Bloech J u.a. (2007) Einführung in die Produktion. 6., überarb. Aufl. Berlin u.a. Brungs F (2012) Der Milkrun in der Produktionslogistik. Aachen

Brungs F (2012) Der Milkrun in der Produktionslogistik, Aachen

Endlicher A (1981) Organisation der Logistik. Untersucht und dargestellt am Beispiel eines Unternehmens der chemischen Industrie mit Divisionalstruktur. Forschungsberichte zur Industriellen Logistik 18. Dortmund

Grundig C-G (2008) Fabrikplanung. Planungssystematik – Methoden Anwendungen. 3., neu bearb. Aufl. München

Günther H-O, Tempelmeier H (2012) Produktion und Logistik. 9. aktual. und erw. Aufl. Berlin u.a.

Hahn D (1989) Prozeßwirtschaft – Grundlegung. Produktionsprozeßplanung, steuerung und -kontrolle – Grundkonzept und Besonderheiten bei spezifischen Produktionstypen. In: Hahn D, Laßmann G (Hrsg) Produktionswirtschaft – Controlling industrieller Produktion. Band 2: Produktionsprozesse – Grundlegung zur Produktionsprozeßplanung, -steuerung und -kontrolle und Beispiele aus der Wirtschaftspraxis. Heidelberg, S. 5-237

Herrmann F (2009) Logik der Produktionslogistik. München

Ihde G B (2001) Transport, Verkehr, Logistik. Gesamtwirtschaftliche Aspekte und einzelwirtschaftliche Handhabung. 3., völlig überarb. u. erw. Aufl. München

Jünemann R, Schmidt T (2000) Materialflußsysteme. Systemtechnische Grundlagen. 2., Aufl. Berlin u.a.

Küpper H-U, Helber S (2004) Ablauforganisation in Produktion und Logistik. 3., überarb. und erw. Aufl. Stuttgart

Meissner S (2013) Schlanke Materialversorgungsprozesse am Beispiel eines Nutzfahrzeugherstellers. In: Günthner W, Boppert J (Hrsg.) Lean Logistics. Methodisches Vorgehen und praktische Anwendung in der Automobilindustrie. Berlin Heidelberg, S. 293-304

Pfohl H-Chr (1994) Interorganisatorische Probleme in der Logistikkette. In: Pfohl H-Chr (Hrsg) Management der Logistikkette: Kostensenkung – Leistungssteigerung – Erfolgspotential. Berlin, S. 201-251

Pfohl H-Chr (2016): Logistikmanagement. Funktionen und Instrumente. 3. neu bearb. und aktual. Aufl. Berlin/Heidelberg

Pfohl H-Chr, Pfohl P (2000) Postponement in der Supply Chain. In: Hossner R (Hrsg) Jahrbuch der Logistik. S. 40-45

Pfohl H-Chr, Yahsi B, Kurnaz T (2015) The Impact of Industry 4.0 on the Supply Chain. In: Kersten W, Blecker Th, Ringle Chr M (Hrsg) Innovations and Strategies for Logistics and Supply Chains. Hamburg, S. 31-58

Salzer J J (1981) Puffer im Materialfluß. In: VDI (Hrsg) Automatisierte Materialflußsysteme. Rationalisierungsreserven in der Produktion. VDI-Berichte 393. Düsseldorf, S. 5-15

Schenk M, Wirth S (2004) Fabrikplanung und Fabrikbetrieb. Methoden für die wandlungsfähige und vernetzte Fabrik. Berlin/Heidelberg

Thonemann U (2015) Operations Management: Konzepte, Methoden und Anwendungen. 3., aktual. Auflage. Halbergmoos

Vahrenkamp R (2008) Produktionsmanagement. 6., überarbeitete Aufl. München

Wäscher G (1993) Logistikorientiertes Layout von Fertigungssystemen. In: Milling P, Zäpfel G (Hrsg) Betriebswirtschaftliche Grundlagen moderner Produktionsstrukturen. Herne/Berlin, S. 77-104

Wäscher G (1998) Layoutplanung für Produktionssysteme. In: Isermann H (Hrsg) Logistik: Gestaltung von Logistiksystemen. Landsberg a. L., S. 321-336

Zäpfel G (1991) Produktionslogistik. Konzeptionelle Grundlagen und theoretische Fundierung. In: ZfB 61 2, S. 209-235

Distribution Logistics 11

11.1 Definition and Conception of Distribution Logistics

Definition
Like procurement logistics, distribution logistics is a *market-linked logistics system*. It combines a company's production logistics with the customer's procurement logistics. Distribution logistics comprises all activities related to the supply of finished goods and merchandise to the customer. Deliveries can be made directly from the production process or from the sales warehouse located at the production site and, if necessary, via other regional delivery warehouses. The *objects* in distribution logistics are not normally changed. Exceptions can occur if complementary services are offered in connection with logistics services when supplying the customer, for example, adapting a product to individual customer requirements (Customizing).[1]

The demarcation between the sales or distribution task area of a company in the sense of the characteristic of a specialized function (sales) marketing[2] and distribution logistics can be appropriately made according to the following view. The *sales task area* of the company has the task of making customer capacities available, maintaining existing customer capacities and developing future ones. It does this by using the instruments of market research to identify the needs (problems) of potential customers in current markets and by using the instruments of marketing policy to generate the various types of utilities that lead to the satisfaction of needs (problem solving). The customer capacity created in this way is to be constantly maintained by the use of the appropriate instruments of marketing policy,

[1] For a more detailed account, see Pfohl, 2016, pp. 110ff. and pp. 146ff. See the reference to the potential to carry out work to "finish" the product in the delivery vehicle in Crowley, 1994, p. 61.

[2] For a distinction between marketing as a corporate principle and marketing as a specialised function, see Part I, Sect. 3.4.

H.-C. Pfohl, *Logistics Systems*,
https://doi.org/10.1007/978-3-662-64349-5_11

in order to keep the customers once won as regular customers. In addition, the sales task area of the company must concern itself with which future products can satisfy needs or solve problems in future markets. In other words, it has to take care of the development of future customer capacities, whereby the close relationship with the research and development area becomes clear with regard to future products.

Distribution logistics uses the existing customer capacities by generating the necessary flows of goods to make the goods purchased by the customer physically available in the desired manner. Thus, it involves making goods available to the customer in the form of finished products. Basically, two types of provision can be distinguished here with the bring or fetch principle.[3] With the *bring principle*, the supplier's logistical span of control extends to the customer. The supplier delivers the goods to the customer via his distribution logistics. With the *fetch principle*, on the other hand, the customer's logistical span of control extends to the supplier. The customer collects the goods provided for him by the vendor himself via his procurement logistics (self-collection).

The provision task of distribution logistics is determined by the delivery service required on the market. As was shown in the characterization of the logistics conception through value and utility-oriented thinking, the supplier can often generate the necessary preferences for his goods with the customer precisely by offering an appropriate delivery service. Distribution logistics is therefore also referred to as an *instrument of marketing policy*, although the term marketing logistics is then more commonly used.[4] However, the customer's preferences are only generated by the delivery service as an output of the logistics system. The customer is therefore only interested in this output, not in the input of the logistics system and the processes that take place within it.

There is a similar relationship between sales and logistics as there is between sales and research and development. The sales area defines the requirements for product creation, and the research and development area attempts to create products that meet these requirements. However, cost considerations on the part of the research and development area may cause these requirements to be put into perspective. It is the same with delivery service. The sales area sets the requirements for the delivery service level. The logistics area tries to create logistics systems that meet these requirements. However, here too, cost considerations can lead to these requirements being put into perspective in the sense of a differentiated delivery service policy.

Conception

The statements on the scope of tasks and the conception of procurement logistics are also relevant for distribution logistics. This is because, depending on the type of division of labor undertaken in the logistics channel—depending on the type of division of the logistical span of control between the supplier and the customer—tasks can fall into the

---

[3] See the corresponding comments on procurement logistics in Sect. 9.1.

[4] See Pfohl, 1972, pp. 44ff.

area of either procurement logistics or distribution logistics. In terms of the marketing conception, distribution logistics must also be informed about the customer's procurement logistics problems in order to be able to support the customer in solving the problem. In the following, therefore, statements that are *mirror images of procurement logistics* will not be made for distribution logistics; only the specific distribution aspects will be discussed.

The starting point for the characterization of the logistics conception is value and utility-oriented thinking.[5] *Customer orientation* is therefore important for all subsystems of logistics. However, it is particularly important in the area of the company that establishes the direct link to the customer on the sales market, i.e., also in distribution logistics. This is clearly expressed in the term marketing logistics, which was used in the past. This is also underlined by the fact that the employees in a distribution warehouse often have much more customer contact than the sales representatives.

Customer orientation results in the great importance of *service thinking* for distribution logistics. It is necessary to constantly search for innovative possibilities that represent better logistical problem solutions for the customer. Two trends in marketing pose special challenges in this respect. Firstly, the classic marketing principle of "market-oriented production" is increasingly being replaced by the future-oriented marketing principle of "sell first, then produce". Secondly, the service to the customer is increasingly being provided in a very differentiated manner according to the J4U-principle (Just for You).[6] Both tendencies demand great speed and flexibility in service.

In addition to the emphasis on service thinking, customer orientation has two consequences for logistics *systems thinking:* First, it must be taken into account that the types of utilities generated by logistics never lead to the satisfaction of customer needs on their own, but only together with the types of benefits generated in the other areas of the company. Consideration of the interrelationships between distribution logistics and the instruments of marketing policy, discussed in detail in the following section, is a prerequisite for the customer's satisfaction with the company's performance. The customer appreciates a perfectly executed order, which includes not only a perfect delivery service but also, for example, perfect product quality or perfect advice. Distribution logisticians must be partners of the customer in the same way as sales representatives. The second consequence for systems thinking is the need for cross-company cooperation in the value chain to satisfy the customer. Only if manufacturers, retailers and logistics service providers do not act in an uncoordinated way or even against each other, but in a coordinated way and with a division of labour according to their competences, can maximum benefit be generated for the customer at minimum cost. This is, for example, the insight that has led to the demand for more cooperation in the consumer goods sector in so-called *ECR systems* (Efficient Consumer Response).[7]

[5] See Part I, Sect. 2.1.

[6] See Crowley, 1994, pp. 60 f.

[7] See also paragraph 11.2.

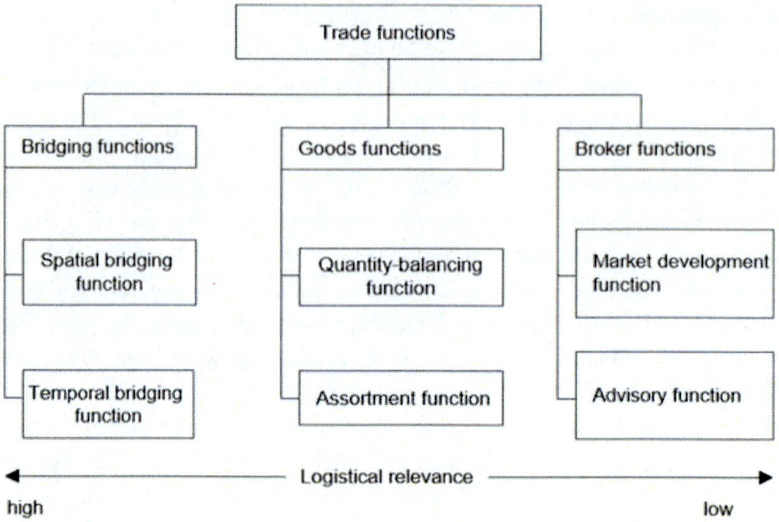

Fig. 11.1 Functions of trade (Source: based on Seyffert, 1972, p. 8 f)

The functions that can be fulfilled by the manufacturer, retailer and logistics service provider on the basis of their competences play a major role for cooperation in the sales channel. The example of retailing will be used to illustrate such functions. From a microeconomic perspective, three groups of retail functions with different logistics relevance can be distinguished, as shown in Fig. 11.1.[8]

The *bridging functions* are logistical functions, while the *goods functions* also have an acquisitive character. Quantity balancing is achieved by regrouping the quantities of goods by combining many small quantities required by individual customers into quantities suitable for production and delivery. The assortment function fulfills customers' desires for multiple choices or "one stop shopping" that a manufacturer alone could not provide. Both sub-functions have a major impact on the logistics of retail companies because the regrouping is generated by picking and packing at upstream stages (usually central or regional warehouses). The direct interdependency between logistics and the assortment function can be illustrated by the fact that a wide assortment of heterogeneous goods also results in higher logistics costs because, for example, the demands on inventory management increase if a consistent delivery service is to be ensured.

The *brokerage functions of* trading are primarily of an acquisition nature. They are classified as the most important for the performance of a trading company. Market development for the manufacturer is essential, since it is often only through the listing of a product, i.e. its inclusion in the product range by retailers, that it can be sold on the market in large quantities. The optimal placement on the shelves can also have a decisive influence

[8] An overview of similar breakdowns is provided by Barth et al., 2007, pp. 25ff.

on success. Accordingly, this question is often the focus of discussions between retailers and manufacturers. However, retailers not only act as brokers vis-à-vis manufacturers, but also in relation to customers, e.g., in the form of purchasing advice. Due to its brokerage function, the retail trade is, therefore, an intermediary between manufacturers and customers and is therefore also referred to as the "gatekeeper" of the sales channel.[9] In this context, the proximity to the customer and the possibility of direct control of the success of the goods are a decisive factor for the power of the trade in the sales channel.

11.2 Distribution Logistics and Marketing Policy Instruments

If the delivery service is seen as an instrument of marketing policy,[10] the interdependencies with the other instruments must be taken into account. This is because the instruments of marketing policy only have an effect on the customer in their combination in the marketing mix. An impression of the existing interdependencies is given in the following, whereby the marketing policy instruments are summarised as instruments of product policy, conditions policy, communication policy and distribution policy.

Product Policy

Components of product policy are the product range, product design, after-sales service and warranty. Decisions regarding the first three components must take into account effects on distribution logistics, while the fourth component is more concerned with spare parts logistics.

In many industries, there is a trend to expand the *product range* through new products or through product differentiation. This is a consequence of the strategy of differentiated market treatment, which requires a specific product and/or other special marketing policy measures for each market segment.[11] However, the effect of an expanded product range on the area of distribution logistics is often overlooked. The larger the product range, the more complex the logistical problems become.

An expansion of the product range results in new problems of order processing, packaging and transport. One major effect is the increase in inventories. Figure 11.2 shows an example of the relationship that experience has shown to exist between the expansion of the product range and the level of stocks: For example, in a company a conventional shampoo (A) is to be replaced by three special shampoos for blond (B), black (C) and brown (D) hair. If we make the pessimistic assumption that sales do not increase compared to shampoo A and are split between shampoos B, C and D in the ratio 60:30:10, then Fig. 11.2 shows that this can increase inventories by about 60%. If one makes the

[9]On the concept of gatekeeper, see also Meffert et al., 2012, p. 145.

[10]See Havighorst, 1980, pp. 96ff.

[11]For strategies of market cultivation see Meffert et al., 2012, pp. 186ff. and pp. 233ff.

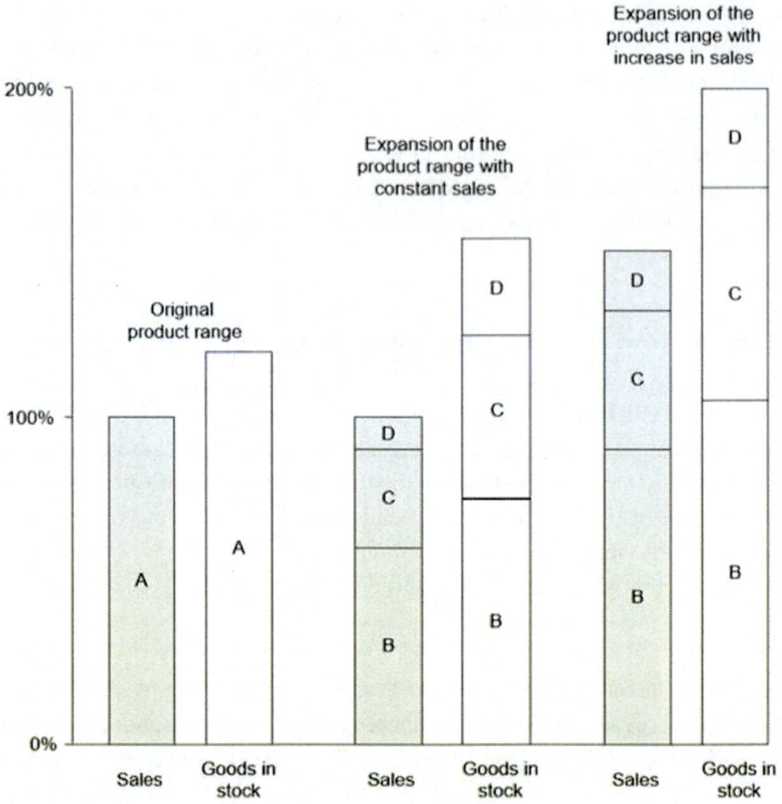

Fig. 11.2 Effect of an expanded product range on inventories

optimistic assumption that sales will increase by 50% compared to product A, then inventories will increase by 100%. The increase in sales achieved by expanding the product line increases the unit cost of inventory. Of course, these empirical values of a business consultant do not always apply. However, they are in any case correct in tendency. For it is generally true for the composition of inventories that the lower the sales of a product, the larger the inventories must be relative to the sales of that product.

If a product is newly included in a company's product range, care must be taken when introducing the product to the market to ensure that the supplier has sufficiently high stocks which can be delivered quickly. This is because experience shows that retailers always have very low stock levels during the introductory period of a new product until, as a result of constant demand, they have become accustomed to taking sufficient account of the product in their orders.[12] During the launch period of a product, retailers must therefore be

[12]See Stackelberg, 1969, p. 62

able to be supplied particularly quickly so that they are in a position to satisfy the demand directed at them.

A product that sells well is not always a product that can be easily moved through the logistical system. A poor *product design* from a logistical point of view, for example, requires oversized packaging, thus increasing the volume and decreasing the density of the units to be transported. This results in increased costs for handling, storage, transport and packaging. Sometimes market requirements make it impossible to sufficiently consider logistical aspects in product design. Often, however, people are completely unaware of the impact of product design on distribution logistics.

When designing a product, one should always consider what problems it may cause for its delivery during transport and in the warehouse. Such problems[13] may arise from the weight or bulkiness of the product, its shape, its fragility and special packaging requirements. For example, in the case of one manufacturer of chairs, it was demonstrated that there could be a doubling of transport costs when the chairs were delivered if they were constructed in such a way that they did not fit together. In another case, a manufacturer of office machines managed to reduce transportation costs for an office machine by 60% by simply redesigning the console of the machine. It also reduced the previously high percentage of damage during delivery to an insignificant level, saving further costs while increasing customer satisfaction.[14]

A major advantage for the logistics system arises if, in product design, it is possible to achieve a certain degree of standardization in the dimensions of the products. This makes the tasks of packaging, storage, handling and transport much easier.

If spare parts are required for the provision of *after-sales services*, then the quality of the after-sales service depends to a large extent on the support provided by appropriate spare parts logistics.[15] One might think, for example, of customer service in the office machine industry or in the automotive industry.

Conditions Policy

The components of the conditions policy are price, financing conditions and leasing. Relationships with distribution logistics exist primarily with regard to price. In the case of international transactions, there are also relationships between distribution logistics and financing conditions, but these cannot be discussed here.

If one assumes that the company must orient itself to a cost price as the lower price limit when *setting prices*, there is always a fundamental relationship between distribution logistics and pricing policy via logistics costs. In two areas of pricing policy, spatial price differentiation and price differentiation according to sales quantities, additional logistical cost considerations must be made. Although price differentiation cannot be

[13] See Bowersox et al., 1968, pp. 147 f.; Magee et al., 1985, p. 34 and pp. 46 f.

[14] For this example, see Smykay/LaLonde, 1967, pp. 36 f.

[15] See Sect. 12.

carried out on the basis of cost considerations alone, since price differentiation can only be successful if there are different demand elasticities in the company's overall market and if submarkets can be delimited, cost considerations are essential in order to decide whether prices are viable for the company.

Quantity discounts are to be set within the framework of price differentiation according to sales quantities. The supplier should try to pursue an *optimal discount policy*, which is also determined by factors influencing distribution logistics.[16] This is because a rebate structure that is graded according to order sizes or purchase quantities results in concentration points in the distribution of order sizes or purchase quantities. Care must be taken to ensure that the transport, handling, storage and packaging requirements associated with this concentration of order sizes or purchase quantities match the capabilities of the logistics system.

Among the questions that need to be answered in the context of *geographical price differentiation* is the extent to which the cost of delivery of the goods is included in the price. If delivery is made free domicile, the supplier bears the entire cost and risk. The greatest advantage is gained by customers whose location is furthest away from the supplier. If the price is understood to be ex works, the customer will generally add the additional costs incurred to the price when calculating the purchase price. This is because, in the end, the only price that counts for the customer is the price at which he actually has the good in his warehouse. It is not possible at this point to discuss the problem of whether the delivery costs should be borne in full by the supplier or by the customer, or shared in some way.[17]

Price reductions from a logistical point of view should also be considered if deliveries are made to a central location of the customer and the customer takes over the supply of its branches or plants itself. The customer will then have to weigh the costs of his procurement logistics against the price discounts granted.

Communication Policy

Components of communication policy are advertising, public relations, sales promotion and personal selling. When using all instruments of communication policy, it must be ensured that it is coordinated in terms of location and time with the necessary measures of distribution logistics in order to be able to satisfy an increase in demand caused by it. Interdependencies exist, however, not so much between distribution logistics and public relations not related to individual products, but above all between distribution logistics and advertising, certain sales promotion measures (e.g. measures to influence consumers at the point of sale or support for special promotions by sales intermediaries) and personal selling.

Advertising campaigns must be carefully backed up by logistical measures, taking into account the time lag between the use of advertising and the change in sales. After all, the

[16] See Heskett et al., 1973, p. 236; Magee et al., 1985, pp. 387ff; Powers/Closs, 1987.

[17] See Heskett et al., 1973, pp. 216ff; Magee et al., 1985, pp. 371ff.

most original advertising is of no use if the demand it stimulates cannot be satisfied in time. An in principle good advertisement can have negative consequences by lack of logistical support, if the customer moved by the advertisement to the purchase is not supplied immediately, turns away therefore annoyed from the enterprise and is lost thus as customer. However, a lack of coordination between advertising and distribution logistics not only has a negative effect on the outside in the form of lost sales and lost customers, but also on the processes taking place within the company.[18] In the area of distribution logistics, for example, high-cost measures suddenly have to be taken in an attempt to respond to the unprepared increase in demand. The working atmosphere for cooperation is certainly not promoted by such events.

Another relationship between advertising and distribution logistics results from the fact that a good delivery service can be highlighted in advertising. An improved delivery service with the help of distribution logistics provides one of the strongest advertising arguments in many industries. In addition, an attempt can be made to develop a better product image through advertising, which allows a higher price and thus compensates for higher costs in distribution logistics.[19]

It is also possible that a psychological effect caused by the location of a delivery warehouse can be exploited through communication policy to increase demand. Thus, there is reason to believe that some customers will purchase their goods from the company that maintains a delivery warehouse in their city due to the psychological effect of the location.[20]

The need to consider the relationship between distribution logistics and *sales promotion* arises in particular when packaging is regarded as a medium of sales promotion.[21] Packaging should then take on the advisory, but also purchase-stimulating function which, in relation to the consumer goods sector, comes into its own above all in the case of impulse buying. Sales-promoting packaging principles are essentially advertising-psychological principles that are based on perceptual-psychological findings. Frequently, sales-promoting and logistical packaging principles will be in conflict with each other, so that sales promotion and distribution logistics will have to make a compromise in packaging design.

The fundamental relationship between *personal sales* and distribution logistics is that the sales representative must be informed about the performance of the logistics system. Otherwise, there is a risk that he will make delivery service promises to the customer in the personal sales discussions that the distribution logistics cannot keep or that cause too high costs.

Distribution Policy

Components of the distribution policy are sales channel, field service and delivery service. All three aspects are related to each other.

---

[18] See Bowersox et al., 1968, pp. 23 f.

[19] See Smykay, 1973, pp. 36 f.

[20] See Constantin, 1966, p. 43.

[21] See the marketing functions of packaging in Part II, Sect. 7.1.

Sales channels differ according to whether they connect the supplier directly or indirectly with the final purchaser and which institutions are involved as sales intermediaries (trading companies) in the second case. In the case of field service, the question is whether sales are organized via the company's own sales organs (members of the management or travelling salesmen) or via external sales organs (commercial agents, brokers). Sales channels and field service are also referred to as selling channels.[22] The *sales channel* occupies a special position among the instruments of marketing policy, since it strongly determines the use of all other instruments. This is because a company decides through the selection of the sales channel which marketing tasks are to be carried out by itself and which are to be delegated to independent market partners.

The close connection between distribution logistics and the sales channel results from the fact that the sales channel decision essentially determines the number of receiving points (a few wholesalers or many retailers) to be served by the logistics system. There is a close connection between *field service* and distribution logistics if the field service employees take over logistics functions of order processing. If the organization of the company assigns them the task of receiving and transmitting orders, they play an important role at the beginning of the order processing process and trigger the flow of information. However, routine order processing in day-to-day business is not one of the original tasks of the field sales force, which is much more concerned with the intensive support of current customers and the acquisition of new customers.

The close connection between distribution logistics and the sales channel results from the fact that the logistics systems of the institutions responsible for the distribution of a manufacturer's goods must be coordinated with each other. In addition, it is always important to bear in mind that lost sales caused by stock shortages—at the retail level, for example—are also lost sales for all upstream stages in the sales channel, right through to the manufacturer. Distribution logistics can do a lot to avoid such shortages by providing a good *delivery service*. However, the frequent shortages of a product caused by poor delivery service can also cause a retailer to pay less attention to selling that product. The head of marketing of an American company in the cosmetics industry clearly expresses the double effect of out-of-stocks when he says: "We lose sales and shelf space if the goods are not there."[23]

To avoid this, integrated information systems are being set up as part of *Efficient Consumer Response* (ECR) programs. Based on the retailer's merchandise management systems and starting from the scanner checkout, sales data is reported to the retailer's central warehouses and, together with the inventory data there, reported to the manufacturers. The manufacturers thus have an overview of the sales as well as the stock levels of their products and can thus continuously adjust their sales forecasts and production.

---

[22] See Meffert et al., 2012, p. 544.

[23] Arbury et al., 1967, pp. 27 f.

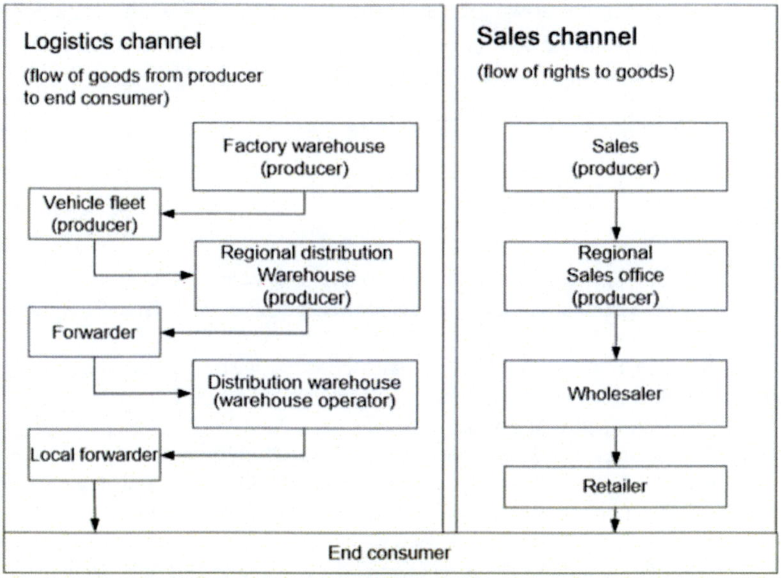

Fig. 11.3 Example of a breakdown of the marketing channel into logistics channel and sales channel (Source: Pfohl, 1975, p. 289, following the example of Bowersox et al., 1968, p. 45)

Another important module of ECR is the Continuous Replenishment Program (CRP). Here, the manufacturer assumes responsibility for the retailer's ability to deliver by taking over inventory control in the central warehouse. ECR requires close cooperation between manufacturer and retailer. Not only the data processing systems have to be coordinated, but also the supply has to be coordinated between both partners. For this purpose, joint teams of retailers and manufacturers are formed, in which not only logistical aspects are dealt with, but also joint marketing strategies are worked out, e.g. product launch or sales promotion strategies. Both parties benefit from this partnership. The manufacturers receive improved information, can manufacture according to demand, increase customer loyalty and can better influence the sales promotion activities of the trade. Retailers benefit from the cost savings, the improved availability of goods and are relieved with regard to storage.[24]

However, when discussing the relationship between the sales channel and distribution logistics, it should also be noted that different *types of flows* can be identified according to the objects flowing in the marketing channel. A distinction is often made between the physical flow of goods, the flow of property (flow of rights to the goods) and the flow of sales promotion.[25] It is by no means necessary that all flow types must pass through the same institutions; on the contrary, they can be decoupled from each other. Figure 11.3

[24]On ECR, see Stock/Lambert, 2001, pp. 40ff; Kotzab/Lienbacher, 2010; Pfohl, 2016, pp. 154 f.

[25]See Kotler et al., 2007, pp. 852 f., who also distinguish between information flow, negotiation flow, order flow, financing flow, risk flow and payment flow. See also Part IV, Sect. 17.3 and Part I, Sect. 2.1.

gives an example of a breakdown of the marketing channel into the logistics channel, which contains the physical flow of goods, and the sales channel, which contains the flow of ownership or flow of rights to goods. However, the interdependencies between the different flow types must be taken into account.

References

Arbury J u.a. (1967) A New Approach to Physical Distribution. New York

Barth K, Hartmann M, Schröder H (2007) Betriebswirtschaftslehre des Handels. 6., überarb. Aufl. Wiesbaden

Bowersox D J, Smykay E W, LaLonde B J (1968) Physical Distribution Management. Logistics Problems in the Firm. Rev. ed. New York

Constantin J A (1966) Principles of Logistics Management. New York

Crowley J (1994) Logistics and Marketing. In: Pfohl, H.-Chr. (Hrsg.): Future Developments in Logistics and the Resultant Consequences for Logistics Education and Training in Europe. Proceedings of the Logistics Educators Conference 1994. European Logistics Association (ELA). Brüssel

Havighorst D (1980) Konzept und Leistungspotential der Marketing-Logistik. Weinheim

Heskett J L, Glaskowsky N A, Ivie R M (1973) Business Logistics. Physical Distribution and Materials Management. 2. Aufl. New York

Kotler Ph, Keller K L, Bliemel F (2007) Marketing-Management: Analyse, Planung, Umsetzung und Steuerung. 12., aktualisierte Aufl. München u.a.

Kotzab H, Lienbacher V (2010) Efficient Consumer Response. Marketinglogistisches Kooperationsmanagement in der Konsumgüterwirtschaft. In: Schönberger R, Elbert R (Hrsg) Dimensionen der Logistik. Funktionen, Institutionen und Handlungsebenen. Wiesbaden, S. 357-371

Magee J F, Copacino W F, Rosenfield D B (1985) Modern Logistics Management. Integrating Marketing, Manufacturing, and Physical Distribution. New York u.a.

Meffert H, Burmann Chr, Kirchgeorg M (2012) Marketing: Grundlagen marktorientierter Unternehmensführung: Konzepte – Instrumente – Praxisbeispiele. 11., überarb. und erw. Aufl. Wiesbaden

Pfohl H-Chr (1972) Marketing-Logistik. Gestaltung, Steuerung und Kontrolle des Warenflusses im modernen Markt. Mainz

Pfohl H-Chr (2016) Logistikmanagement. Funktionen und Instrumente. 3., neu bearb. und aktual. Aufl. Berlin u.a.

Powers Th L, Closs D J (1987) An Examination of the Effects of Trade Incentives on Logistics Performance in a Consumer Products Distribution Channel. In: Journal of Business Logistics 8 2, S. 1-28

Seyffert R (1972) Wirtschaftslehre des Handels. 5., neu bearb. Aufl., Opladen

Smykay E W (1973) Physical Distribution Management. 3., Aufl. New York

Smykay E W, LaLonde B J (1967) Physical Distribution. The New and Profitable Science of Business Logistics. Chicago/London

Stackelberg K G v (1969) Marktstrategie ohne Geheimnisse. Düsseldorf/Wien

Stock J R, Lambert D M (2001) Strategic Logistics Management. 4. Aufl. Boston u.a.

Spare Parts Logistics 12

12.1 Definition and Conception of Spare Parts Logistics

Definition
In *DIN 24 420*, spare parts are defined as *objects* of spare parts logistics as parts (also called individual parts), groups (also called assemblies and groups of parts) or complete products which serve to replace damaged, worn or missing parts, groups or products.[1] A characteristic feature is that they are not independent components of systems.[2]

For further considerations, it is first necessary to define the *task* of spare parts logistics more precisely. A distinction must be made between the spare parts logistics of the supplier (producer) and the spare parts logistics of the customer (user). Whereas the scope of spare parts logistics for the customer covers procurement, storage and the use of spare parts in the context of *maintenance*, the task of spare parts logistics for the manufacturer of spare parts is the supply of spare parts to customers in line with their requirements as part of *after-sales service*. This requires a flow of goods from the provision of goods to the distribution of goods to the use of goods. Figure 12.1 shows the relationship between the spare parts logistics systems of the supplier and the customer.

Since the tasks of spare parts logistics at the customer are largely the same as those of general procurement logistics, only the deviating special features of procurement logistics for spare parts will be presented here. Further consideration will then be given to the design of spare parts logistics at the manufacturer, as this has some special features compared to the logistics of the primary product.

[1] See German Institute for Standardization, 1976.

[2] See Wohinz, 1974, p. 200. For further characteristics, see Arnold et al., 2008, pp. B7–10.

H.-C. Pfohl, *Logistics Systems*,
https://doi.org/10.1007/978-3-662-64349-5_12

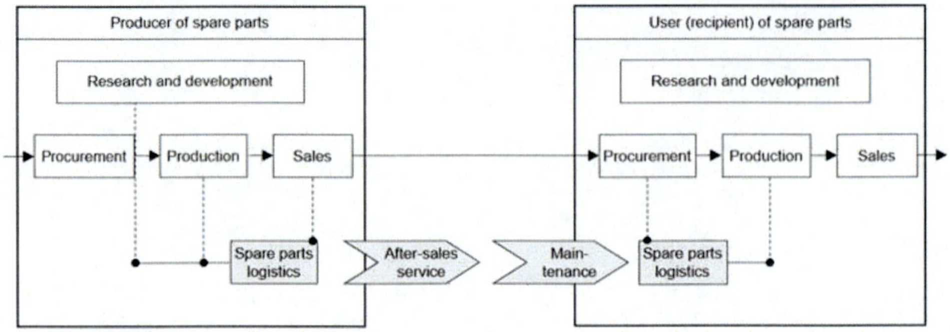

Fig. 12.1 Spare parts logistics at the producer (supplier) and user (customer) of spare parts (Source: with minor changes taken from Pfohl, 1991, p. 1033)

Conception of Spare Parts Logistics at the Customer

Due to the increasing complexity and networking of production plants and the associated rise in *downtime costs*, the user of spare parts requires fast and reliable maintenance and repairs of the failed production facilities. In addition to sufficient maintenance capacity, which is usually maintained in-house but can also be provided by a service provider or the supplier of the primary products/spare parts itself, this immediate repair also requires a reliable and fast supply of spare parts by the supplier. The scope of tasks of spare parts logistics here relates primarily to the procurement of spare parts, storage and deployment planning for maintenance.[3]

Several operational functional areas are involved in the spare parts *procurement* of the industrial customer, in addition to maintenance, e.g., plant management, logistics and controlling. The goal of the joint activities is the provision of spare parts in line with demand at minimum costs of spare parts logistics. The requirements on the efficiency of the spare parts logistics at the customer are influenced by the design of the procurement marketing.[4]

In the context of stockpiling, one of the most important issues is to determine the *type and number of spare parts to be stockpiled*. The determination of which spare parts to stock should be based on an economic cost-benefit assessment (cost of stocking, availability of the spare part on the market, downtime costs of the installation concerned, etc.).[5] In

[3]See Biedermann, 1995, p. 9. For different maintenance strategies and their possible effects on the requirements for spare parts supply, see ibid. p. 19ff. According to DIN 31 051, maintenance is the generic term for servicing (measures to maintain the target condition of the technical resources of a system), inspection (measures to determine and assess the actual condition of the technical resources of a system) and repair (measures to restore the target condition of the technical resources of a system), see Deutsches Institut für Normung, 1985.

[4]See Sect. 9.2.

[5]For the investment-related benefit components and their characteristics as well as the cost components, see Hug, 1986, pp. 100ff., pp. 147ff. and pp. 172ff.

determining the number of spare parts to be stockpiled, reference should be made to the requirements planning for spare parts. While it is generally possible to use mathematical forecast models to plan requirements for primary products, which determine future requirements based on analyses of past requirements and taking into account known expected changes, it is more difficult to plan requirements for spare parts.[6] In this case, it is advisable to use causal analytical procedures for requirements planning, in which the spare parts requirement itself is derived from the expected development of the influencing variables on the spare parts requirement.[7]

Conception of Spare Parts Logistics at the Supplier

Basically, spare parts are secondary products whose turnover and supply requirements must always be considered in conjunction with the primary product. This results in *special features of the spare parts turnover*, which have a significant influence on the requirements and the design of the spare parts logistics. The following are worth mentioning at this point[8]:

- The purchase of a replacement part is preceded by a negative experience on the part of the customer in the form of a failure of the primary product.
- The demand for spare parts can only be planned to a limited extent.
- The demand for spare parts depends, among other things, on the number of primary products sold, the maintenance measures carried out and the service life of the parts used. This means that spare parts sales can only be increased to a limited extent by marketing measures. The only possibilities here are to gain sales shares from competing suppliers for these spare parts or to expand sales through marketing for preventive maintenance policies, for example.
- Spare parts often require a great deal of explanation and are frequently sold in connection with an after-sales service.
- Assortments of spare parts—e.g. depending on their necessity for the functionality of the primary product—cannot always be formed solely on the basis of economic considerations.
- Product innovations lead to the replacement of discontinued primary products. However, spare parts must be provided for old and new primary products, which leads to a constant expansion of the product range.
- End users are often not spare parts customers, but often repair shops.

[6] It should be noted that not only the purchaser of machines and systems carries out requirements planning as part of his inventory management, but that the supplier of spare parts also needs requirements planning in order to be able to guarantee sufficient availability for the purchasers during ongoing series production and to have sufficient quantities in stock for future requirements when production of the primary product and the associated spare parts is discontinued.

[7] On causal-analytical procedures, see Meidlinger, 1994; Loukmidis/Luczak, 2006, pp. 258ff.

[8] See Baumbach, 2004, pp. 169ff.

These characteristics of spare parts turnover result in special features in the spare parts logistics system in contrast to the logistics system for primary products. *Systems thinking* therefore demands not only a close connection with after-sales service, but also the view of spare parts logistics as a part of the overall logistics system that is closely related to the other logistics subsystems. The spare parts logistics system should contribute to the achievement of the goals of the overall system. The goal of spare parts sales can either be defined as supporting primary product sales, or spare parts sales can be managed as a separate sales object (profit center).[9] In any case, an optimization between delivery service and logistics costs is to be aimed for here as well, whereby the compound effects to the primary product must be taken into account.

For the requirements on the *delivery service* for spare parts the following special influencing factors are to be mentioned[10]:

- The required delivery time is an important component for spare parts, as it helps to avoid unnecessary downtimes.
- Delivery reliability is particularly important if maintenance tasks can only be performed at certain times (for example, at weekends) or if maintenance is not performed regularly as a preventive measure, but only when necessary.
- With regard to the condition of the delivery, it should be noted that the customer reacts particularly sensitively after a negative experience with the primary product and therefore the flawless condition of the spare parts delivery is necessary.
- As spare parts orders are often extremely urgent orders, delivery flexibility must be given with regard to order transmission, transport routes and means of transport.

The delivery service offered is compared with the *costs of supplying spare parts*, the main components of which are storage costs and transport costs. The storage costs are comparatively high because the spare parts inventories generally represent a large assortment of which only small quantities are needed at any one time. In addition, there is generally a high concentration of sales on a small number of spare parts in this assortment. Therefore, special attention must be paid to the assortment formation and the stock policy for individual products. With regard to transport costs, it should be noted that the possible inclusion of additional nodes in the logistical network in the form of repair workshops and the servicing of rush orders increases the complexity of the logistical network and the transport processes that take place within it, and thus the costs. The question then arises to what extent such costs can be passed on to the customer.

---

[9] See Ihde et al., 1999, pp. 9ff.

[10] On these influencing variables, see Thurow, 1977, pp. 116ff. See also the customer requirements for delivery service, communication, quality of spare parts and the costs of spare parts supply in Ester, 1997, pp. 138ff.

12.2 Activity-Specific Subsystems of Spare Parts Logistics

Order Processing

In the area of order processing, a distinction must first be made between rush orders (reactive processes) and replenishment orders (anticipatory processes). The former arise due to stock shortages at the customer, the workshop or the delivery warehouse. The latter are used to replenish and secure stock at the customer, in the workshop or in the delivery warehouse. During order acceptance and preparation, the customer often needs expert advice to identify the failed parts on the production facility and to order the correct spare part. This advice is particularly necessary for primary products that have already been discontinued. In addition, orders for spare parts are often placed by a customer service representative or repair shop rather than the end customer. In such cases, it is advisable to standardize the order transmission process or to install a direct EDP link.

Warehouse

For the warehouse subsystem, the focus is on determining the number of stocking echelons as well as the number and locations of warehouses. The number of stocking echelons depends, among other things, on the design of the sales channel (e.g. in the case of sales via workshops, these usually also represent a stocking echelon). The number and location of warehouses should ensure an optimal supply of spare parts to customers. To this end, a sufficient number of delivery warehouses or workshops should be set up in the vicinity of the customer. The location of the central warehouse, on the other hand, is usually determined production oriented. Proximity to fast transport links is also important when deciding on a location. For example, spare parts deliveries to non-European countries require an airport connection.

Inventory Management

Within the framework of inventory management, it must be decided in which quantity and at which stocking echelon the parts to be stocked are to be held. With regard to the quantity, it is necessary to refer once again to the questions of requirements planning for spare parts, which have already been addressed under the spare parts logistics of the customer. Within the framework of a causal analytical demand forecast for spare parts, which has already been addressed under the spare parts logistics of the customer, the influencing factors on the spare parts demand shown in Fig. 12.2 must be taken into account, among other things. Reference should also be made here to the product life cycle as the basis of the demand forecast with the points in time "Start of Production, End of Production, End of Delivery Obligation, End of Service and End of Life".[11]

[11] See Voss, 2006, p.109.

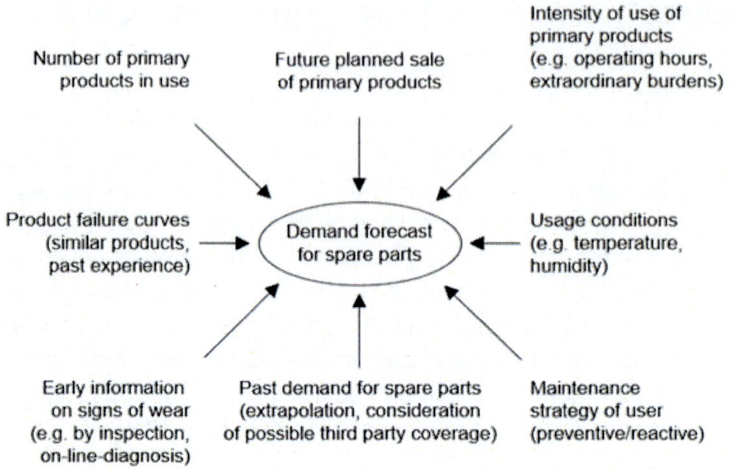

Fig. 12.2 Factors influencing spare parts requirements (Source: Pfohl, 1991, p. 1038)

The level of inventory costs can be influenced by the application of selective storage. For example, all parts that are used frequently and are indispensable for the functioning of the primary product should be kept in stock close to the customer, i.e. in delivery warehouses or in workshops. It must be ensured that after-sales service staff can be supplied with these parts as quickly as possible, e.g., delivery of the spare parts to the staff's customer service vehicle parked at night, or that they carry a basic assortment with them.[12] Parts that are used less frequently and/or are not absolutely necessary for the functionality of the primary product can be stocked in central distribution warehouses, since a somewhat longer delivery time is tolerated by the customer. It would also be possible to select the parts to be stocked according to the maintenance policy of the customer: all parts that are only required as part of preventive maintenance or that the customer stocks himself could be located at central warehouse echelon, as it is possible to work with less urgent replenishment orders here. The importance of selective storage also arises from the fact that the concentration curves for spare parts are very pronounced. Only a few spare parts contribute to the largest share of sales.

Packaging

In the context of packaging, the protective and warehouse function as well as the information function must be emphasized. Since spare parts may be stored for a relatively long time, the packaging must provide sufficient protection against mechanical (manipulation processes) and chemical-physical (heat, moisture, etc.) effects for this period and allow measures to be taken for maintenance and inspection. Due to the high number of spare part

[12]See Patton, 1984, p. 281, who proposes the grouping of the spare parts required for a repair order together with the corresponding equipment, tools and documentation into so-called kits.

items and the need for explanation of spare parts, the packaging must contain sufficient information to identify the spare part as well as necessary technical details. In addition, the packaging is important for many suppliers in their marketing function, identifying the spare part as an original spare part and thus differentiating it from replica spare parts.

Transport

When transporting, it should be noted that the order volume per customer (even for workshops) is small. For this reason, very many points in the logistics system must be approached during delivery, each of which only receives a relatively small number of parts. Sufficient utilization of the supplier's own fleet of vehicles can thus hardly be achieved. Often, outsourcing the transport is the more economical alternative.

A new technology that is completely changing the design not only of transport but of the entire spare parts logistics system is 3D printing. With this additive production technology, components are printed in layers at the customer's site or at a logistics service provider near him. The focus of transportation then shifts from transporting the spare parts to transporting the powders as raw materials. High-quality slow-moving parts, which incur high logistics costs due to their long storage times, are considered particularly suitable for 3D printing.[13]

12.3 Importance of Spare Parts Supply as a Competitive Instrument

Criterion for Supplier Selection

For the competitive situation of the company, it is decisive that primary product and spare parts supply are not perceived separately by the customer. The market requirements present themselves as the *purchase of the functionality of* an aggregate and can also include the tasks of maintenance and technical advice in addition to the supply of spare parts.[14]

The increasing importance of spare parts supply is already shown by a survey of 354 capital goods suppliers conducted in 1992 for the sale of capital goods.[15] It reflects the assessment of the success factors for the sale of capital goods from the point of view of the companies surveyed. The respondents considered the second most important factor in capital goods competition to be the services associated with the primary product. The second most important after-sales service is the supply of spare parts. The speed of spare parts supply has a major influence on the attractiveness of the offer from the suppliers of capital goods and thus also on the willingness to pay prices.[16]

[13] See Bottler, 2015, p. 27. On technology, see Fastermann, 2012.

[14] See Belz, 1991, p. 53.

[15] See Burkhardt, 1992, pp. 26 f.

[16] See Diller, 2004, p. 962.

Sector Dependency

The realization of the importance of spare parts supply leads to the fact that the service policy is considered as an independent component of the marketing instruments.[17] Particularly in the capital goods market, supplying the customer in the after-sales phase is important with regard to follow-up orders. Here, "after the purchase is before the purchase" applies.[18] With regard to the importance and the design of the spare parts supply in the context of capital goods marketing, a differentiated consideration must be carried out for different business types of capital goods, which require different marketing measures. HERMANNS/FLORY, for example,[19] define—according to the level of consideration of value creation—the types series manufacturer, variant supplier, single item producer, process manufacturer, system supplier and frame order supplier. In the context of series production (e.g. manufacturers of ball bearings or measuring instruments) and variant production (e.g. manufacturers of forklift trucks, geared motors), great importance is attached to the repair service and the supply of spare parts, while in the case of process production (e.g. basic chemical materials, sheet metal), for example, no supply of spare parts is required.

In sectors such as the automotive industry, where there is a market for parts removed from old or damaged primary products, possibly reconditioned, there is a special connection between spare parts logistics and the disposal logistics to be dealt with in the following section. In order to prevent these parts from being offered on the market in an uncontrolled manner as competition to the original spare parts, the manufacturer must pay particular attention to the control of the disposal logistics channel!

References

Arnold D u.a. (Hrsg) (2008) Handbuch Logistik. 3., aktualisierte und korr. Aufl. Berlin/Heidelberg

Baumbach M (2004) After-sales-Management im Maschinen- und Anlagenbau. 2., überarb. Aufl. Regensburg

Belz Chr (1991) Mit Problemlösungspaketen Kunden behalten und dazugewinnen. In: io Management-Zeitschrift 60 11, S. 53-58

Biedermann H (1995) Ersatzteil-Logistik. Beschaffung. Disposition. Organisation. Düsseldorf

Bottler S (2015) Die Revolution aus dem Drucker. In: VerkehrsRundschau 28, S. 26-27

Burkhardt R (1992) Ein neues Marketing-Gefühl. Investitionsgüter-Studie. In: Top Business o. Jg. 9, S. 18-34

Deutsches Institut für Normung e.V. (DIN) (Hrsg) (1976) Deutsche Norm, DIN 24 420, Ersatzteillisten – Allgemeines, Teil 1. Berlin/Köln

Deutsches Institut für Normung e.V. (DIN) (Hrsg) (1985) Deutsche Norm, DIN 31 051, Instandhaltung: Begriffe und Maßnahmen. Berlin/Köln

[17] On the development of spare parts supply and on sectoral differences, see Ester, 1997, pp. 128ff.

[18] See Thurow, 1977, pp. 41ff; Pfohl et al., 1995.

[19] See Hermanns/Flory, 1995.

Diller H (2004) Preisstrategien im Industriegütermarketing. In: Backhaus K, Voeth M (2004) Handbuch Industriegütermarketing. Wiesbaden, S. 947-968

Ester B (1997) Benchmarks für die Ersatzteillogistik: Benchmarkingformen, Vorgehensweise, Prozesse und Kennzahlen. Berlin

Fastermann P (2012) 3D-Druck/Rapid Prototyping. Eine Zukunftstechnologie – kompakt erklärt. Heidelberg/Berlin

Ihde G B, Merkel H, Henning R (1999) Ersatzteillogistik: theoretische Grundlagen und praktische Handhabung. 3., völlig neu bearb. Aufl. München

Hermanns A, Flory M (1995) Typologisierung der Wertschöpfungsstrukturen im Investitionsgütermarketing. In: ZfB 65 1, S. 49-67

Hug W (1986) Optimale Ersatzteilwirtschaft. Köln

Loukmidis G, Luczak H (2006) Lebenszyklusorientierte Planungsstrategien für den Ersatzteilbedarf. In: Barkawi K, Baader A, Montanus S (2006) Erfolgreich mit After Sales Services: Geschäftsstrategien für Servicemanagement und Ersatzteillogistik. Berlin u.a., S. 251-270

Meidlinger A (1994) Dynamisierte Bedarfsprognose für Ersatzteile bei technischen Gebrauchsgütern. Frankfurt a. M.

Patton Jr J D (1984) Service Parts Management. North Carolina Pfohl H-Chr (1991) Ersatzteil-Logistik. In: ZfB 61 9, S. 1027-1044

Pfohl H-Chr, Ester B, Jarik J (1995) Qualitätsmerkmale der Ersatzteilversorgung – Ergebnisse einer Kunden- und Anbieterbefragung. Arbeitspapiere zur Logistik. Nr. 18. Fachgebiet Unternehmensführung, Institut für Betriebswirtschaftslehre. Technische Hochschule Darmstadt. Darmstadt

Thurow W (1977) Allgemeingültiges Modell für die Konzeption integrierter Marketing-Logistik-Systeme der Ersatzteilbewirtschaftung. Diss. Technische Universität Graz

Voss H (2006) Life Cycle Logistics. Der Weg zur produkt-lebenszyklusorientierten Ersatzteillogistik. Bern/Stuttgart/Wien.

Wohinz J W (1974) Morphologie des Ersatzteilbedarfes. In: io Industrielle Organisation 43 4, S. 199-202

Disposal Logistics 13

13.1 Definition and Conception of Disposal Logistics

Definition
Disposal logistics can be defined[1] as the application of the logistics concept to residues in order to design an economically and ecologically efficient flow of residues with all activities of spatio-temporal transformation, including changes in quantity and type. Differentiation criteria of disposal logistics from procurement, production, distribution and spare parts logistics are both the *objects* of disposal logistics and their flow direction. In accordance with the goal content reference of the objects, a distinction can be made between target products and residues as the output of companies. While target products represent objects of supply logistics, the spatial-temporal transformation of *residues* is the task of disposal logistics. According to the type of reusability, residues can be subdivided into secondary raw materials and waste, which can also be differentiated according to their aggregate state into solid, pasty and liquid substances.[2] This broad interpretation of the term "residues" also allows used and consumed products, exchange aggregates, returns, obsolete stock as well as empties, containers and packaging to be assigned to residues. An overview of the objects of disposal logistics is given in Fig. 13.1.

The *flow direction* of the residues in the disposal area is opposite to the flow direction in the supply area. A distinction is made here between different types of residue flows.

[1]See Pfohl/Stölzle, 1992, pp. 573 f.

[2]The use of the terms residue, secondary raw material and waste deviates from previously common definitions and replaces the terms residual material and recyclable material. On these terms see Stölzle, 1993, p. 163ff. The new definitions are based on §2 of the Circular Economy and Waste Management Act (KrW/AbfG). According to the law, residues are secondary raw materials if they are to be recycled in accordance with the law or are waste if they may not be recycled as secondary raw materials.

H.-C. Pfohl, *Logistics Systems*,
https://doi.org/10.1007/978-3-662-64349-5_13

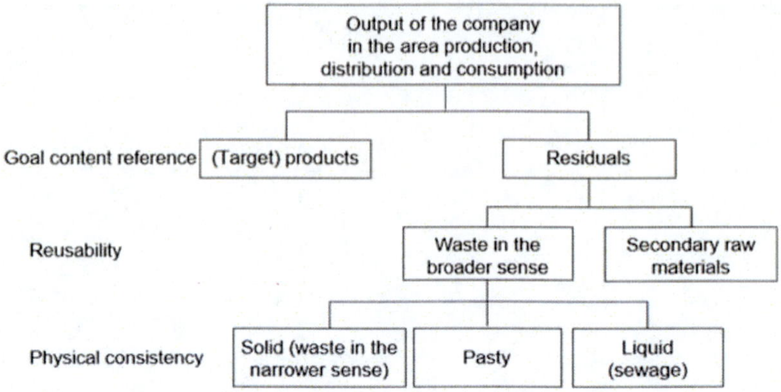

Fig. 13.1 Objects of disposal logistics (Source: Based on Stölzle, 1993, p. 167)

Depending on the type of company involved and the echelon of the disposal logistics chain, there are single-echelon and multi-echelon redistribution channels, residue cycles and disposal channels.[3] *Redistribution channels* are identical to distribution channels for target products and involve distribution intermediaries and helpers in the physical transfer of residues. The source of the residue flow corresponds to the sink of the target product flow, so the residue flow occurs in the opposite direction of the target product flow.[4] *Residue cycles* are intended to ensure the reuse of secondary raw materials and link the place of residue generation with the manufacturing companies of used or consumed target products, whereby the residue flow can be interrupted by the involvement of recycling companies. *Removal channels* finally realize the residue flow between residue generators and the institutions responsible for the proper disposal of waste (e.g. incinerators or landfills). Selection criteria for the structure of the disposal logistics systems are the type purity of the residues, their reuse possibilities, the residue quantities as well as special legal regulations, such as hygienic requirements of the transport or collective loading prohibitions of the Dangerous Goods Ordinance Road (GGVS). The logistical processes in the disposal channels include collection, transport and handling, storage, exploitation, thermal treatment and orderly removal (see Fig. 13.2).[5]

In addition to these mesologistics systems, which form the basis of the development of a national economy towards a *circular economy*, there are also micrologistics systems of disposal. They refer to the transformation of residues within public and private institutions. Objects can be, for example, the return of lubricants or production waste for direct reuse in the company.

[3] See Pfohl/Stölzle, 1995, pp. 2234ff. and the literature listed there.

[4] On the design and evaluation of redistribution channels, see Haasis, 1999. In English, therefore, disposal logistics is often referred to as "reverse logistics".

[5] See Arnold et al., 2008, p. 488. See also Ivisic, 2002.

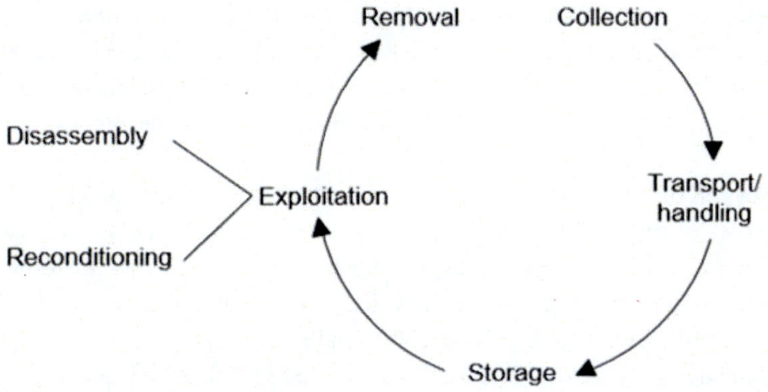

Fig. 13.2 Logistical processes in the disposal channels (Source: based on Arnold et al., 2008, p. 489)

Conception

The definition shows that disposal logistics pursues both economically and ecologically oriented *objectives*. The economic objective consists of a reduction in logistics costs and an improvement in the service level of disposal logistics (acceptance of residues at the points of origin in line with requirements and delivery of secondary raw materials to the sources of reuse in line with the type, quantity, space and time requirements). The ecologically oriented goal aims at the conservation of natural resources and the reduction of emissions from disposal logistics processes.[6] In some cases, ecological and economic objectives complement each other, e.g., in the conservation of natural resources. Their finite availability means that less landfill space is available and the costs of disposal are constantly increasing, so that a reduction in the volume of waste disposed of also leads to a slower increase in disposal costs.[7] The inclusion of ecological targets in the target system of disposal logistics illustrates the interactions between logistics and the natural environment. While the *ecological aspects of supply logistics* are limited to the reduction of process-dependent input and output-side environmental impacts,[8] disposal logistics actively contributes to the solution of ecological problems by taking on specific tasks of disposal and demonstrates the role of logistics in environmental protection.

In accordance with the objectives of disposal logistics, ecological approaches are also taken into account in the conception of disposal logistics, which is integrated into the overall conception of logistics, i.e., in system, total cost, service and efficiency thinking. The basis of planning, monitoring, realization and control of disposal logistics processes is *systems thinking*. On the one hand, this way of thinking makes it possible to avoid suboptimal solutions in the area of disposal logistics, such as a suboptimal design of the

[6] See Pfohl/Stölzle, 1992, p. 575f; Emmermann, 1996, pp. 62 f.

[7] See Emmermann, 1996, pp. 62 ff.

[8] On the possibilities of reducing the environmental impact of logistics ("green logistics"), see Bretzke and Barkawi, 2010; Rausch et al., 2010; Gregori and Wimmer. 2011; McKinnon et al., 2012.

storage of residues, in favor of overall optimal disposal logistics solutions. On the other hand, the consideration of the entire logistics chain allows the exploitation of cross-divisional optimization potentials, such as the integration of disposal processes into distribution logistics activities, which is characteristic for the design of redistribution channels.

The application of *total cost thinking* to disposal logistics pursues the goal of disclosing all relevant costs of disposal logistics decisions. In this context, ecological cost variables should also be included in the decisions, insofar as they can be determined on a reliable basis. Approaches to this include depreciation for specific environmental protection equipment, such as air filter systems in transshipment halls.

Service thinking in the field of disposal logistics distinguishes between an in-put and an output-oriented approach to service. While the input-oriented service level refers to the delivery of residues to reconditioning, treatment and removal plants as well as the place of reuse, the output-oriented service concentrates on the disposal of the residues at the points of occurrence. The service components of delivery or collection time, reliability and flexibility apply to both subsectors. In contrast, the delivery quality refers solely to the input-oriented view, for example by describing the purity of recyclable residues.

In accordance with the target definition of disposal logistics, *efficiency thinking* encompasses the aspects of economic and ecological efficiency. The economic dimension is oriented towards the total costs of disposal logistics and the input- and output-related service level facing them. One possible starting point for determining ecological efficiency is the recycling rate.[9] When considering overall efficiency, it should be borne in mind that despite conflicting relationships between the two types of efficiency, in the short term there are generally congruent target relationships in the long term. For example, by switching to ecology-oriented production processes in the area of disposal logistics, competitive advantages can be gained in the long term and thus economic goals can also be pursued.

13.2 Acticity-Specific Subsystems of Disposal Logistics

Corresponding to the differentiation of activity-specific functional subsystems in the area of supply-oriented logistics, the subsystems order processing, inventory management, warehouse, packaging and transport can also be delineated in disposal logistics. Specific to disposal logistics is the task area of collection and separation. The special features of these subsystems compared to supply logistics will be outlined below.

[9] It should be noted that the recording of the recycling rate does not take into account the target hierarchy of avoidance before reduction of environmental impacts and the logistics-related environmental impacts, see Stölzle, 1993, p. 187.

Order Processing

Compared to supply logistics, a large number of legal requirements have to be observed, which are based on legal decrees on labelling obligations for residues.[10] The aim of these decrees is to document the flow of environmentally hazardous residues as comprehensively as possible. In the area of advance information flow, these decrees regulate, for example, the obtaining of permits for the collection and transport of hazardous residues. Accompanying documents and warning signs on the truck identify the residues matching the flow of residues. The forwarding of waste disposal certificates is specific to the follow-up flow of information in waste disposal logistics.

Inventory Management

Comparable to the supply sector, the storage of residues fulfils various functions, the characteristics of which, however, are weighted differently in disposal logistics. For example, the time bridging function of inventories is of less importance in disposal logistics than in the supply sector. This is because only aspects of the economic transport lot size are to be taken into account, since the capital commitment costs are to be assessed as less significant due to the generally lower value of residues. The safeguarding function of stockpiling, on the other hand, is of greater interest with regard to the reuse and recycling of secondary raw materials. Temporal uncertainties in the type and quantity of secondary raw materials can thus be compensated for and a continuous utilisation of the recycling plants can be ensured.[11] The importance of the specialization function of secondary raw material inventories results from the establishment of secondary raw material-specific recycling plants and the associated division of tasks in the waste management industry. Finally, the speculation function and the acquisition function of stockpiles have negligible importance in residue storage. According to the functions of residue inventories, the objectives of inventory management in disposal logistics focus on the provision of sufficient capacity and the proper timing of residues to realize the most efficient recovery or removal. The objective of maintaining a certain service level is limited to the input storage facilities of recycling plants.[12]

Warehouse

In the context of disposal logistics, storage warehouses serve to provide residues at the right time and in the right quantities as well as to receive processed secondary raw materials. Transshipment warehouses enable the establishment of transport chains by realizing transshipment operations between different modes of transport. Distribution warehouses have the task of concentrating residues from different collection regions and distributing

[10] For an overview of the laws and regulations in waste management logistics in Europe, see Nguyen, 2012, pp. 72ff.

[11] See Pfohl/Stölzle, 1992, p. 581.

[12] See Stölzle, 1993, pp. 225 f.

them to specific recycling and removal facilities. Residue-specific collection warehouses can hold waste and secondary raw materials at the place where they are generated until they are transported away. Due to the high safety requirements in the waste management sector, the tasks in the warehouse subsystem are essentially limited to ensuring the necessary safety standards. This includes, in particular, compliance with specific legal requirements, such as bans on combined storage, fire protection and immission control regulations, as well as a suited technical design of the warehouse.[13]

Packaging

Requirements for packaging of residues predominantly relate to their protective function towards the environment. Particularly in the case of packaging for hazardous residues, the focus is on the requirements for safety and durability.[14] Whereas in supply logistics the communication function is of particular importance with regard to marketing, this tends to take a back seat in disposal logistics and is limited to the labelling of the residues. Factors influencing the packaging design are, in addition to special legal framework conditions, essentially residue-specific and interface-related requirements. In the case of residues, environmentally relevant properties as well as the weight, quantity, form and type of residues must be included in the packaging decision. In order to avoid technical problems in the design of interfaces in linked disposal logistics systems, the packaging should also be standardised, automatically manageable, universally transportable on different modes of transport and usable as intermediate storage. These properties also make it easier to set up interconnected systems between disposal and supply logistics, such as those implemented in redistribution channels.

Transport

Special features of the disposal logistics subsystem transport result primarily from the inhomogeneity of the transport objects, the high risk of accident-related environmental pollution and the problem of avoiding empty runs. The inhomogeneity of residues and the specific requirements that each type of residue places on transport lead to a high complexity of the transport service, which in turn causes high transport costs. In contrast, pure residues generally occur only to a small extent, so that transport specialisation only makes sense in isolated cases for economic reasons. An increased environmental risk due to accidents exists in particular in the case of hazardous goods transports. The importance of this aspect results from the increasing transport performance and the rising accident frequency in this area,[15] which has also led to an increase in legal requirements. The problem of avoiding empty runs arises in waste disposal logistics in particular due to residue-related contamination of the transport containers. On the one hand, these contaminations limit the selection

[13] See Becker/Hüning, 1993, pp. 74ff.

[14] See Pfohl/Stölzle, 1992, p. 587.

[15] See Bilitewski/Härdtle, 2013, p. 107.

of residue-specific return loads, on the other hand, they generally prevent the transport of supply goods and residues with the same vehicle due to hygiene regulations, as would be possible, for example, within the framework of redistribution channels.[16]

Collection and Separation

Closely linked to the demand for residue flows that are as homogeneous as possible is the task of collection and separation. The aim of this specific disposal logistics subsystem is to increase the purity of the residues in accordance with the requirements of the residue sinks. Since the processes of residue separation are often combined with collection, the subtasks can be combined in one task area.[17]

The organisation of collection and separation represents an essential decision element. Residues can be separated at different echelons of the disposal logistics channel. First of all, separate collection at the point of origin is a suitable option. This leads to a comparatively high degree of purity of the residues, but requires additional effort in the provision of containers and in the transport of the residues. In contrast, the reuse of residues in a mixed collection is dependent on subsequent handling processes. Economic advantages result here in particular from the lower space requirement of the collection containers, which are however offset by additional process costs for storage, transport and handling processes. Decisions on the organizational form of collection and separation must therefore be made on a case-by-case basis, taking into account overall cost considerations and the reusability of the residues.

Another decision element of collection and separation is the degree of involvement of the waste producer in the disposal. Whereas in the case of fetch systems the residues are merely made available at the place where they occur, bring systems require the involvement of the residue producer in the transport of the residues. In this case, collection takes place at central collection points, such as recycling centres. The choice of an alternative depends essentially on the willingness of the waste producer to become involved in disposal.[18] In multi-echelon residue cycles, the fetch and bring systems can be combined according to the respective requirements, as shown in Fig. 13.3.[19]

13.3 Technical Design of Disposal Logistics Processes

Special features of the technical design of disposal logistics processes can be found above all in collection and transport technology. Specific technical facilities and equipment features in the areas of warehousing and transshipment result from substance-dependent

[16] See Pfohl/Stölzle, 1992, p. 584.

[17] See Pfohl/Stölzle, 1992, pp. 586 f.

[18] See Stölzle, 1993, p. 240.

[19] See Arnold et al., 2008, p. 515.

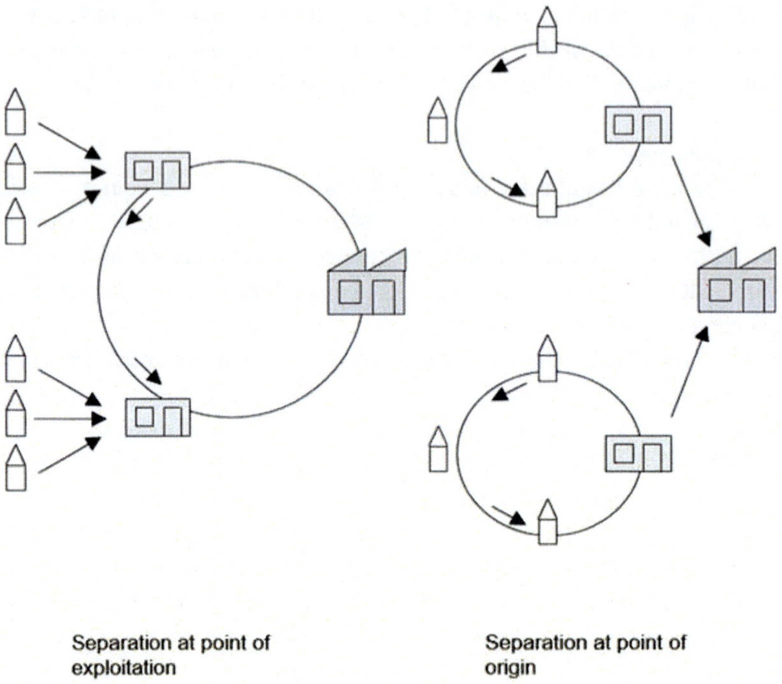

Fig. 13.3 Combinations in multi-echelon residue cycles (Source: Arnold et al., 2008, p. 516)

requirements that also exist to a lesser extent in the supply area, e.g., in the handling and storage of hazardous substances.

Collection Technology

Decisions on collection technology must be made taking into account the type and quantity of residues, the organisation of collection and separation (separate or mixed collection and bring or fetch system) and the organisation of transshipment (emptying or swap procedure). The type of residue determines the need to consider special environmental protection facilities in the area of collection technology. The quantity of residues has an influence on the container size. The smallest unit is the system waste bin (SME) with a capacity of 35 or 50 L. The next larger unit is the system waste bin (SMT) with a capacity of 70 or 100 L. Frequently used collection containers are large refuse containers (MGB) with a capacity of 120, 240 or 360 L, which are used, for example, in household refuse collection. Since 2005 there are the multifunctional containers (MFB), which are suitable for front, rear and side loader use. For larger quantities of residue, 660, 1100 or 5000 L units or larger swap bodies are suitable.[20]

[20]See Arnold et al., 2008, pp. 501 f.

The choice of separate or mixed collection is decisive for the number of containers or the number of separate compartments in the containers. Multi-chamber systems are mainly used for the separate collection of used glass, paper and batteries in the household refuse sector.[21] Decisions regarding bring or fetch systems as design alternatives for collection and separation influence the choice of containers in terms of manageability.

Taking into account the involvement of residue producers into the transport in bring systems, the initial collection at the place of occurrence takes place in small, manageable containers that can be transported to the central collection points without any problems. Due to the design of these collection points (e.g. provision of sufficient manoeuvring space for trucks) and existing connections to the transport infrastructure network, central residue collection can generally take place in large residue troughs or containers. Fetch systems, on the other hand, require containers which, on the one hand, have sufficient volume for the collection of residues, but, on the other hand, also meet the requirements of manual handling and the technical handling systems. Emptying or swap procedures as design options of the handling organisation[22] finally require a coordination of the containers with the respective technical equipment of the vehicles. Since decisions regarding the factors influencing the choice of container are not made independently of one another, but are usually combined systems, e.g. fetch system in the emptying process with mixed residue collection, several decision parameters must be taken into account in each case when choosing a container type, which are also subject to economic consideration.

Means of Transport

Special features in the technical design of means of transport arise from the compaction of residues often associated with transport and the need to maintain the separation of different residue fractions. Equipment for compacting residues is used in particular for large-volume residues, such as bulky waste. The compaction devices can be differentiated according to their technical design into press plate, rotary drum and press screw principle. Compaction enables more efficient utilisation of the collection vehicles, which can, however, be offset by a reduction in the recyclability of the residues.

In the waste management industry, trucks are most often used because waste management systems are often complex networks and road transport is particularly advantageous in terms of network formation.[23] However, due to the nature of waste—bulk, low capital commitment, low requirement for delivery time—rail or water transport is a good alternative solution.[24] Collection vehicles with swap bodies represent a variant for combined modes of transport between road and rail.

[21] See Becker/Lenz, 1993, pp. 35ff.

[22] See Stölzle, 1993, p. 238.

[23] See Arnold et al., 2008, p. 505.

[24] See Meyer/Rauh, 2003, p. 19.

References

Arnold D u.a. (Hrsg) (2008) Handbuch Logistik. 3., neu bearb. Aufl. Berlin/Heidelberg

Becker T, Hüning R (1993) Lagertechnik. In: Wehking K-H, Rinschede A (Hrsg) Entsorgungslogistik II. Entwicklung und Bewertung neuer Konzepte und Technologien. Berlin, S. 67-85

Becker T, Lenz G (1993) Sammeltechnik. In: Wehking K-H, Rinschede A (Hrsg) Entsorgungslogistik II. Entwicklung und Bewertung neuer Konzepte und Technologien. Berlin, S. 35-48

Bilitewski B, Härdtle G (2013) Abfallwirtschaft. Handbuch für Praxis und Lehre. 4., aktual. u. erw. Auflage. Berlin/Heidelberg

Bretzke W-R, Barkawi K (2010) Nachhaltige Logistik. Antworten auf eine globale Herausforderung. Berlin/Heidelberg

Emmermann M (1996) Managementorientierte ganzheitliche Entsorgungslogistik. München

Gregori G, Wimmer Th (Hrsg) (2011) Grünbuch der nachhaltigen Logistik. Handbuch für die ressourcenschonende Gestaltung logistischer Prozesse. Wien/Bremen

Haasis H-D (1999) Produktkreislauflogistik. In: Pfohl H-Chr (Hrsg) Logistikforschung: Entwicklungszüge und Gestaltungsansätze. Berlin, S. 253-277

Ivisic R-A (2002) Management Kreislauforientierter Entsorgungskonzepte. Bern/Stuttgart/Wien

McKinnon A, Browne M, Whiteing A (Hrsg) (2012) Green Logistics. Improving the environmental sustainability of logistics. 2nd. Ed. London

Meyer P, Rauh T (2003) Reisen auf Gleisen. In: Entsorga-Magazin, 11-12, S. 18- 22

Nguyen T V H (2012) Development of Reverse Logistics. Adaptability and Trasferability. Dissertation Technische Universität Darmstadt.

Pfohl H-Chr, Stölzle W (1992) Entsorgungslogistik. In: Steger U (Hrsg) Handbuch des Umweltmanagements. Anforderungs- und Leistungsprofile von Unternehmen und Gesellschaft. München, S. 571-591

Pfohl H-Chr, Stölzle W (1995) Stichwort „Retrodistribution“. In Tietz B, Köhler R, Zentes J (Hrsg) Handwörterbuch des Marketing. 2. Aufl. Stuttgart, S. 2234-2247

Rausch K-F, Kadow M, Elbert R (2010) Grüne Logistik. Handlungsfelder und – strategien für Logistikdienstleister am Beispiel von DB Schenker. In: Schönberger R, Elbert R (Hrsg) Dimensionen der Logistik. Funktionen, Institutionen und Handlungsebenen. Wiesbaden, S. 681-707

Stölzle W (1993) Umweltschutz und Entsorgungslogistik. Theoretische Grundlagen mit ersten empirischen Ergebnissen zur innerbetrieblichen Entsorgungslogistik. Berlin

Part IV
Institutional Aspects of Logistics Systems

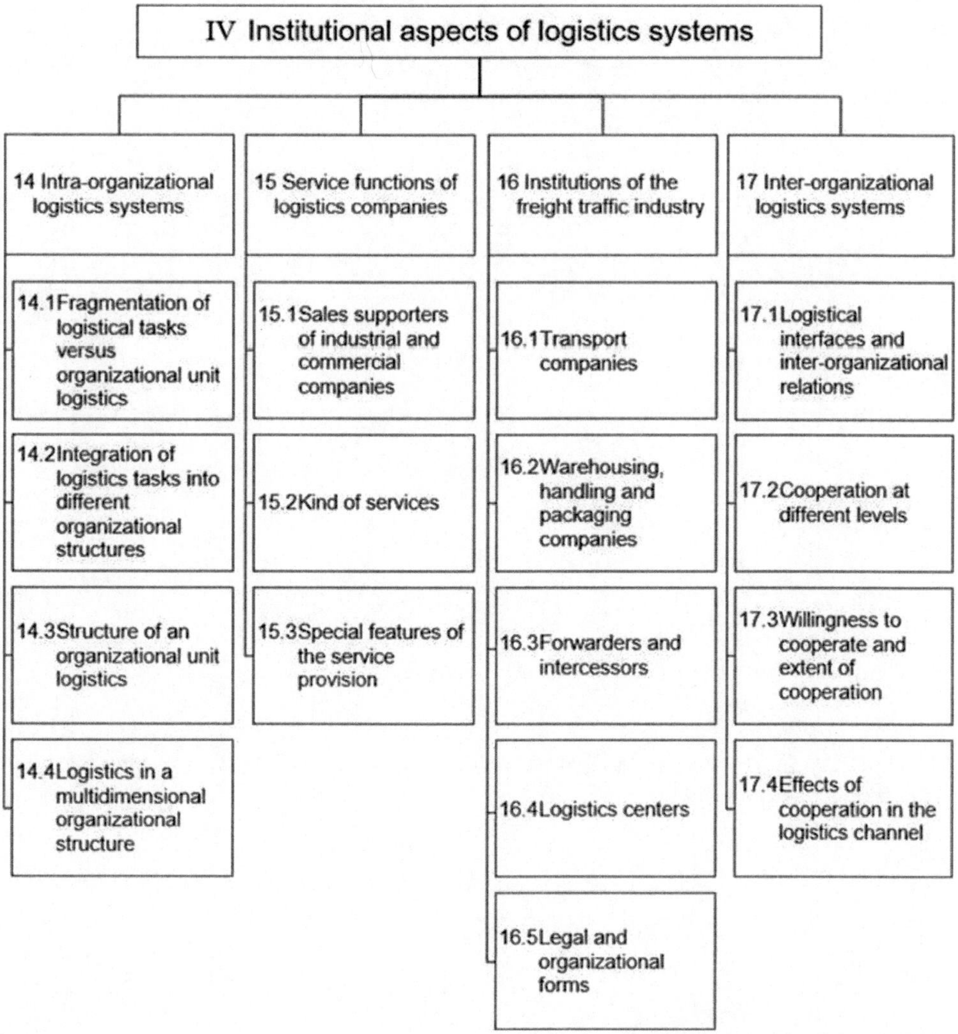

The functional subsystems of logistics presented in Parts II and III are institutionally anchored in corporate practice in very different ways. This applies both to the performance of logistics tasks in companies and to the distribution of logistics tasks among the companies cooperating in the value chain. The related questions will be dealt with in Part IV. Thus, on the one hand, the *intra-organizational problem of* the institutional anchoring of *logistics in the company* and, on the other hand, the *inter-organizational problem of* the institutional anchoring of *logistics in the value chain or in the supply chain* are addressed.[1]

The first chapter deals with the organizational form of intra-organizational logistics systems. The focus here is on industrial companies, in which the organization of logistics is usually very complex. The second and third chapters deal with logistics companies that specialize in different ways in providing logistics services in the value chain. The fourth chapter deals with inter-organizational logistics systems that are created through the cooperation of different institutions in the value chain.

[1]For a distinction between different levels of logistics systems, see Part I, Sect. 1.4.

14 Intra-organizational Logistics Systems

14.1 Fragmentation of Logistics Tasks Versus Organizational Unit Logistics

Basic Theses for the Organizational Implementation of the Logistics Conception

The starting point for statements about the integration of logistics tasks into the organizational structure of companies is the *logistics conception*.[1] It can be deduced from the logistics conception that logistics tasks can only be performed efficiently if all the processes required for this purpose are viewed as a whole and performed in a correspondingly coordinated manner. The demand for coordinated performance of individual logistical tasks directly addresses the organizational structure of logistics, because the organizational structure of a company is formed by the rules that govern the division of labor and coordination.[2]

However, the demand for an organizational unit logistics cannot be derived directly from the necessity of coordinating the performance of logistics tasks. Rather, two fundamentally different views can be distinguished. One thesis states that coordination is facilitated by the *integration of logistics tasks* in an organizational unit that specializes in them.[3] This is because in these organizational units, in which task bearers specialized in logistics tasks work together, the need for coordination and also the resistance to coordination are lower than in the case of organizational fragmentation of logistics tasks. From this, the demand for the organizational concentration of logistical tasks in an organizational unit logistics that is as comprehensive as possible can be derived. The opposing thesis is that the necessary coordination of logistical tasks, especially in the case of small companies, can

[1] See for the following Pfohl/Large, 1998, pp. 91ff.

[2] See Kieser/Walgenbach, 2010, pp. 18 f.

[3] See Pfohl, 1980, p. 1207; Pfohl, 1992, pp. 1255ff.

H.-C. Pfohl, *Logistics Systems*,
https://doi.org/10.1007/978-3-662-64349-5_14

also be achieved *without structural-organizational integration.*[4] The prerequisite for this, however, is the use of a comprehensive set of coordination instruments.

It is difficult to answer which of the two theses is more likely to be the truth, because to date there are *hardly any theoretical statements* on the efficient integration of logistics into the organizational structure of companies in a particular situation. KIRSCH et al. already point out that there is no clear answer to the question of the correct organizational anchoring of logistics.[5]

Problems Caused by the Fragmentation of Logistical Tasks

If the coordination of logistical task fulfilment is not achieved either by means of an appropriate set of coordination instruments or, in accordance with the first thesis, by means of structural-organizational integration, coordination problems are to be expected which prevent the realisation of the logistics conception. The problems arising from the fragmentation of logistical tasks can be roughly divided into three groups:

- Conflicting goals between organizational units.
- Conflicting goals within organizational units.
- Communication problems.[6]

Goal conflicts between organizational units: In the performance of logistics tasks, strong goal conflicts occur between the organizational units to which logistics tasks are assigned.[7] Organizational conflicts can have their *causes* in the socio-emotional behavior of organizational members and accordingly express themselves in status struggles or antipathies and mistrust. However, they can also have their cause in the task-oriented behavior of the organization members and then express themselves in different value concepts and problem views. Different, task-oriented behavior is particularly a consequence of the fact that specialists work in the individual organizational units who develop behavioral and thought patterns that correspond to the requirements of their work and their training. Therefore, the different specialists have different perceptions of how a particular task should be performed. Individual logistical decisions are then not made on the basis of logistical objectives and logistical thinking, but on the basis of the objectives of the organizational units in which logistical tasks are performed and according to the prevailing way of thinking there.

Goal conflicts within the organizational units: Within the organizational units, conflicting goals occur between the fulfillment of the main tasks and the *logistics tasks* that arise, which are considered *auxiliary services* and neglected as *secondary tasks.* In the

[4] See Ihde, 1985, p. 726.

[5] Kirsch et al., 1973, p. 343.

[6] See Pfohl, 1980, pp. 1205 f.

[7] For an overview of possible conflicts between organizational units, see Kotler et al., 2007, pp. 888ff.

organizational unit production, for example, the main focus is on the production activity and it is very likely that little attention is paid to the logistics tasks besides. Nevertheless, the logistics tasks burden the employees in the organizational units and keep them from fulfilling their actual tasks. The actual tasks of the organizational units of procurement, production and sales were elaborated in Part III in the sections on procurement, production and distribution logistics. In the conventional organizational form, they are the bearers of a significant proportion of the logistics tasks that arise.

Communication problems: Due to the demarcation of organizational units from one another, communication problems occur that prevent consistent application of logistical systems thinking and thus lead to wrong decisions, delay the flow of logistical processes and reduce the flexibility of logistical systems. In a company in which, for example, external transportation tasks are performed by both the sales area and the procurement area, an unreliable forwarding agent was terminated by the sales area. Since the procurement area performs transport tasks without coordination with the sales area, the forwarding agent was given new orders by the procurement area.

Possibilities for the Coordination of Fragmented Logistics Tasks

A number of coordination tools are available to handle the coordination problem.[8] One can divide the available coordination instruments into structural and non-structural coordination instruments. The structural coordination instruments are defined by organizational rules.[9] The *non-structural coordination instruments* are based on the basic idea that the activities of the organizational members can be most easily aligned with the organizational goal if all organizational members identify with the organizational goal. A differentiated set of *indoctrination instruments* is available for this purpose, which cannot be discussed in detail here. For logistics, however, *personnel policy measures* are of particular importance which are aimed at increasing the knowledge of logistical interrelationships among those members of the organization who perform logistical subtasks. In addition, the motivation and attitude of organizational members can be influenced by incentives in such a way that they actually take the recognized relationships into account in their decisions.

Structural coordination instruments are divided into coordination by personal instruction, coordination by self-coordination, coordination by programs and coordination by plans.

Coordination by *personal instruction* is based on the hierarchical nature of coordination. In this case, coordination takes place by, for example, the heads of procurement, production and sales instructing their employees who deal with logistical sub-problems how to take into account the existing interdependencies.

In the case of *self-coordination*, coordination is achieved by those members of the organization who have to deal with logistical sub-problems and who belong to different

[8] See for the following Pfohl, 1980, pp. 1206 f.

[9] See Kieser/Walgenbach, 2010, pp. 100ff.

areas coordinating with each other themselves. This self-coordination needs not be limited to informal contacts between organizational members and must be left to the members' own initiative and discretion. Issue-specific interaction, for example, involves determining which logistical problems require one entity to coordinate with which other entities. For coordination, institutionalized interaction is also possible, in which communication between the bodies that are to coordinate with each other becomes even more structured. Colleges (committees, commissions, working groups, meetings, conferences, etc.) are then set up for self-coordination.

In the case of coordination through *programs*, procedural guidelines or manuals are usually formulated in writing. They specify how logistical tasks are to be performed. For example, programs can be defined for the ordering situation, for transport processing, etc.

In the case of coordination through *plans*, coordination takes place through the specification of objectives, measures and resources. In contrast to programmes, plans are not general specifications. They only apply to the planning period and to the logistical planning problems that arise in it.

However, it is questionable whether these coordination instruments are sufficient to realize the logistics concept, i.e., to overcome the coordination problems. In the following, therefore, empirical studies on the existence of an organizational unit of logistics will be considered first.

Existence of an Organizational Unit Logistics

In accordance with the first thesis, the coordination problems identified can be effectively reduced by *integrating logistics tasks* in an organizational unit specialising in logistics. Accordingly, organizational units that concentrate a large number of logistical tasks in their area of responsibility would have to become established and empirically proven in operational practice.

There are a *large number of empirical studies* on the organizational anchoring of logistics in Europe and the USA.[10] *The* core of these studies was often the question of the *existence of an organizational unit for logistics*.

However, the existence of an organizational unit bearing the name Logistics is not yet a clear indication that organizational integration of logistics tasks has taken place. The prerequisite for this is that the logistical subtasks are largely concentrated in this unit. Based on company surveys, Fig. 14.1 shows the proportion of companies whose logistics organization can be assigned to one of the three degrees of centralization.

Although at present a large proportion of industrial companies obviously have an organizational unit for logistics with a high scope of tasks, it cannot be clearly clarified whether such a unit will continue to be the efficient way to realise the logistics conception in the future. Certainly, a compelling development towards a comprehensive organizational

[10] See Bowersox et al., 1994; Bowersox/Closs, 1996, p. 596; Pfohl/Large, 1998, pp. 97ff; Straube et al. 2005, p.27; Straube and Pfohl, 2008, pp. 20ff; Handfield et al. 2013, pp. 42ff.

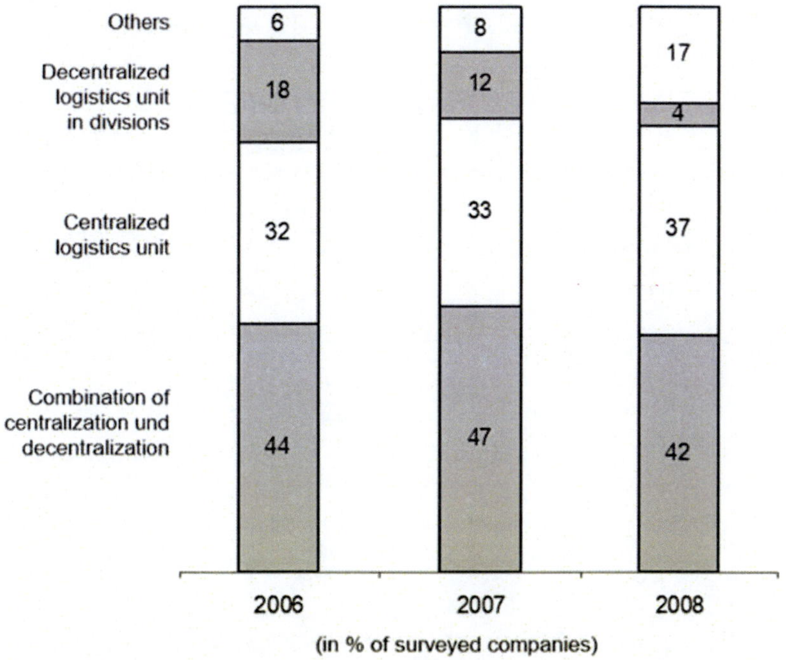

Fig. 14.1 Organizational anchoring of logistics (Source: own representation based on LaLonde/Ginter, 2006, p. 7; LaLonde et al., 2007, p. 7; LaLonde/Ginter, 2008, p. 7)

unit logistics cannot be assumed. Rather, the adequate organizational form depends on situational influencing factors.

However, if one assumes the necessity of an organizational structure grouping due to the characteristics of these factors, this cannot be reduced to the aspect of the existence of an organizational unit logistics. Other important issues are the integration into an existing organizational structure and the internal structure of the organizational unit logistics.[11]

14.2 Integration of Logistics Tasks into Different Organizational Structures[12]

Functional Organizational Structure

The organizational units for the fulfillment of logistical tasks can be integrated centrally or decentrally into a functional organizational structure according to the two theses presented. In the case of decentralized integration, the logistics tasks remain organizationally divided

[11] See Kirsch et al., 1973, p. 344; Ihde, 1980, pp. 1228 f.; Pfohl/Large, 1998, pp. 92 f.

[12] For a discussion of the incorporation of logistics into various organizational structures, see Pfohl, 1980, pp. 1208ff.; Felsner, 1980, pp. 64ff.; Large, 2016, pp. 186 ff.

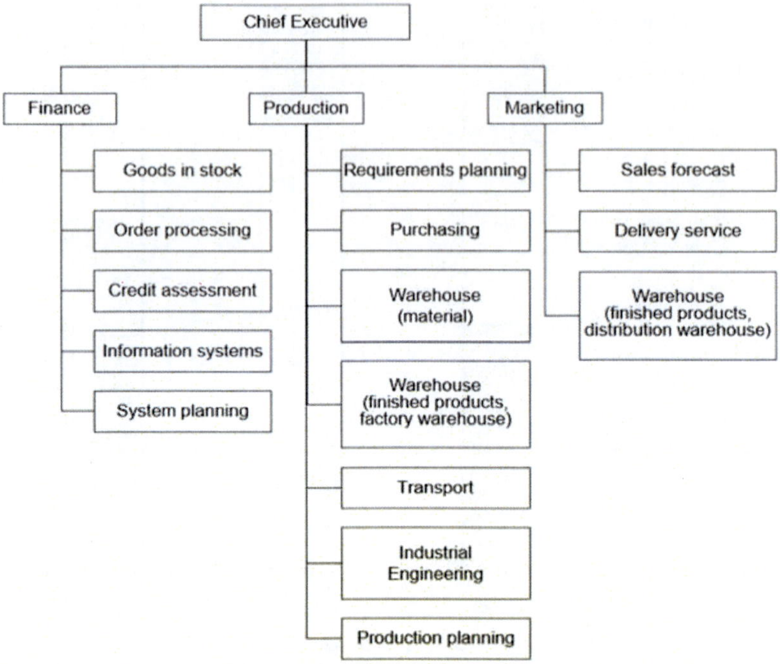

Fig. 14.2 Example of fragmentation of logistics tasks (Source: Adapted from Bowersox/Closs, 1996, p. 599)

among different functional areas. Figure 14.2 gives an example of such a fragmentation of logistics tasks.

In the case of *central* task grouping, the logistics tasks are integrated into a single functional area. In principle, there are two possibilities for this:

- Grouping of tasks in an independent logistics functional area alongside the other functions or, in the sense of a cross-sectional function, overlapping the basic business functions,
- Consolidation of logistical tasks under an existing functional area.

Logistics can therefore be equated with areas such as production, marketing/sales—these two terms are used synonymously in the following to designate organizational units—, procurement or administration, or logistics can be subordinated to these areas. It depends on the scope of the operating logistics system, seen as an institutional unit, how logistics is to be classified in the functional organizational structure.

BOWERSOX/CLOSS[13] distinguish between the three types of integration of logistics into a functional organizational structure outlined in Fig. 14.3. All three types can be found

[13]Bowersox/Closs, 1996, pp. 599ff.

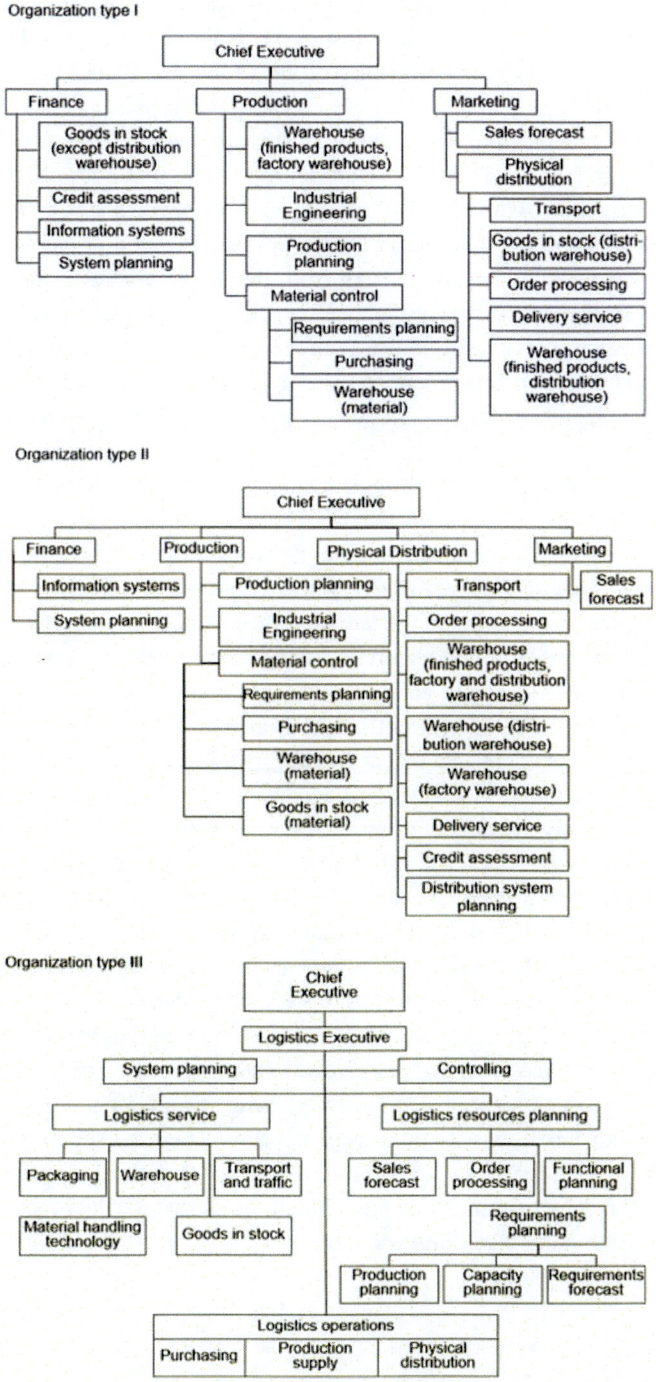

Fig. 14.3 Degree of centralization of logistics tasks in functionally structured companies (Source: Adapted from Bowersox/Closs, 1996, p. 600, p. 601 and p. 603)

in companies today. However, the different types also reflect the previous development of the organizational implementation of the logistics concept over time. Type I organization is characteristic of logistics-oriented companies of the 1950s and 1960s. Type II is typical of the 1970s and 1980s. Type III is increasingly encountered at present.[14]

Characteristic of *organization type I*, although still completely separate from each other, are Materials Management and Physical Distribution as the first crystallization points for combining logistics tasks. If the need to reorganize the logistics area is recognized in a company, it is to be expected that one or two such crystallization points will arise. In the marketing area, this usually happens starting from the delivery service and in the production area starting from material procurement. However, since logistics tasks continue to be highly fragmented in the organizational structure, organizational type I provides comparatively little support for the management of logistics system interrelationships. This is also underlined by the fact that logistics tasks do not receive any significant hierarchical upgrading.

In *organization type II*, some of the logistics tasks are outsourced from the other areas and centralized in a separate area. In the USA at least, this is very often the physical distribution area, starting from the focal point of delivery service. Such an area is typical for the food industry, for example, where delivery service problems and distribution costs play a major role. However, depending on the industry, for example, in the automotive industry, where supply service problems and materials management costs play a major role, this can also be the Materials Management area. In both cases, the centralized logistics tasks are upgraded hierarchically. The other areas begin to accept Physical Distribution/Materials Management as an independent area that acts on its own initiative and does not just react to the demands of the other areas. However, organization type II does not yet provide support for managing the synergy effects between material logistics and distribution logistics.

Finally, in *organization type III*, all logistics tasks are centralized under uniform management. The logistics department is accepted by the other departments as having full equal rights and is also given influence on the strategic management of the company. The System Planning staff supports the logistics management in the performance of strategic tasks, while the Controlling staff supports the planning and control of costs and services in the overall logistics system. For the remaining logistics tasks, Fig. 14.3 distinguishes between three exemplary organizational task areas. The logistics operations are performed by the organizational units Purchasing, Production Supply and Physical Distribution, each of which has its own responsibility. They are supported by organizational units of the Logistics Service unit. Planning and coordination is carried out jointly for all logistics operations by Logistics Resource Planning, so that the interdependencies of the overall logistics system are taken into account.

[14] For statements on the life cycle approach, see Pfohl/Large, 1998, p. 92; Pfohl, 2016, p. 26.

Divisional Organizational Structure

If the organizational structure is divisional (divisional organization), the basic forms for the integration of logistical tasks are central integration or decentralized integration into the divisions, as shown by alternatives a and b in Fig. 14.4.

The central function, in which all logistics tasks are grouped together centrally above the divisions, is the most suitable form of *central integration*. This form is suitable if one or more of the following conditions exist:

- a small number of divisions with a relatively low turnover of the divisions,
- a small number of production sites, or
- a comparatively low degree of decentralization of functional tasks to the divisions.

An alternative to the formation of a central division would be the formation of a logistics division, which would, however, pose the problem of being treated as a profit center.

The proposal to have the logistics tasks for all divisions performed by a separate, legally independent logistics company goes one step further.[15] Such a company can market externally the logistics services they offer to intra-group customers.

A *decentralized integration* of logistics into the divisions is suitable for relatively large companies with largely independent divisions or with divisions whose logistical problems differ greatly due to the characteristics of their products or special requirements of the markets. With such a decentralized classification, all logistics tasks that occur in a division are combined in this division.

However, decentralization can have a major disadvantage for the economic solution of logistical problems, as the profitability of a flow of goods tends to depend on its volume. Centralization offers the advantage of combining the various flows of goods in the divisions and using a single logistics system for all products. This results not only in synergy, specialization and standardization effects, but also in greater purchasing power for the purchase of logistical services. Therefore, even in companies that in principle have a decentralized organizational structure, a centralization of logistics tasks can be observed.[16] The possibility for centralization is limited by the individual requirements placed on transport, handling, storage and packaging by the characteristics of the product or the market. In addition, new coordination problems arise from the fact that the logistics function is centralized but the other functions remain decentralized in the divisions.

Possible Combinations

One way of combining centralization and decentralization in a *functional organizational structure* is to combine logistical staff tasks centrally and to classify logistical line tasks decentrally. The distinction between the two types of tasks for logistics is made in the next

[15] See Bowersox et al., 1986, pp. 317 f.

[16] See Magee et al., 1985, pp. 407 f.

section. If the central staff is to be able to actually influence logistics decisions, it must be a strong staff equipped with the necessary competencies in the decision-making process.[17]

One possibility for combining centralization and decentralization in a *divisional organizational structure* is the centralization of planning (system planning and specialized planning), coordination and control tasks as shown in Fig. 14.4 under alternative c, and the decentralization of the operational tasks of steering and execution. In practice, various forms can be implemented, each of which differs according to how the competencies are divided in detail between central logistics and the decentralized logistics organizational units of the divisions. In essence, this is expressed in the extent to which central logistics is given functional (specialist) authority to issue instructions to decentralized divisional logistics. Two extremes can be distinguished here[18]:

- A *central logistics department with specialist expertise* is responsible for system planning (i.e. setting up and implementing a functioning logistics system and developing the logistics methods and instruments to be used) and specialist planning (planning logistics objectives, logistics measures and logistics resources) as well as coordination and control. The decentralized divisional logistics are only assigned the tasks of steering (e.g. the physical flow of goods for a week is planned by the head office, but control within the week is left to the decentralized logistics) and the tasks of execution.
- In contrast, a *central logistics department with the authority to issue guidelines* can only define the framework within which the decentralized system logistics must operate by means of rough planning. Otherwise, the head office only has advisory tasks. In principle, the essential line tasks of logistics remain in the divisions.

Figure 14.1 showed the organizational integration of logistics into the divisional organizational structure in the USA. In the companies surveyed, the logistics function is predominantly implemented as a central department or as a combination of central department and integration into the divisions. It should be noted that these surveys are not a longitudinal study, as the composition of the companies that provided an evaluable questionnaire changed. Nevertheless, it can be seen that the type and manner of organizational integration changed only insignificantly in the period under consideration.

The explanations about the integration of logistics into a divisional organizational structure apply analogously to companies with *several plants*. Figure 14.5 shows an example from the automotive industry of a combination of centralization and decentralization. The combination problem becomes more complex when a company has several divisions, several plants and several functionally organized central units.[19]

---

[17] On the demand for a strong staff, cf. Heskett and others, 1973, pp. 685 f. On the possible combinations, see Bowersox et al., 1968, pp. 379 f.; Magee et al., 1985, p. 401. On this, see also the "hybrid" organisational units in Pfohl, 2016, pp. 285 ff.

[18] See in general Pfohl/Stölzle, 1997, pp. 192 f. and pp. 200 f.

[19] See Endlicher, 1981, pp. 209ff.

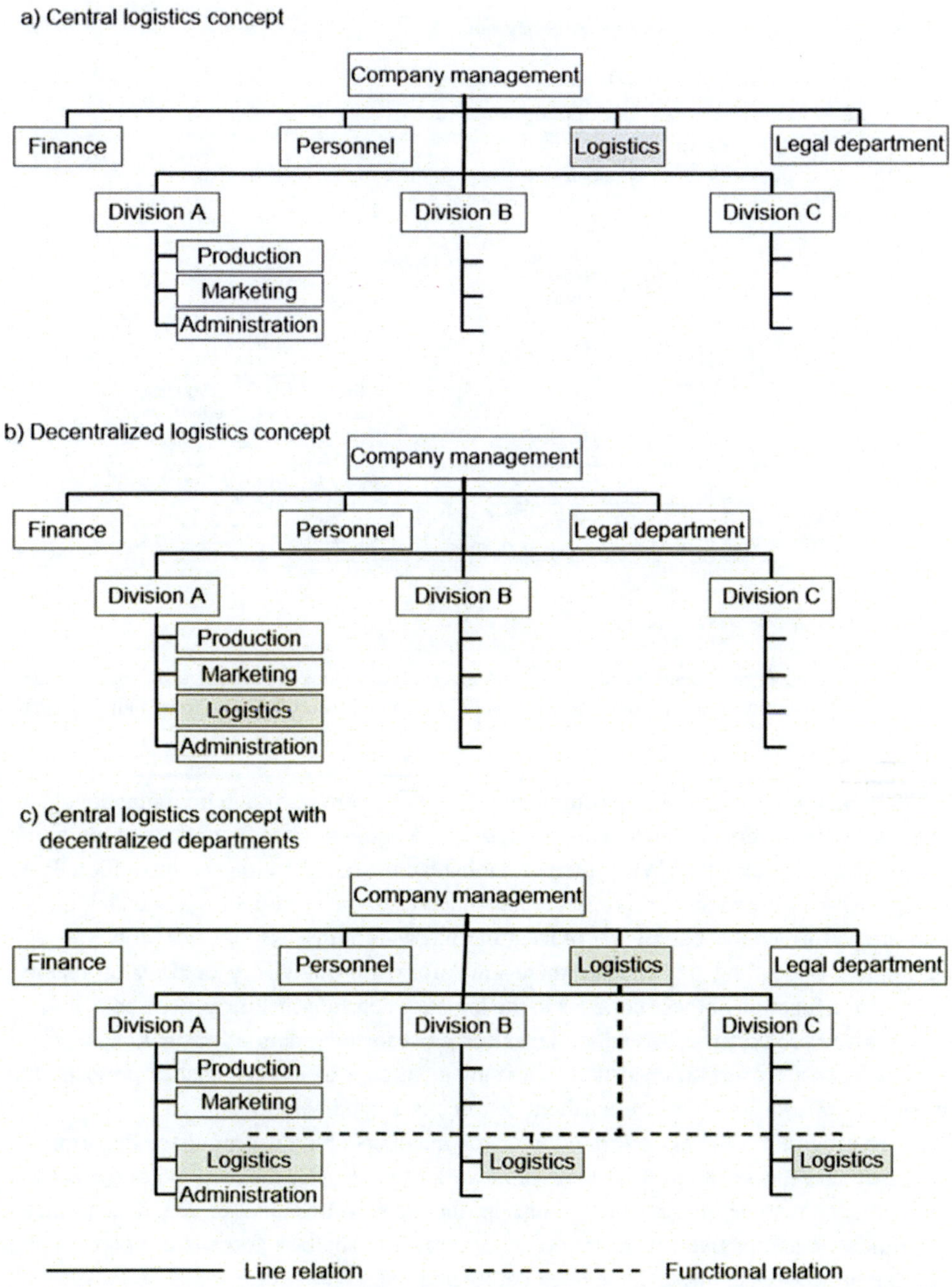

Fig. 14.4 Degree of centralization of logistics tasks in divisional companies (Source: Felsner, 1980, p. 72)

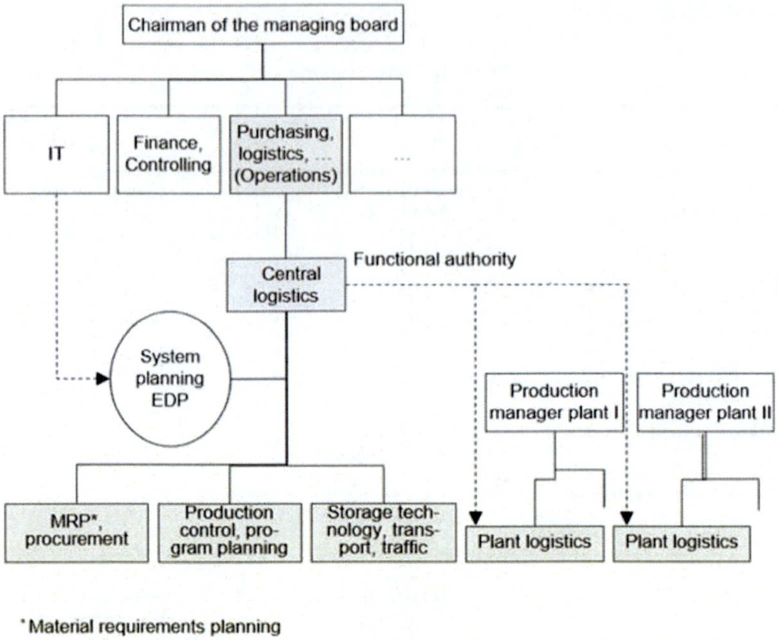

Fig. 14.5 Combination of centralized and decentralized performance of logistics tasks in a company with several production sites (plants) (Source: taken with minor modifications from Felsner, 1980, p. 70)

One possibility for combining centralization and decentralization in both functional and divisional organizational structures is offered by the *colleges* already mentioned under the coordination instruments. With their establishment, coordination is institutionalized through self-coordination. In this case, representatives of the departments in which logistics tasks are performed meet at regular or irregular intervals to discuss logistics problems with each other. Depending on whether there are information, advisory or decision-making colleges, the function of these colleges in the logistics decision-making processes will vary. In contrast to project groups, such colleges can exist for long periods of time.

If colleges are overburdened with the implementation of larger logistics projects, the *project organization* in the form of the task force is a good solution.[20]

A project group is therefore formed in which members of the decentralized departments concerned and specialists from headquarters work together. For the duration of the project work, the corresponding positions both in the head office and in the decentralized departments are kept free for these project employees. The task force is dissolved again after the project-related tasks have been completed.

[20] On project organization, see Pfohl/Stölzle, 1997, pp. 206ff.

In addition to the organization of project execution, e.g., in the form of the task force, the organization of the project sponsorship must also be taken into account in the project organization. This determines who is the responsible client for the project and, as such, has to exercise the supreme supervisory and decision-making function. An essential aspect for the organization of the project sponsorship is its composition of persons who can promote the project in the necessary way as expert and power promoters.

Hierarchy Level

The problem of the organizational grouping of logistics tasks can only be solved in connection with their hierarchical classification in the organizational structure. The hierarchical classification of the logistics department is an indication of the importance that management attaches to the logistics concept. The lower the institutionalization of logistics in the corporate hierarchy, the greater the danger that the logistics tasks will be regarded merely as a secondary function and subordinate to the objectives of the organizational unit to which logistics has been assigned. Furthermore, the hierarchical position influences the enforceability of logistics-specific objectives because of the power relationships within the company. Basically, the following hierarchy levels are available for selection:

- Board level,
- Business unit level,
- Main department level,
- Department level,
- Group level.

Since the *board level* is not differentiated according to function in the case of undivided management, this level may be omitted for the hierarchical classification of a logistics department. In the case of multipersonnel management, however, this restriction no longer applies. The implementation of the logistics concept in the company is supported most effectively by assigning the Logistics department to the board level.

Classification at *group level* is of secondary importance, as the following empirical studies show, since it is only conceivable at the beginning of a reorganization process, at which point a logistics group in a department begins to work on logistics problems. The business unit, main departmental and departmental levels are thus still of importance and will be discussed first in connection with the *functional organizational structure.*

Classifying logistics at *business unit level* has the great advantage that the logistics function is visibly given the same weight as the traditional functions of procurement, production and sales. The integration of logistics tasks into a specially created unit should have the effect that logistics problems are not merely dealt with in passing in another unit, but that they are solved with the attention and expertise they deserve.

Three main objections were raised early on against the integration of logistics tasks into another unit, e.g. distribution logistics into the sales unit[21]: In the sales unit, turnover and not cost thinking prevails, so that an excessive delivery service is offered disregarding the cost situation. The choice of location is often based on the location of existing sales offices, although such a warehouse location can be very unfavourable from a logistical point of view. In addition, people in the sales unit are often entrusted with logistical tasks that are not suitable for them.

However, these objections are no longer valid if the tasks of distribution logistics are integrated into the sales unit in such a way that they are combined in this unit in a *main department*. In this way, distribution logistics is given the necessary independence and weight in the sales unit. It then ultimately depends on the ability of the head of this main department whether or not the aspects of distribution logistics are given due consideration in the overall sales unit. This also ensures that a salesperson does not have to perform logistical functions that he or she is not normally equipped to perform because of his or her training. In this way, the salesperson can be relieved of any logistical activity so that he can devote himself exclusively to his actual task of selling and promoting sales. The same arguments also apply to the incorporation of procurement logistics alongside purchasing at main department level in the procurement unit.[22]

Thus, if the typical logistics system is procurement or distribution logistics based on the nature of the company's operations, it is perfectly possible to integrate the corresponding logistics tasks at the main department level. More comprehensive logistics systems, on the other hand, generally require their independent hierarchical anchoring at the *business unit level*.

The integration of logistics tasks at *departmental level* under a main department generally does not do justice to the importance of logistics. If, for example, distribution logistics is subordinated to a main sales department, there is a very great danger that logistics tasks will not be performed with the attention and expertise they deserve. This then manifests itself, for example, in the fact that salespersons who no longer perform their acquisition function to the satisfaction of the head of the main department are assigned logistics tasks that are considered to be of lesser value.

If, in a *divisional organizational structure*, all logistics tasks are combined in a central unit, no problems arise with regard to hierarchical integration, since this is the highest hierarchical level for logistics. In the case of a decentralized integration of logistics, the considerations made for the functional organizational structure apply because the individual divisions are normally organized functionally. If the two options are combined, it depends on the scope of the tasks and competencies that are assigned to central logistics. If

[21] See Reese, 1967, pp. 58 f. There is therefore a trend to establish logistics as an independent organisational unit during a reorganisation. See Handfield et al., 2013, p. 43.

[22] For the organizational classification of purchasing and procurement logistics, see Large, 2013, pp. 264ff.

this is large, a central logistics unit should be set up. Otherwise, a main logistics department in another central unit is sufficient. In the case of decentralized divisional logistics, if the central logistics department is equipped with the appropriate competencies, a classification at department level is also conceivable.

There are a large number of empirical studies on the *hierarchical integration* of logistics in the organizational structure of industrial and retail companies.[23] An empirical study from 2013 determined the highest logistics position of the companies surveyed as an indicator for the hierarchical integration of logistics (see Fig. 14.6). It can be determined that there are different emphases in the hierarchical level due to situational influencing factors in different countries and different industries.

14.3 Structure of an Organizational Unit Logistics

Descriptive and explicative statements about the internal structure of an organizational unit logistics are problematic; because this depends not only on the assigned scope of tasks, but also on the form and hierarchical level of integration.[24] This results in a large number of possible structure types. For this reason, data on the *internal structure of an organizational unit* logistics are rarely collected by empirical studies (see, however, Fig. 14.7). Detailed information on the more precise internal structure of a logistics organizational unit can be obtained primarily through case studies.

Staff and line positions can be found within the organizational unit logistics. The importance of line units has increased over time. It also became clear during the discussion of the integration of logistics into the divisional organizational structure that combinations of central departments and decentralized line departments can be found in the divisions, with staff positions often being located in the central departments.

Another indication of the internal structure of logistics is the *proportion of time* that managers and employees in an organizational unit logistics spend performing their individual tasks, used as indicators of the task profile. Figure 14.8 shows the results of annual surveys of American companies on the time spent by logistics managers in performing their assigned tasks. As expected, the classic logistics tasks of transportation, warehousing, and storage dominate in terms of time shares. This suggests a high importance within the tasks performed in the organizational unit logistics. However, it is also clear that non-logistical activities take up a noticeable proportion of time. This is an indication of the performance of interface tasks (e.g. sales forecasting), which require coordination with other organizational units. The importance of such interface tasks is typical of a multidimensional organizational structure, which will be presented below.

[23] See LaLonde/Ginter, 2006; LaLonde et al., 2007; LaLonde/Ginter, 2008; Straube/Pfohl, 2008, p. 23; Handfield et al., 2013, p. 42.

[24] See Pfohl/Large, 1998, p. 96.

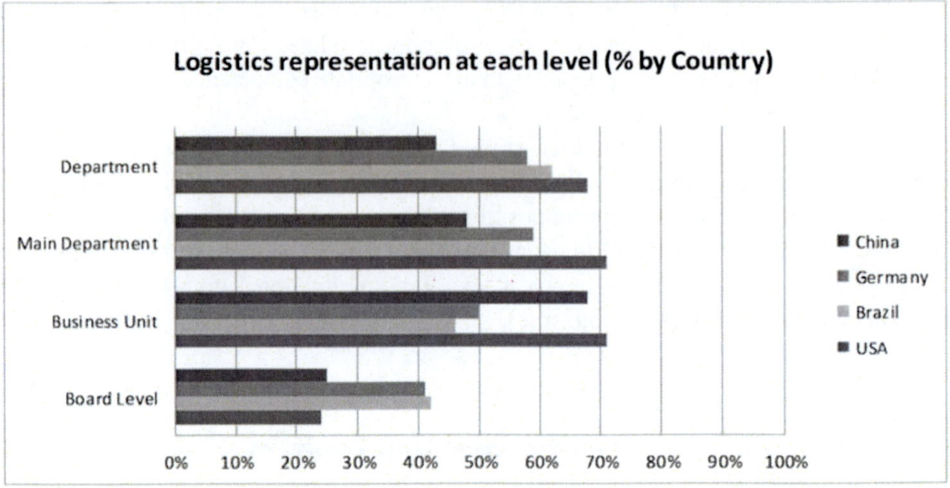

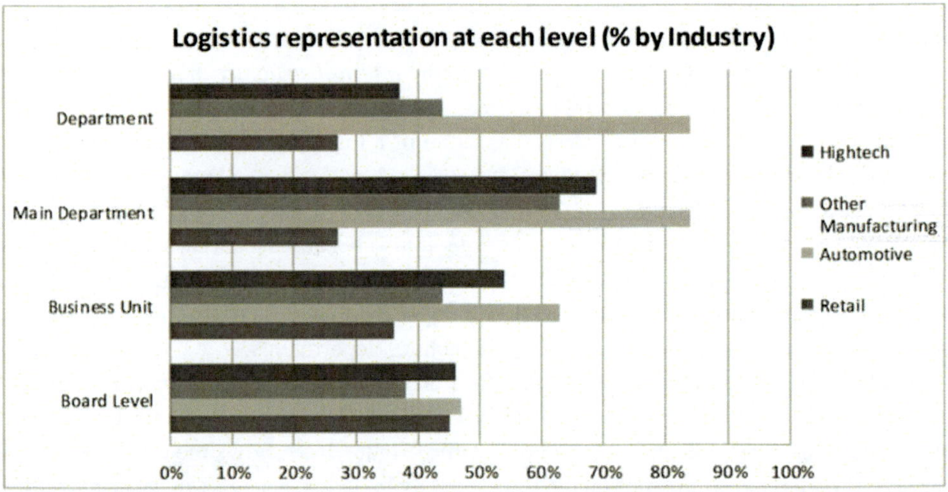

Fig. 14.6 Hierarchical anchoring of logistics in the company. Results of a survey conducted in 2013 in Brazil, Germany, China and the USA among 1757 people (Source: Hanfield et al., 2013, p. 42)

14.4 Logistics in a Multidimensional Organizational Structure

One- or Multidimensional Organizational Structure

The problems to be solved in a company generally have several dimensions, i.e., technological, economic and social problem dimensions must be taken into account, for example. *Onedimensional organization models* correspond to this multidimensionality of the problems by using different criteria for the centralization of tasks in organizational units

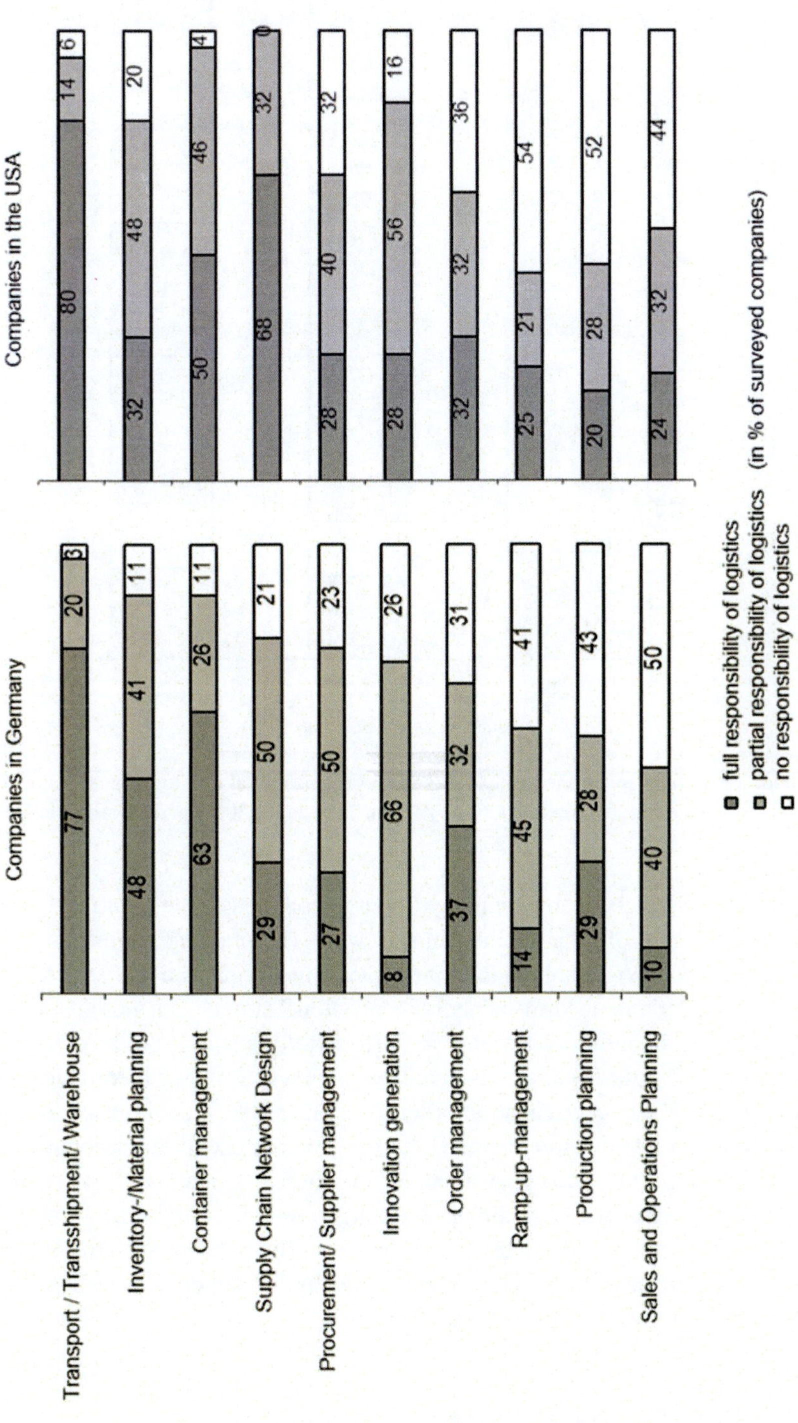

Fig. 14.7 Assignment of company tasks to an organizational unit logistics. Results of a survey conducted in 2008 in Germany, China and USA among 1189 companies (Source: Straube/Pfohl, 2008, pp. 25ff)

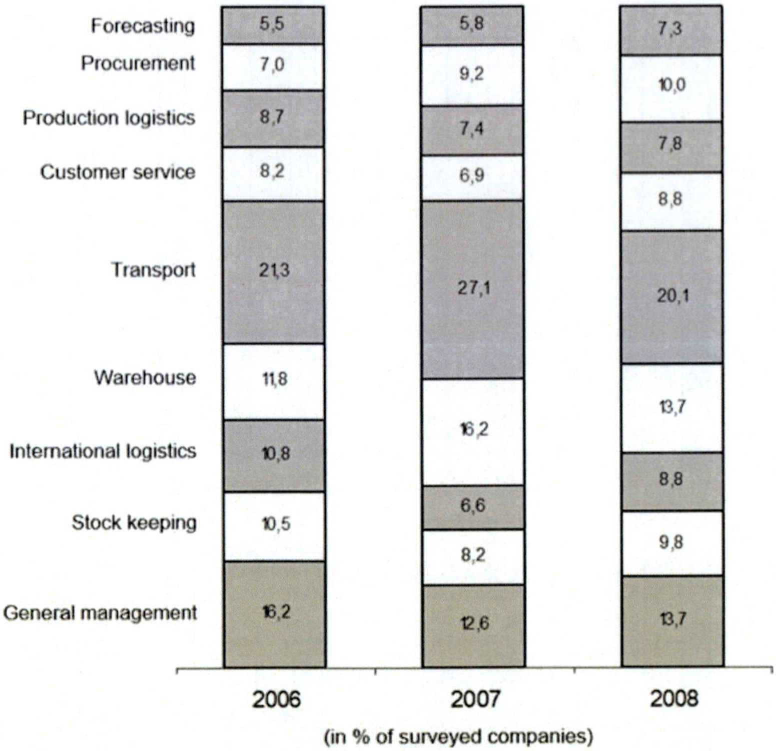

Fig. 14.8 Proportion of time spent on various tasks in an organizational unit logistics (Source: own representation based on LaLonde/Ginter, 2006, p. 12; LaLonde et al., 2007, p. 13; LaLonde/Ginter, 2008, p. 13)

at different hierarchical levels of the company. In large companies, the criteria of region, product and execution are often combined in this way. The *multidimensionality* of the problems is thus *successively* dissolved in a one-dimensional organizational structure.[25]

A balance of the criteria in onedimensional organizational structures is not given due to their hierarchical order. In *multidimensional organizational structures*, on the other hand, different *dimensions of a problem* are considered *simultaneously* on one hierarchical level with equal weighting. If two centralization criteria are taken into account, we speak of a matrix organization. If three or more centralization criteria are taken into account, this is called a tensor organization. Onedimensional organizational structures are suitable for simple and stable environments. Multidimensional organizational structures, on the other hand, are suitable for complex and dynamic environments. They are designed to counteract the danger of suboptimal problem solving. This is done by forcing differently specialized organizational units to work together on a problem.

[25] See Bleicher, 2004, pp. 338 f.; Schulte-Zurhausen, 2014, pp. 277ff.

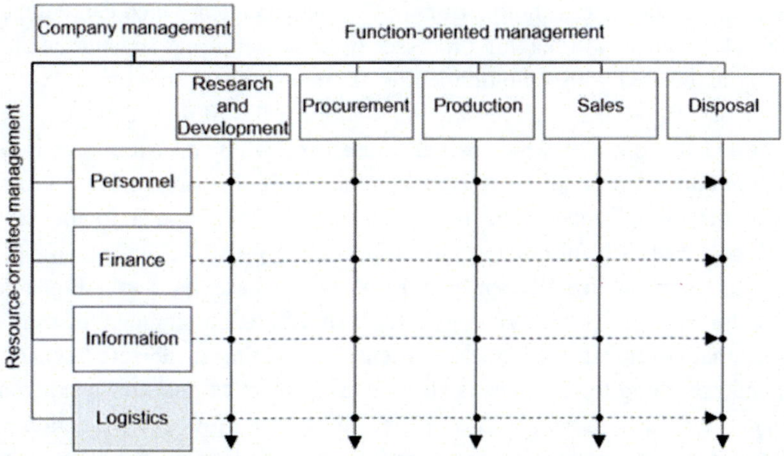

Fig. 14.9 Integration of logistics in a function- and resource-oriented matrix organization

Cross-Sectional Function Logistics in a Multidimensional Organizational Structure

The cross-sectional function of logistics[26] suggests that the logistics conception should be implemented in a multidimensional organizational structure. One step in this direction is the combination of central units with divisions or plants. If a central unit logistics has sufficient competencies, this combination already achieves a multidimensional organizational structure.

Figure 14.9 shows a *matrix organization in* which logistics is anchored as part of resource-oriented management alongside traditional function-oriented management. The horizontal specialization in the resource-oriented organizational units is on an equal footing with the vertical specialization in the traditional function-oriented organizational units. The consequence of this way of performing tasks is that, as can also be seen from Fig. 14.9, interfaces arise at which an overlap-free delimitation of competences is practically no longer possible. Multidimensional organizational structures are generally associated with overlapping competences or responsibilities.

Overlapping Competences and Conflict Institutionalisation

The conflicts of competence resulting from overlapping competences are not only accepted in the multidimensional organisational structure, but are also deliberately institutionalised. This *institutionalisation of conflict* is intended not only to ensure that the organizational units concerned think multidimensionally, but also that conflicts are dealt with openly and not in secret or suppressed. Then there is also the possibility of exerting influence to ensure that conflicts develop as few harmful, person-related forms as possible. Conflict institutionalisation does not mean letting conflicts run wild. There are possibilities for

[26]See Part I, Sect. 2.6: In Fig. 2.9, the term service is used instead of resource.

regulating competences and techniques of conflict management which can counteract the dangers of delay in decision-making processes or of power struggles and the shifting of responsibility associated with a multidimensional organizational structure.

By means of differentiated *competence regulations* (rules of precedence), it is possible to determine, at least formally, which partial competences—e.g. decision-making, advisory or information competence—an organizational unit has in the joint fulfilment of tasks. In this context, the competences need by no means remain assigned to the same units throughout the entire task fulfilment process. It is quite possible, for example, to assign decision-making competence differently in the planning phase than in the implementation phase (structural change). A well-known example of different assignment of authority in a multidimensional organizational structure oriented according to objects (e.g. projects or products) and according to functions is also the assignment of decision-making authority over the What and When to the object manager and over the How to the function manager. In the real task performance process, however, it will also depend in particular on the power that is not only tied to formal positions—e.g. expert power—who influences it the most.

The techniques of *conflict management* can be assigned to three areas, namely, corporate philosophy, personnel development and organization.[27] The corporate philosophy, in which the fundamental value system of the company is expressed, to which each individual manager can orientate his decisions, facilitates conflict management. This is done, on the one hand, by reducing the pluralism of values in the company and, on the other hand, by including in the corporate philosophy principles of conduct for the regulation of conflicts. In the field of human resources development, techniques can be used that can change people's attitudes towards conflict. Sensitivity training and the confrontation technique are particularly worthy of mention here. Their aim is to make the people involved aware of latent conflicts, to lead them to a constructive way of dealing with them and also to increase their ability to tolerate conflicts. Finally, the organizational area includes measures such as the selection of members for a group, the regulation of coordination processes with the aim of carrying these out as regularly and as early as possible, or also the definition in the form of a subsidiary decision-making process of which other body makes the decision in the event of disagreement.

Interface Tasks

The previous explanations of the multidimensional organizational structure make it clear that the tasks at the interfaces require *close cooperation* between the logistics organizational unit and *other organizational units.*[28] The importance of this cooperation can in turn be measured by the time spent on non-logistics tasks by the managers of a logistics organizational unit. As Fig. 14.8 already illustrates, the logistics managers surveyed spend a considerable proportion of their working time on general management tasks or

[27] See Krüger, 1973.

[28] See for the following Pfohl, 1980, pp. 1216 f.

interface tasks. This indicates that *logistics managers* represent logistics interests to a significant extent vis-à-vis managers from other functional areas.

In addition to the tasks of production planning and sales forecasting, the typical interfaces with different organizational anchoring are considered to be the determination of the delivery service level as well as the supplier selection.[29] It is important to ensure the cooperation of the organizational units affected by these interface tasks. It seems less important in which organizational unit the interface is initially integrated. This can be decided differently from case to case.

Thus, it may make organizational sense to place *production planning*—which is primarily about process and requirements planning—primarily with production or primarily with logistics. In companies in which production is not controlled on the basis of lot sizes but on the basis of material flows due to the production strategy pursued and the resulting production technology, or in which lot size changes only have an insignificant effect on manufacturing costs, production planning should be combined with the other logistics tasks from an organizational point of view. If, on the other hand, low cost of goods manufactured is decisively dependent on large lot sizes, production planning should be organizationally anchored in the production area. In both cases, it goes without saying that both areas should interact in order to take into account the interdependence of lot-size planning and material requirements planning.

A similar approach can be taken, for example, in the organizational integration of the tasks of *setting the delivery service level* and *sales forecasting*. *Setting the delivery service level* is part of the marketing policy and has a significant impact on logistics costs. Therefore, cooperation between logistics and marketing is necessary in any case. In order to ensure the connection of the delivery service with the other instruments of marketing policy, it will generally make sense to anchor the setting of the delivery service level primarily in the area of marketing. With regard to the primary anchoring of the *sales forecast*, on the other hand, it will depend on who is the source of information for the forecast. If, for example, salespeople are the main sources of information for the sales forecast, it makes sense to anchor the sales forecast primarily in the marketing area and at the same time ensure close cooperation with the logistics area.

Supplier selection is part of the purchasing department's area of responsibility. Since the geographical spread, the number and the size structure as well as the delivery service of the suppliers have a significant influence on the logistics costs, the cooperation between purchasing and logistics must also be ensured at this interface. This applies in particular because opportunities for interorganizational cooperation between supplier and buyer will play a greater role in the logistics area in the future. If supplier selection is also understood to mean the selection of suppliers of logistical services—i.e. logistics companies—then this

[29] See Pfohl/Large, 1998, p. 95 and the literature listed there. See also the phase-specific subsystems of logistics in Part III, Chaps. 9–11.

naturally belongs to the area of responsibility of logistics. The following section provides information on the range of logistics services.

References

Bleicher K (2004) Das Konzept Integriertes Management: Visionen - Missionen – Programme. 7., überarb. und erw. Aufl. Frankfurt/Main

Bowersox D J, Closs D J (1996) Logistical Management. Überarb. Ausg. der 3. Aufl. New York u.a.

Bowersox D J, Closs D J, Helferich O K (1986) Logistical Management. A Systems Integration of Physical Distribution, Manufacturing Support, and Materials Procure-ment. 3. Aufl. New York/London

Bowersox D J, Frayer D J, Schmitz J M (1994) Organizing for Effective Logistics Management. In: Robeson J F, Copacino W C (Hrsg) The Logistics Handbook. New York u.a., S. 773-784

Bowersox D J, Smykay E W, LaLonde B J (1968) Physical Distribution Management. Logistics Problems in the Firm. Rev. ed. New York

Endlicher A (1981) Organisation der Logistik. Untersucht und dargestellt am Beispiel eines Unternehmens der chemischen Industrie mit Divisionalstruktur. Forschungsberichte zur Industriellen Logistik 18. Dortmund

Felsner J (1980) Kriterien zur Planung und Realisierung von Logistik- Konzeptionen in Industrieunternehmen. Bremen

Handfield R, Straube F, Pfohl H-Chr, Wieland A (2013) Trends und Strategien in Logistik und Supply Chain Management. Hamburg

Heskett J L, Glaskowsky N A, Ivie R M (1973) Business Logistics. Physical Distribution and Materials Management. 2. Aufl. New York

Ihde G B (1980) Stichwort "Logistik, Organisation der". In: Grochla, E. (Hrsg.): Handwörterbuch der Organisation. 2., völlig neu gest. Aufl. Stuttgart, Sp. 1224- 1234

Ihde G B (1985) Die organisatorische Handhabung der Logistik. Stellungnahme zum Beitrag von Lutz J. Heinrich und Elisabeth Felhofer. In DBW 45 6, S. 725-727

Kieser A, Walgenbach P (2010) Organisation. 6., überarb. Aufl. Stuttgart

Kirsch W u.a. (1973) Betriebswirtschaftliche Logistik. Systeme, Entscheidungen, Methoden. Wiesbaden

Kotler Ph, Keller K L, Bliemel F (2007) Marketing-Management: Analyse, Planung, Umsetzung und Steuerung. 12., aktualisierte Aufl. München u.a.

Krüger W (1973) Grundlagen und Instrumente der Konflikthandhabung in der Unternehmung. München

LaLonde B J, Ginter J L (2006) The Ohio State University 2006 Survey of Career Patterns in Logistics

LaLonde B J, Ginter J L (2008) The Ohio State University 2008 Survey of Career Patterns in Logistics

LaLonde B J, Ginter J L, Stock J R (2007) The Ohio State University 2007 Survey of Career Patterns in Logistics

Large R (2013) Strategisches Beschaffungsmanagement. Eine praxisorientierte Einführung. Mit Fallstudien. 5. vollst. überarb. Aufl. Wiesbaden

Large R (2016) Logistikmanagement. Betriebswirtschaftliche Logistik. Band 2. Berlin/Bosten

Magee J F, Copacino W F, Rosenfield D B (1985) Modern Logistics Management. Integrating Marketing, Manufacturing, and Physical Distribution. New York u.a.

Pfohl H-Chr (1980) Aufbauorganisation der betriebswirtschaftlichen Logistik. In: ZfB 50 11-12, S. 1201-1220

Pfohl H-Chr (1992) Stichwort "Logistik, Organisation der". In: Frese, Erich (Hrsg.): Handwörterbuch der Organisation. 3., völl. neu gest. Aufl. Stuttgart, Sp. 1255-1270

Pfohl H-Chr (2016) Logistikmanagement. Funktionen und Instrumente. 3., neu bearb. und aktual. Aufl. Berlin/Heidelberg

Pfohl H-Chr, Large R (1998) Eingliederung der Logistik in die Aufbauorganisation von Unternehmen. In: Isermann H (Hrsg) Logistik: Gestaltung von Logistiksystemen. 2., überarb. und erw. Aufl. Landsberg a. L., S. 91-105

Pfohl H-Chr, Stölzle W (1997) Planung und Kontrolle. 2., neu bearb. Aufl. München

Reese B (1967) Physical Distribution the Neglected Marketing Function. In: Marks N E, Taylour R M (Hrsg) Marketing Logistics. Perspectives and Viewpoints. New York/London/Sydney, S. 56-61

Schulte-Zurhausen M (2014) Organisation. 6., überarb. und aktual. Aufl. München

Straube F, Pfohl H-Chr (2008) Trends und Strategien in der Logistik - Globale Netzwerke im Wandel. Umwelt, Sicherheit, Internationalisierung, Menschen. Hamburg

Straube F, Pfohl H-Chr, Günthner W A, Dangelmaier W (2005): Trends und Strategien in der Logistik. Ein Blick auf die Agenda des Logistik-Managements 2010. Hamburg

15 Service Functions of Logistics Companies

15.1 Sales Supporters of Industrial and Commercial Companies

Institutions in the Marketing Channel

Each *marketing channel* can be described as a system in which different groups of active elements (institutions) assume sales management functions.[1] The *primary elements are* the production or extraction company supplying the merchandise, the trading company acting as sales intermediary and the final buyer of the sales channel purchasing the merchandise. *Secondary elements* do not act as buyers or sellers of the merchandise, but are involved in the sales process as sales supporters. In addition to logistics companies, they include, for example, commission agents, advertising agencies or credit institutions. They offer services—and are therefore also referred to as service providers—which serve to initiate or carry out sales in the channel.

Logistics companies thus play the role of *sales supporters in the marketing channel.* Their primary performance is logistics services so that they do not have to be provided as a secondary service by other institutions in the marketing channel.[2] Their service offering has to be done mainly under four conditions by which their marketing activities are decisively influenced[3]:

First, the demand for the services of logistics companies is not original, but *derived demand.* Thus, it does not occur alone, but only in conjunction with the demand for the

[1] See Meffert et al., 2012, pp. 544 f. Further explanations to the distribution policy see Part III, Sect. 11.2. The following explanations apply mirror-invertedly also to the procurement. For explanations on procurement policy, see Part III, Sect. 9.2.

[2] For the definition and meaning of this service, see Part I, Sects. 2.1 and 2.2. For the logistics conception as basic for the marketing of logistics companies, cf. Pfohl, 1980.

[3] See Pfohl, 1980, pp. 423 f.; Zöllner, 1990, pp. 7 f.

H.-C. Pfohl, *Logistics Systems*,
https://doi.org/10.1007/978-3-662-64349-5_15

merchandise of the marketing channel. This does not exclude the possibility that it is only through the provision of a specific, logistics service that demand for a product can be created. For example, the supply of flowers and fruits from Africa or the Middle East has only become possible through the provision of corresponding logistical services by air freight companies.

Second, logistics companies always have to deal with *two market partners* when *selling their service*, namely the consignor and the consignee of the goods flowing in the marketing channel. The producers of the good and the intermediaries involved in its distribution, on the other hand, generally have to deal with only one market partner, the respective buyer of the good.

Thirdly, it is often typical for logistics companies to offer only part of the service demanded by the shipper themselves and therefore have to *coordinate* their *service offering* with that of other service providers.[4] For example, the offered service might include only local transport but not long-distance transport.

Fourthly, the process of creating logistics services is characterised by a *combination of* internal production factors, i.e., factors that are obtained from the procurement market or for which there is a right of use (transport routes, stations) and external factors, i.e., in particular tangible goods on which logistics processes are carried out.

Systems Thinking

All four conditions make it clear that a *holistic view of the marketing channel* is necessary for offering logistics services. The focus must be on concepts of the flow of goods through the entire system of the marketing channel. Taking advantage of the division of labor, these concepts must solve the problems associated with bridging spatial and temporal distances in an optimal way for the entire marketing channel, whereby the flow of goods may well pass through other institutions than, for example, the flow of property.[5] This underlines the importance of *cooperation* for logistics companies. This is because the holistic approach requires a rethinking of the traditional roles of institutions in the marketing channel. The functions to be performed in the flow of goods must be shifted to the points in the marketing channel where they can be optimally performed.

However, systems thinking is also important for the provision of logistics services from another point of view, which has already been addressed in the discussion of the importance of logistics.[6] Due to the increasing spread of the logistics conception in the shipping industry, integrated logistics services are increasingly in demand. In order to be able to satisfy this demand, a transport and forwarding company, for example, must not only offer transport services, but demand-oriented, *logistical service packages* through which all or at

[4] On the creation by division of labor, cf. Pfohl, 1993, pp. 123 f.

[5] See Fig. 11.3 in Part III, Sect. 11.2.

[6] See Part I, Sect. 3.3.

least a large part of the service needs can be satisfied. They then no longer offer partial solutions to logistics problems, but complete problem solutions.

Purchase of Logistics Services

Logistics services are purchased to a considerable extent. This applies in particular to transport services, but increasingly also to services of other kinds.

In the discussion of demand pressure as a factor influencing the importance of logistics, empirical studies on the importance of various criteria for the selection of logistics companies were presented in Figs. 3.5 and 3.6. If one disregards company-specific factors, it can be summarised that the type of goods plays a major role when it comes to the *importance of decision criteria* for the selection of logistics companies when purchasing logistics services.

Basically, the tendency can be seen that in industries with mass transport volumes (e.g. mining and basic industries), price is the most important decision criterion, while the quality of the delivery service components becomes a priority as the goods become more mature for consumption. The importance of the quality of the logistics services offered is also reflected in the fact that poor quality cannot simply be compensated with a lower price. Moreover, individual delivery service components are not easily substitutable. This is because the overall high demands that are ultimately placed on all delivery service components mean that good fulfillment of a criterion that ranks high in the hierarchy of decision factors does not easily permit poor fulfillment of a lower-ranking criterion.

Whether and, if so, which decision criteria are used for supplier analysis when purchasing logistics services can vary from decision process to decision process. Theoretical approaches from business-to-business marketing can be used to *analyse the purchasing behavior* of shipper institutions when purchasing logistics services.[7] This refers to all sales processes directed at companies or other organizations, i.e., including the sale of logistics services. Fundamental to these approaches is, on the one hand, the recognition that the purchasing decision is not an isolated act, but that a multi-phase decision-making process—with the phases of stimulation, search, evaluation and selection—takes place. On the other hand, different purchasing situations are assumed to influence the course of this decision-making process.

The situation in the purchase of logistics services is characterized by an asymmetric distribution of information between the market partners involved. Therefore, both the buyer and the supplier strive to compensate for the existing information deficits. Since the demander, due to his subjective perception and due to his limited ability to judge—which is caused, for example, by a lack of know-how or a lack of willingness to acquire this know-how—cannot judge all services equally, different characteristics of services can

[7] See Homburg, 2017, pp. 1053ff. On questions of organizational buying and interaction behavior and on problems of product and distribution policy as well as pricing and contract design in capital goods marketing, cf. Backhaus/Voeth, 2014.

be distinguished from the point of view of information economics.[8] These performance characteristics can be classified according to their assessability and the time at which they can be assessed:

- A service has *search* qualities if it can be fully assessed by inspecting the service offering or by searching for information *before purchase.*
- A service has *experience* qualities if it can only be assessed *after purchase.*
- A service has *credence* qualities if it is *not* possible to assess it *either before or after purchase.*[9]

Since the purchase of logistical services is usually not made on the basis of a finished, inspectable product, but the provider can initially only make a promise of performance, experience and trust characteristics of the service offer play an important role. They are of particular importance when purchasing situations are classified on the basis of the characteristic of novelty of the problem for the buyer.[10] According to this characteristic, *three purchasing decision types* can be distinguished for which the purchasing decision process will be completely different.

Pure repeat purchase: The purchasing decision situation is characterized by a problem that arises in the same way again and again. In order to solve such a problem, the phases of searching for alternatives and evaluation in the decision-making process does not take place at all. New alternatives are not seriously considered in the decision, but the known solution alternative is routinely taken. A pure repeat purchase occurs, for example, when a planned transport is carried out with the same transport or forwarding company that was previously awarded the contract.

Modified repeat purchase: The problem definition of this decision situation is not new, but deviates in partial aspects from the previous problem definition. This requires the search for previously unused alternative solutions and their evaluation. A modified repeat purchase occurs, for example, when the shipper is dissatisfied with his previous transport or forwarding company and wishes to place the order with another company.

First purchase: The problem of this decision situation is completely new. Previous experience plays only a very minor role. To solve such problems, new information must be obtained at all stages of the decision-making process. An example of a first purchase is the decision between different previously unused modes of transport.

While for the pure repeat purchase the experience characteristics of the service are of outstanding importance, in the modified repeat purchase both the search characteristics

[8] See Weiber/Adler, 1995, pp. 58ff.

[9] An example of this is taxi journeys in a city unknown to the taxi user, where the taxi driver knows the exact route to the destination, but the taxi user does not. See Balafoutas et al., 2013, pp. 3ff.

[10] See Meffert et al., 2012, p. 144. Further characteristics can be, for example, the organizational change associated with the use of the service as well as the investment value of the service for the demander, see Wagner, 1978, pp. 272ff.

(e.g. if the services of two transport companies are very similar) and the experience characteristics (e.g. if a transport or forwarding company already provides similar services for the buyer) can be of primary importance.[11] For the first purchase, on the other hand, the trust properties of a service are relevant. This applies in particular to those services that are to a large extent individually created.[12]

Which decision criteria are used in the modified repeat purchase or in the first purchase will depend on the company-specific characteristics of various *variables*, which can be divided into four *classes*. These are the "environmental variables" (e.g. the transport technology, the transport infrastructure or the competitive situation in the markets of logistics or freight forwarding companies), the "organizational variables" (e.g. the objectives pursued by the shipper organization, the logistics tasks derived from these, their formal integration into the organizational structure and the technological and human resources available), the "interpersonal variables" (e.g. the perception of specific roles such as that of the user, the buyer, the influencer, the gatekeeper and the decision-maker in the purchasing process of logistics services) and the "intrapersonal variables" (e.g. training, information, motives, attitudes and expectations of the persons involved in purchasing).[13] The logistics company has to match the wishes of the shippers in the purchasing of logistics services with a corresponding range of services. The next section gives an impression of the kind of such services.

15.2 Kinds of Services

Service Program

The range of services offered by logistics companies includes, first of all, the provision of services directly related to the *realization of the distribution of goods*. They refer to the guarantee of all service components and the related performance of logistics tasks. Another service provided by logistics companies is *logistics consulting*. In this case, logistics companies assume the function of a management consultant specializing in logistics. Finally, the range of services offered by logistics companies also includes *non-logistical services* that are provided in connection with the distribution of goods. These include, for example, shelf services or the take over of collections. Figure 15.1 gives an overview of the

[11] For the importance of evaluation criteria in supplier selection, see Fig. 3.4 in Part I, Sect. 3.3.

[12] See the reference to the importance of psychographic targets in the provision of services in Part I, Sect. 2.1.

[13] A survey conducted in Germany in 2016 among 132 sellers of logistics services on the involvement of organizational units with their customers revealed that the logistics department is involved in defining requirements in 82.6% and the purchasing department in 71.8%. In contrast, the logistics department dominates in shaping the relationship after the contract is signed, with 92.9% compared to 28.3% of the purchasing department. See Large, 2016.

Functions	Institutions
Main functions	
• Disposition function - Choice of means of transport, route, tariff - Conclusion of freight contracts - Issue of shipping documents - Freight control	Forwarder Carrier Agent
• Transportation function - Short-distance traffic: Groupage and distribution traffic, transport to long-distance traffic with other means of transport - Long-distance traffic: national and international	Carrier
• Logistics consulting - Advisory, analysis, planning and organization	Forwarder
Supplementary resp. complementary functions	
• Handling function - Organization and realization of the handling - Operation of stations/terminals	Carrier Handling company
• Warehousing function - Take into and out of stock, storage - Order picking - Warehouse management (e.g. inventory management)	Forwarder Warehousing company
• Groupage function (piece good) - Groupage and distribution of piece goods - Consolidation of load units	Forwarder Carrier
• Packaging function - Advisory and selection of transport packing - Transport auxiliary means - Pack, unpack	Forwarder Packaging company
• Manipulation function - Shipment manipulation (e.g. labeling) - Goods manipulation (e.g. security measures)	Carrier Packaging and handling company
• Information function - Installation of information chains to plan, coordinate, control and check the transport process, the storage and consignment status	Forwarder Carrier
• Internal transport, handling and storage tasks with the customer	Forwarder
Special functions	
• Merchandising function - Taking over merchandising measures by order of sender in favor of recipient (e.g. rack jobbing)	Forwarder Carrier
• After-sales service function - Taking over after-sales service functions by order of sender in favor of recipient (e.g. storage of spare parts, maintenance)	Forwarder Carrier Warehouse company
• Assembly function	Forwarder
• Transport insurance function - Risk management in the field of transport insurance - Conclusion of insurance contracts - Handling of claims	Forwarder Agent
• Customs clearance function - Customs declaration, -booking and -dispatching	Forwarder Customs agent
• Credit function - Freight and customs loans	Forwarder

Fig. 15.1 Services provided by logistics companies (Source: based on Stabenau, 1994, pp. 15 f. and Aberle, 2009, pp. 532ff.)

types of services provided by logistics companies.[14] The importance of supplementary and complementary functions and also of special functions—today often referred to as value added services—as a competitive tool will increase in the future. However, it can be assumed that small companies in particular do not have the resources to follow this development.

Not listed in Fig. 15.1 are the event logistics services offered by specialized logistics companies. The importance of event logistics is increasing in the context of event management in recent times.[15] A market survey shows a trend for logistics companies to develop a new market of professional management of events, which is predicted to grow rapidly in turnover through coordinated logistics management of major meetings, events and occasions.[16]

Demarcation Criteria

Four criteria can be used to assess the similarity of goods distribution operations and thus specialization in the service program of logistics companies: Scope of service, spatial dimension, goods dimension and qualitative dimension.[17]

According to the *scope of services,* the service programs of logistics companies differ in the extent to which logistics tasks of the shipper are taken over, that is, in the shipper's share of the logistics system. For example, the scope of services of a logistics company that distributes goods that have already been picked by the shipper for the customer is smaller than the scope of services of a logistics company that carries out picking for the shipper.

The *spatial dimension* refers to the geographical location and size of the source area (delivery points) and destination area (receiving points) for which a logistics company performs goods distribution tasks. From this point of view, the service program can be delimited, for example, according to economic areas or national borders.

The *goods dimension* refers to the type and quantity of goods to be distributed. Here it is important to what extent different goods also place different demands on the logistics systems. Possible classification criteria are, for example, volume, weight, aggregate state or sensitivity with regard to temperature, odour or impact, etc.

The *qualitative dimension* refers to the level of service guaranteed in various service components. An example of this is the provision of express freight services by transport and forwarding companies. The conditions under which logistics services can be offered by

---

[14] On the changes with regard to the tasks of logistics service providers and the influence of ecological aspects on logistics services, see Göpfert/Wehberg, 1995; Stabenau, 1999, pp. 92 f. On the increasing importance of value added services, cf. European Logistics Association/A.T. Kearney, 2009a, p. 15 and 2009b, p. 14.

[15] See Bobel, 2009, p. 8.

[16] See Klaus/Kille, 2008, p. 133.

[17] See for the following Krass, 1984, pp. 220ff. Another structuring possibility of the service program of logistic service companies can be found in Zöllner, 1990, pp. 60ff. who distinguishes the three dimensions geographical areas, customer groups and customer functions.

logistics companies depend on their capabilities to create (produce) such services. In this regard, some special features can be cited that distinguish logistics companies from other companies.

15.3 Special Features of the Service Provision

There is a comprehensive discussion in the business management literature on the extent to which the production of transport services exhibits typical peculiarities that justify a special business management discipline, namely *traffic management.*[18] It is not necessary to go into this discussion in detail here. It will suffice to briefly describe the special features of service production with which logistics companies are confronted and which do not occur, at least to this extent or in this combination, in other industries. These include the intangible nature of services and the external factor that has to be brought into the service process. This has already been discussed in Part I, Sect. 2.1. Specific to logistics services are the co-production in transport and the differentiated production processes of logistics companies.[19]

Co-production During Transport

Among the logistic services, the transport service has another peculiarity in its creation. This is because the creation of transport services does not have a fixed location. It usually proceeds as shown in Fig. 15.2.

A transport service is requested from B to C, a vehicle for this is not available at the delivery point (loading location) B, but must be brought from its location A to the loading location B in *staging traffic.* After loading, the goods are transported to the receiving point

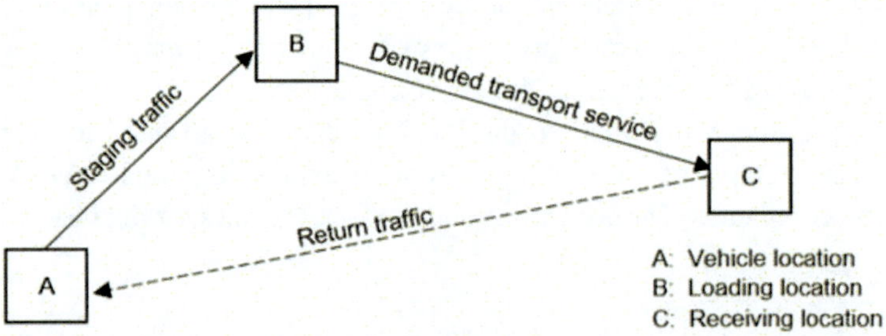

Fig. 15.2 Organizational co-product in the production of transport services (Source: Stabenau, 1994, p. 53)

[18]On the special features of traffic, cf. Ihde, 2001, pp. 101ff.

[19]See for the following Stabenau, 1994, pp. 51ff; Isermann, 1998, pp. 33ff; Aberle, 2009, pp. 230ff.

(receiving location) C according to the demanded transport service. Especially for road freight transport, the vehicle usually has to be returned to the location for organizational reasons, resulting in a *return traffic* from C to A. In addition to the demanded transport service, the staging and return traffic therefore arise as a co-product.

This is also referred to as an *organizational co-product*. This is because the staging and return transports generally do not arise for technical reasons, but for organizational reasons together with the production of the transport service demanded. Of course, these co-products can be marketed by not carrying out staging and return transports as empty runs, but by trying to find at least partial loads for these transports at points A and C. This is the only way to market the co-product. The limits of the marketing of the co-product are due to the *unpaired nature of the traffic flows*. This means that a traffic flow from A to B is not matched by a corresponding traffic flow from B to A, so that in return traffic the vehicles do not find a corresponding return load. The reasons for the fact that the demand for transport services in region A does not correspond to that of region B are the transport volumes resulting from the different economic structure of the regions as well as administrative reasons resulting from state interventions in the national and international transport markets.

Differentiated Production Processes

In terms of transport service, there is hardly any other sector in which there is such a great difference in production processes as in the case of logistics companies. For example, the transport services of railways, road haulage, shipping, air transport or pipeline are produced using completely different production processes, which inevitably lead to very different cost structures. For example, the share of fixed costs in rail or inland waterway transport is much higher than in long-distance road haulage. The proportion of personnel costs, for example, is higher for rail and lower for inland waterways than for long-distance road haulage. The depreciation period of the means of transport for railways and inland waterways is much longer than for trucks. These differences in the cost structures resulting from the different ways in which services are provided are essential for the calculation of the transport services to be provided by different institutions in the transport industry.

References

Aberle G (2009) Transportwirtschaft. Einzelwirtschaftliche und gesamtwirtschaftliche Grundlagen. 5., überarb. und ergänzte Aufl. München

Backhaus K, Voeth M (2014) Industriegütermarketing. 10., überarb. Aufl. München

Balafoutas L, Kerschbaumer R, Sutter M (2013) Second-Degree Moral Hazard in a Real-World Credence Goods Market. JZA Disscussion Paper No. 7714. Bonn

Bobel T (2009) Logistikorientiertes Management von Events. Grundlagen und Handlungsempfehlungen für die Eventlogistik. Berne

Göpfert I, Wehberg G (1995) Ökologieorientiertes Logistik-Marketing. Konzeptionelle und empirische Fundierung ökologieorientierter Angebotsstrategien von Logistik-Dienstleistungsunternehmen. Stuttgart/Berlin/Köln

European Logistics Association (ELA), A T Kearney (2009a) Supply-Chain-Excellence in der globalen Wirtschaftskrise

European Logistics Association (ELA), A T Kearney (2009b) Supply Chain Excellence amidst the Global Economic Crisis

Homburg C (2017) Marketingmanagement. Strategie - Instrumente - Umsetzung - Unternehmensführung. 6., überarb. u. erw. Aufl. Wiesbaden

Ihde G B (2001) Transport, Verkehr, Logistik. Gesamtwirtschaftliche Aspekte und einzelwirtschaftliche Handhabung. 3., völlig überarb. u. erw. Aufl. München

Isermann H (1998) Grundlagen eines systemorientierten Logistikmanagements. In: Isermann H (Hrsg) Logistik: Gestaltung von Logistiksystemen. 2. überarb. und erw. Aufl. Landsberg a. L., S. 21-60

Klaus P, Kille C (2008) Die Top 100 der Logistik – Marktgrößen, Marktsegmente und Marktführer in der Logistikdienstleistungswirtschaft, Ausgabe 2008/2009, Hamburg

Krass R (1984) Kooperation zwischen Verlader und Spedition. Konzept zur Effizienzsteigerung der Marketing-Logistik. Darmstadt

Large R (2016) Einkauf von Logistikleistungen. In: Beschaffung aktuell 19, S. 14-16

Meffert H, Burmann Chr, Kirchgeorg M (2012) Marketing: Grundlagen marktorientierter Unternehmensführung: Konzepte – Instrumente - Praxisbeispiele. 12., überarb. und erw. Aufl. Wiesbaden

Pfohl H-Chr (1980) Die Logistikkonzeption als Grundlage für ein modernes Marketing von Transport- und Speditionsunternehmen. In: DBW 40 3, S. 423-434

Pfohl H-Chr (1993) Logistische Dienstleistungen im Zusammenwirken von Industrie, Handel und Verkehr. In: Simon H (Hrsg) Industrielle Dienstleistungen. Stuttgart, S. 109-132

Stabenau H (1994) Verkehrsbetriebslehre. Betriebswirtschaftliche Grundlagen für eine langfristig orientierte Unternehmenspolitik in strukturell sich ändernden Verkehrsmärkten. 3. Aufl. Düsseldorf

Stabenau H (1999) Wohin entwickelt sich die Transportwirtschaft? In: Faller P (Hrsg) Transportwirtschaft im Umbruch: Strukturwandel, Anpassungserfordernisse, Gestaltungsaufgaben. Wien, S. 91-94

Wagner G R (1978) Die zeitliche Desaggregation von Beschaffungsentscheidungsprozessen aus Sicht des Investitionsgütermarketings. In: Zeitschrift für betriebswirschaftliche Forschung 30 4, S. 266-289

Weiber R, Adler J (1995) Informationsökonomisch begründete Typologisierung von Kaufprozessen. In: ZfbF 47 1, S. 43-64

Zöllner W (1990) Strategische Absatzmarktplanung. Kunden- und Wettbewerbsanalyse für Logistikunternehmen. Berlin u.a.

Institutions of the Freight Transport Industry 16

The following section provides an overview of the various specialized institutions that provide services for the shipping industry in the logistics channel. *Shippers* are all industrial, commercial and service companies that act as demanders and thus as clients of or for logistics services. These can be both the supplier and the customer of a particular good. With regard to the transport service, it is, for example, the supplier in the case of delivery free domicile, and the recipient in the case of delivery ex works. In this case, the recipient is also considered to be the principal with regard to the transfer of risk and the assumption of transport costs if the transport company is commissioned by the supplier. The term freight transport economy[1] is chosen to describe the institutions that provide logistics services as primary services because a term logistics economy has not yet become established and the term transport economy is now used in the sense of a logistics economy. Within the freight transport industry, whose primary function is the spatial movement of goods from a shipping point to a receiving point, three service areas can be distinguished[2]:

- Services based on the *change of location of the means of transport*, i.e. transport services in the narrower sense.
- Services provided at a *fixed location*, e.g. handling, storage, packaging and commissioning, i.e. transport services in the broader sense.
- Services consisting of *advice, intercession, organization* and *sale of* transport services.

Various institutions in the freight transport industry can be described according to these service areas. However, a clear demarcation of specialized logistics companies is not

[1] In order to distinguish the freight transport considered here from passenger transport, the term freight transport economy is used instead of transport economy.

[2] See Stabenau, 1994, pp. 13 f.

H.-C. Pfohl, *Logistics Systems*,
https://doi.org/10.1007/978-3-662-64349-5_16

possible, as companies are adapting to changing customer requirements and industry and trade expect logistics service providers to take on increasingly comprehensive functions, including the management of entire logistics systems.[3] As a result, in addition to standardized transport, handling and warehousing services that take on the character of commodities, individual special goods that are difficult to substitute are increasingly appearing in the form of system offerings and complex logistics service packages. They are offered by *logistical system providers,* which are to be distinguished from *logistical component providers.* Finally, *logistics centres* should be mentioned as institutions of the freight transport industry, in which various logistics companies are grouped together locally.

16.1 Transport Companies

An overview of the different means of transport[4] was given in the presentation of the logistical subsystem transport. In this section, the institutions that operate these means of transport will be considered. The operators of means of transport usually act as carriers. In contrast to the term shipper, this is a *legally defined* term: A carrier is someone who is obliged by a contract of carriage to transport goods to a destination and to hand them over to the consignee there. The prerequisite is that the carrier carries out this business commercially.[5] This includes, for example, trucking companies, inland shipping companies, railway companies or airlines.

Road Freight

In all industrialised countries, road freight transport has a very high share in the *modal split*, i.e., in the distribution of freight transport volume among the individual modes of transport. Figure 16.1 illustrates this dominant position with figures on the volume of freight transport in Germany. If the weight of the transported goods is used as a measure, approx. 83% of freight transport is by road. If, on the other hand, the weight of the goods multiplied by the transport distance is taken as the measure—the more meaningful tonne-kilometres relative to freight transport, the so-called freight transport performance—the road share is approx. 12% lower. This is due to the fact that in road freight transport, short-distance transport exceeds long-distance transport many times over, although long-distance road freight transport has recorded very high growth rates in recent years.[6]

[3] Thus, no uniform typology of transport companies can be given, see Stabenau, 1994, p. 35.

[4] See Part II, Sect. 8.3.

[5] See §407 HGB. Sea shipping companies are not carriers in the sense of the HGB, but special carriers. Special provisions apply to them, which are regulated in the fifth book of the HGB on maritime trade.

[6] The share of long-distance road freight transport (including foreign trucks) in total freight transport performance, measured in tonne-kilometres (tkm), rose from 31.6% (1980) to 52.3% (1997) and to 70.3% (2007), see Federal Ministry of Transport, Building and Urban Affairs, 2008, pp. 236 f.

Means of transport	Volume of freight transport		Performance of freight transport	
	in mio. t	share in %	in billion tkm	share in %
Rail road [a]	361,3	7,9	116,0	17,6
Inland waterway	219,0	4,8	53,2	8,1
Ocean shipping	292,6	6,5	x	x
Road haulage [b]	3593,3	78,7	471,8	71,5
Pipelines long distance [c]	92,4	2,0	18,8	2,8
Air transport [d]	4,5	0,1	x	x
total	4563,9	100,0	659,9	100,0

[a] without road haulage; including internal charges DB
[b] german and foreign trucks
[c] crude oil and mineral oil pipelines
[d] freight and air mail; including double counts in transshipment

Fig. 16.1 Freight transport volume and performance in 2016 in the Federal Republic of Germany (Source: Data taken from: Federal Statistical Office, 2017, p. 1)

About half of road freight transport is operated as *works transport*, primarily as local transport. In Germany, the consolidation processes of recent years have changed the structure of the transport and logistics industry, but the transport industry as a whole still consists predominantly of small and medium-sized enterprises.[7] Small companies often enter into contractual commitments with certain clients. In this way they secure a certain volume of transport, but become heavily dependent on one or a few principals. In addition, in the road freight transport sector, there is considerable interconnection between transport companies and other companies, both freight forwarders and industrial and commercial companies.

Rail Freight

In the Federal Republic of Germany (as in most other European countries), the vast majority of rail freight transport is handled by a public company, Deutsche Bahn AG. These public companies are usually the result of the conversion of former state-owned companies into public limited companies under private law, whose sole or main shareholder in most cases, however, continues to be the respective state. Up to now, in addition to Deutsche Bahn AG, there have been a large number of smaller, non-state-owned railway companies, which are responsible for fulfilling special tasks, particularly in distribution and feeder transport, as well as parts of regional transport. However, the complete separation of infrastructure and rail transport operations is intended to give new transport providers, in addition to foreign railway companies, increasing non-discriminatory access to the railway network, so that a change in the structure of the

[7] See Deutscher Speditions- und Logistikverband e. V., 2005, p. 28; IKB Deutsche Industriebank AG, 2007, pp. 8 f.

rail transport market is to be expected.[8] Works transport is of much less importance in rail freight transport than in road freight transport, although large companies in the coal, iron and steel industry and the chemical industry maintain large works railways.

The *decline in the share of rail transport* in freight transport in Germany observed so far is partly attributed to the fact that the Bundesbahn was a state-owned company until December 31, 1993. As such, it was not sufficiently flexible to adapt to changes in the transport market. Moreover, the Bundesbahngesetz had entrusted it with public service tasks which had prevented it from competing more effectively. The structural reform of the railways in Germany has improved market opportunities for the railways. The transfer of traffic to the railways is very important from the point of view of environmental policy and because of the highly congested road transport capacity.[9]

Shipping

The institutions that operate *inland navigation* can be divided according to the type of service they provide into commercial transport by shipping companies and particuliers, as well as works shipping companies and charterers.[10] *Shipping companies* are large companies that undertake the commercial execution of transports with a centrally managed fleet of inland waterway vessels or with external shipping space and acquire cargo usually via several land offices. Shipping companies perform both the tasks of a forwarder and a carrier and generally also perform warehousing and transhipment tasks. In some cases, they also offer shippers all-inclusive packages that additionally include depot maintenance, trucking, container repairs, etc.[11] *Particuliers* (small, private or individual skippers) are usually skippers of their own vessel. They use up to three vessels and do not own any offices for cargo acquisition. Particuliers act as carriers for both shippers and shipping companies, to some of which they are bound by long-term contracts. A large number of them have also joined together in enterprises that are run in the legal form of a registered cooperative or a limited liability company. On the one hand, these enterprises serve to acquire cargo, and on the other hand they take on other functions, such as the business management of the participants. It is difficult to distinguish *works transport* in inland navigation from the forms of commercial transport described above. Only transport carried out with the company's own personnel and vehicles for the own purposes of an industrial or commercial company and where the goods are owned by the parent company is considered as works transport. If, in addition to works transport, companies also carry out commercial transport, as is the case for the majority of own-account shipping companies, they are classified as commercial transport. Consequently, there are only a few companies in Germany that meet the narrow definition of works transport. Another institution in inland

[8] See Aberle/Hedderich, 1993, p. 15.

[9] See Aberle, 2009, p. 148.

[10] See Brandenburg et al., 2006, p. 207ff.

[11] See Bartsch, 1994, p. 36.

navigation is the *charterer*, who concludes contracts of affreightment without owning shipping space. In order to provide the promised transport service, he must have the cargo taken over transported by ship owners. On the basis of the contract of affreightment, the charterer faces the principal with the rights and obligations of a carrier.

In *maritime shipping*, *a* distinction is made between coastal and deep-sea shipping. *Coastal shipping* is mainly operated by small transport companies, where the coastal skipper is at the same time the owner, skipper and carrier, who hires out his ship on a one-off or time charter basis. In *deep-sea shipping*, on the other hand, a distinction must be made between the shipowner, who finances the construction or purchase of a ship, and the shipping company as the equiper and operator. The industry has been experiencing a significant concentration process for years, mainly as a result of the high capital requirements for adapting equipment to new developments in shipping technology.[12] It has resulted in the ten largest shipping companies in Germany accounting for more than half of the total tonnage.

Air Freight

Air freight transport has grown steadily in importance in recent years, and the transport companies operating air freight services are showing high growth rates in turnover. This is also evident from the average annual increase in air freight volumes of around 10% between 2004 and 2007.[13] On the one hand, this is due to the fact that it is primarily goods that are insensitive to transport costs and often have a high value with a low volume or weight that are transported. However, typical air freight goods also include perishable and short-lived goods, such as foodstuffs and last-minute consignments. Basically, it can be stated that due to the price development on the freight transport market and the required transport quality in the international economy, more and more types of goods are being shipped by air freight.[14] *Aircraft* are used to transport goods in *three different ways*. Either passenger aircraft take the cargo in underfloor cargo holds or aircraft with upper deck loads are used (combination aircraft). Cargo-only aircraft are also used.

The largest share of airfreight volume worldwide (over 75% based on revenue tonne-kilometres) is accounted for by the internationally active scheduled airlines, in which the state generally holds a stake in Western Europe. If such airlines use both passenger and cargo aircraft to transport freight, they are also known as combination carriers. In some cases, the scheduled airlines have established subsidiaries that operate exclusively air-freight services. LUFTHANSA AG, for example, has such a subsidiary, LUFTHANSA CARGO AG. Freight carriers (market share approx. 10%), on the other hand, specialise in freight transport, making freight capacities available for charter flights and covering seasonal

[12] See Schieck, 2008, p. 210.

[13] Between 1990 and 2007, transport performance rose from 439.5 million tkm to 1249.4 million tkm, see Federal Ministry of Transport, Building and Urban Affairs, 2008, pp. 236 f.

[14] See Arnold et al., 2008, pp. 757 f.; Pompl, 2008, p. 6.

peaks in demand for air services. Finally, there are the so-called integrators. These companies provide all service processes in the air freight transport chain between consignor and consignee through their own company.[15]

Air freight transport also includes *trucking*, i.e., air freight replacement transport between individual airports by road. Such transports may only take place if no suitable flight connections exist. They are used by the airlines to make greater use of the freight capacity available on long-haul routes by bundling the goods at individual airports by road transport.[16]

16.2 Warehousing, Handling and Packaging Companies

Warehousing Company

Commercial warehousing, which is operated either by independent warehousing companies or by *freight forwarding companies*, is *regulated* by *law*: "A warehouse keeper is anyone who undertakes the storage and safekeeping of goods on a commercial basis."[17] This commercial nature is only not given if the storage is an ancillary duty of another trade, such as transport-related interim storage or if storage and warehousing are not operated on a sustainable basis. It is not uncommon for *independent warehousing companies* to be state-owned or at least to have state involvement. It is common to find warehousing companies specialising in certain types of goods, as these very much determine the technology to be used and the storage space. For example, there are warehousing companies that specialize in bulk goods, general cargo or the storage of special goods, e.g., refrigerated and frozen goods.[18] Similarly, depending on the sensitivity of stored goods, adjustments to storage facilities may be required, e.g., in hazardous goods or valuables warehouses. In some cases, additional services are also provided within the scope of commercial warehousing, e.g. labelling, finishing or order picking of the stored goods.[19]

Handling Company

Handling companies offer their services to other logistics companies or to the shipping industry. They include both *specialized handling companies* and *freight forwarders* offering transhipment services. Companies specialising in cargo handling are either privately or publicly owned. Handling companies include, above all, container terminals operated by the railways and port authorities, air freight terminals operated by airlines or airport companies, groupage transhipment centres operated by groupage freight forwarders, ore

[15] See Beder, 1998, p. 129.

[16] See Mosler, 1993, pp. 507ff.

[17] §416, HGB.

[18] See Jünemann/Schmidt, 1999, p. 335.

[19] See Bjelicic, 1990, p. 12.

transhipment companies as a joint facility operated by several steelworks, and parcel transhipment centres operated by the postal service or other parcel service companies.[20]

Packaging Company

Companies that specialize in the service of packaging are called *wages packers* or *contract packers*.[21] The distinction between these two types is fluid. However, the main distinction is that wages packers offer their customers only personnel, machine and space capacity, but not the packaging materials. If, on the other hand, the packaging companies provide the principals with all or at least a substantial part of the packaging materials and other services, they are referred to as contract packers. The other services may include, for example, advice on the choice of packaging, but also the mixing, granulating, suspending etc. of goods. Few contract packers offer a range of packaging in all areas. Most packers specialize in, for example, export packaging or special packaging for bulky goods. They provide packaging at shippers' plants, either through their own specialists or under the direction and supervision of contractors. Shippers who carry out routine packaging tasks themselves also use the services of packaging companies for special tasks (e.g. when introducing a new product or for sales promotions).

16.3 Forwarders and Intercessors

Forwarders

The concept of a freight forwarder is *regulated by law*[22]: A freight forwarder is entrusted with organizing the carriage of goods for third parties. The forwarder's task is to select the means of transport and the performing companies as well as to secure the sender's claims for damages. To this end, he concludes the necessary contracts with other service providers, e.g., carriers, packers or warehouses in his own name or, if he is authorised to do so, in the name of the consignor.

In addition, the Freight Forwarder's duties include, if so agreed, the performance of other services related to the carriage. The HGB lists, for example, insurance and packaging of the goods, labelling and customs clearance. In fact, however, the offers go beyond this and often include the storage and picking of goods and now even include the configuration of end products or assembly activities at the recipient of the goods.

In practice, the field of activity of freight forwarders comprises the entire range of logistics services as well as other ancillary services, which were also referred to in Fig. 15.1. Freight forwarders purchase national and international logistics services and related ancillary services, supplement them to the necessary extent with services they have

[20] See Gudehus, 2010, p. 979.

[21] See Jünemann/Schmidt, 1999, p. 335.

[22] §§453 et seq, HGB.

produced themselves and sell them together as a total service to the customer at a profit.[23] The freight forwarder can be seen as the nucleus for logistics companies that are able to offer complete logistics service packages. In Anglo-Saxon usage, the term *Third Party Logistics Service Provider* (3PL) is commonly used for such forwarders. The service packages they offer are referred to as contract logistics. Important characteristics of contract logistics are the services that are individually adapted and designed to meet the needs of the shipper.[24]

For the realization of the offer the forwarder has two basic possibilities. It can either limit itself to the actual forwarding activity, which consists of planning, organizing and controlling the flow of goods and information associated with the shipment of goods, and commission other forwarding companies or specialized logistics companies such as transport, handling, warehousing or packaging companies to carry it out. They are also referred to as *Fourth Party Logistics Service Providers* (4PL) in a network of relationships between shippers, their customers as well as executing logistics companies.[25] Or the freight forwarder can carry out these activities himself. This is then referred to as the freight forwarder's *self-entry*.[26] However, since the use of own vehicles is unprofitable, especially for simple transports, a cost-saving reduction of the self-entry by outsourcing the freight carrying to subcontractors is to be expected. As a result, a polarization is emerging among freight forwarders, with suppliers of high-quality system services and logistics components standing out against pure carriers.[27] Thus, *forwarding pyramids* for logistics services are emerging, with individual forwarders with integrating system logistics services at the top, followed by suppliers of logistics subsystems in the middle and many component forwarders at the bottom of the pyramid.[28]

Irrespective of the way in which the logistics tasks are carried out, freight forwarders often specialize in certain means of transport as motor, rail, sea and air freight forwarders, in certain groups of goods as furniture, clothing or heavy goods forwarders, in certain transport relations or in domestic, border or international transports, as well as in individual functional areas, such as dispatching, receiving, place, clearance or transshipment forwarders.[29]

---

[23] See Eßig, 2007, pp. 425ff.

[24] See Klaus/Kille, 2008, pp. 115 f.

[25] See Klaus/Kille, 2008, p. 115. The pure 4PL service provider is "asset free". It does not have its own transport and storage capacities. "The service spectrum of the 4PL includes classic value-added services, network planning and the coordination of logistics service providers." Pfohl/Wagner, 2015, p. 43.

[26] §458, HGB.

[27] See Florian, 1995, p. 55.

[28] See Haubold/Stahl, 1994, pp. 321 f.

[29] See Aberle, 2009, pp. 270ff.

Intercessors

In addition to freight forwarders, who also conclude contracts for the carriage of goods in their own name, there are a large number of providers who act as freight intercessors. They broker contracts between shippers and transport companies, particularly in maritime and air transport, and operate chartering, declaring, agency and/or buying and selling businesses, e.g., for ocean-going vessels.[30] In recent years, a whole series of intercessors have established themselves on the *Internet* in the form of so-called electronic freight exchanges which take over the brokerage of orders, whereby the focus here is frequently on the brokerage of free capacities and thus the avoidance of empty runs.

A distinction is made between brokers and agents. *Brokers* arrange freight for occasional transport in the form of space and time freight contracts on a case-by-case basis without being permanently entrusted with this task by their principals. *Agents*, like commercial agents, are entrusted on a permanent basis by individual shipping companies or airlines with representing them in order to acquire cargo.[31]

16.4 Logistics Centers

In connection with the discussion of *hub systems in transport chains*,[32] logistics centers are among the most frequently discussed planning projects in the transport sector. A logistics center is understood to be an economic center in which one or more companies offer a wide range of logistics services in addition to transport services, which are often supplemented by additional services. A whole range of different forms of logistics centers can be distinguished, the most important of which are described below[33]:

Freight center: A freight center is a facility operated by DEUTSCHE BAHN AG or Deutsche Post which is used for the central collection and distribution of goods.

Transport trade area: The formation of a transport trade area is based on the targeted settlement of logistics companies with the focus on forming an interface between local and long-distance transport. Although there is no central body in a transport business area that coordinates the settlement and activities of the companies, such an area can be characterised by cooperative service offerings and the cooperative use of infrastructure.

Freight distribution center (GVtZ): A freight distribution center, in some cases also referred to as a *goods distribution hub*, primarily serves the distribution of goods, i.e., the provision of distribution services is in the foreground. A GVtZ "is generally understood to be the facility of a larger freight forwarder which performs transportation, warehousing and

[30] See Biebig/Althof/Wagener, 2008, p. 13.

[31] See Kummer, 2006, pp. 283 f.

[32] See Part II, Sect. 8.4.

[33] For a list of the various forms of logistics centres, see Gareis, 2002, pp. 16ff; Vahrenkamp, 2007, pp. 400ff; Gudehus, 2010, pp. 25ff. See also Part V, Sect. 19.1.

transshipment tasks as its primary function and various service tasks as its secondary function. Cooperative groupings can also be described as freight distribution centers. However, a GVtZ cannot be defined by the integration of several modes of transport."[34] Instead, like a transport trade area, a GVtZ is geared towards the formation of transport chains of one mode of transport.

Logistics park: A logistics park can be defined as a collection of freight distribution centers of logistics service companies that are often located near a system node of a courier, express and parcel service provider (CEP service provider). This results in short transit times between the freight distribution centers and the system node of the CEP service provider, which in turn leads to shorter loading cut-off times at the freight distribution center and allows the cut-off time for orders to be extended. Thus, the resident logistics service providers can improve their delivery service and at the same time increase their competitiveness. In addition, logistics service providers can also cooperate by helping each other out with existing capacity bottlenecks.

Industrial park: An industrial park is a joint settlement of several suppliers of a customer and/or the service providers involved, close to the customer. The location of an industrial park is determined by an overall development and settlement planning. In the course of the operation of the site, on the one hand the common buildings, areas and infrastructure facilities are provided, and on the other hand the companies located in the industrial park carry out customer-specific logistics and manufacturing processes. With the establishment of an industrial park, both costs can be reduced and service improved in the context of procurement logistics. Furthermore, close business relationships can be established and secured.

Freight transport center (GVZ): A freight transport center is primarily characterised by its interface function. It serves as a system changeover point between the system-compliant use of different modes of transport (rail, road, inland waterway) and as a changeover point between local and long-distance transport. In a freight transport center—usually located in a region with good transport links and a given infrastructure—transport and service companies are brought together to offer the widest possible range of services in local and regional cooperation.[35] The services of transshipment and reloading are regarded as indispensable components; the core function of a GVZ is generally intermodal transport.

City logistics: Today, freight distribution centers are also discussed in connection with the buzzword city logistics. Although it has already been possible to bundle consignments of goods from a large number of different suppliers by setting up central warehouses for the retail chain stores, inner cities are still under considerable pressure from delivery traffic. The main problems with deliveries to inner-city shops are[36]:

---

[34] Glaser, 1993, p. 215.

[35] See Arnold et al., 2008, p. 778. For further information on combined transport, see Part II, Sect. 8.4 and on the combination of transport networks, see Part V, Sect. 19.1.

[36] See Vahrenkamp, 2007, pp. 408 f.; on the urban logistics dilemma, see also Browne, 2010, pp. 241 f.

- Negative consequences such as the risk of congestion, noise and exhaust fumes as a result of increased delivery traffic.
- Overlap of customer traffic and delivery traffic.
- The logistics service providers supply the individual shops independently of each other, so that the shops are approached at irregular times by a large number of vehicles with low capacity utilization.
- The delivery vehicles often cannot drive up to the unloading point at the shops.
- Delivery restrictions result from time windows for delivery in the city centers

One possible solution is offered by the city logistics concept, which is based on *a holistic view of commercial transport* in cities. It includes considerations of supply and disposal as well as the problems caused by air pollution, noise, accidents and traffic jams.[37] The aim is to achieve a reduction in the volume of traffic in the inner cities by means of spatial rather than goods-related bundling, while maintaining the same transport volume.[38] Participants in such concepts are, on the one hand, retail and logistics service companies and, on the other hand, the municipalities, which set the framework conditions.[39] Such city logistics concepts can be implemented by setting up GVtZs in convenient suburban locations, which are supplied by various manufacturers and transport service providers and whose goods are destined for different customers. The GVtZs can be operated by several service providers in cooperation.[40] Thanks to the bundling effect, the retail companies located in the inner cities are supplied more efficiently with fewer vehicles, which also reduces the burden on the environment.

An example of the positive effects of a cooperative city logistics concept can be found in Regensburg (see Fig. 16.2).[41] More than 100 retail companies were involved in the project and at least 25 shops are supplied daily. Whereas in the past 7–8 poorly utilized vehicles drove through the city center, today, depending on demand, only one or, in rare cases, two well utilized vehicles deliver the tonnage of all partners. Through the cooperation of the forwarding agencies, truck deployments are saved in the old town of Regensburg. Compared to the situation before the introduction of the concept, the cooperative city logistics

---

[37] See Wittenbrink, 1995, p. 5; Berg, 1999, p. 135 et seq. where it is shown that the implementation of a city logistics concept in Munich can reduce traffic congestion (in terms of kilometres travelled) by 27%, which also leads to a significant reduction in noise and harmful substance pollution as well as fuel consumption.

[38] See Hatzfeld/Hesse, 1994, pp. 647 f.; Kaupp, 1998, p. 24.

[39] On cooperation between freight forwarders in city logistics, see Eberhart, 1995, pp. 116ff.; Kaupp, 1998. See also the reference to inter-company cooperation in Sect. 17.2. On the different interests of the actors in city logistics, see Hatzfeld/Hesse, 1994, pp. 648 f.

[40] See Pfohl, 1993, pp. 117 f.; Kaupp, 1998, pp. 23ff.

[41] See Bottler, 2008, p. 6. Other examples of the positive effects of a city logistics concept can be found in Freiburg and Munich, see Berg, 1999.

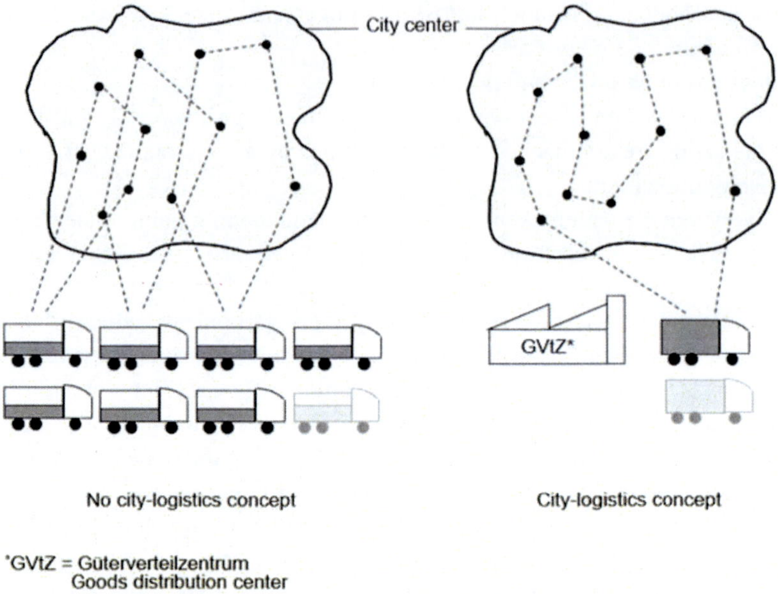

Fig. 16.2 Positive effects of the city logistics concept using the example of the city of Regensburg

concept saves an average of 4300 truck kilometers per year in the city center, which covers about 1 km^2.

However, it is not only a GVtZ that lends itself to efficient inner-city commercial transport. In addition, the following possible solutions are proposed[42]:

- Promoting the consolidation of transports without a GVtZ.
- Cooperation between shops and/or logistics service providers.
- Use of vehicles with suitable dimensions.
- Use of vehicles based on alternative energies and with low noise levels.
- Information and communication systems for traffic management.
- Efficient urban infrastructure.
- Restrictions in the time of delivery and vehicle size.
- Nightly supply and disposal of the shops.

16.5 Legal and Organizational Forms

Legal Forms

The legal forms of logistics companies are basically the same as for other companies. However, a number of special features have emerged that are due to the involvement of the

[42] See Browne, 2010, pp. 143ff.

public sector, the links with other companies and the particularly large capital requirements of maritime shipping. These special features will be discussed below, starting with a breakdown by ownership structure[43]:

- *Public enterprises as enterprises under state control or as special assets*: Such enterprises are fully owned by regional, federal, state or local authorities. They usually also have public service objectives to pursue. Examples of this are the federal railway assets or local traffic companies.
- *Public enterprises as AG or GmbH*: They are also fully or predominantly owned by the public sector. They have to fulfil the requirements of their public owners with regard to the range and provision of services, which are also based on public service objectives. In return, they may receive a certain degree of market protection from the public owner. This legal form is used, for example, by Deutsche Bahn AG, Deutsche Lufthansa AG, airport companies, seaport handling companies and some public traffic companies.
- *Corporate companies*: Such companies have the legal form of a corporation, whereby the capital is held exclusively or predominantly by a group of the shipping industry. This results in the objectives of the logistics company being integrated into the objectives of the group as a whole. In the case of such links between the shipping industry and logistics companies, the transport services they provide are also referred to as non-genuine works transport, which expresses the influence of the shipping industry.[44]
- *Logistics companies as corporations*: These are larger or medium-sized companies whose capital is in family and/or free float.
- *Logistics companies as partnerships*: This includes the many, mostly small companies owned by a single person or a small group of persons.

Among the public companies, Deutsche Bahn AG and Deutsche Lufthansa AG are of particular importance. The privatization of the former, initiated by the railway reform, has been largely completed in organizational terms. The former state-owned company, whose special legal status was regulated in Article 87 of the Basic Law, was transferred to a holding company with Deutsche Bahn AG as the parent company.

As shown in Fig. 16.3, the Group is divided into eight business units[45]:

- In the DB Nets Track business unit, DB Nets AG is a service provider for railway undertakings (RUs). DB Nets is responsible for operating the high-performance railway infrastructure (long-distance/urban network, regional network, train formation and treatment facilities).

[43] See Stabenau, 1994, p. 31.

[44] Conversely, it can also be observed that intra-group service providers are changing into service providers operating openly on the market, see Kowalski, 1999, pp. 42 f.

[45] See Deutsche Bahn AG, 2017.

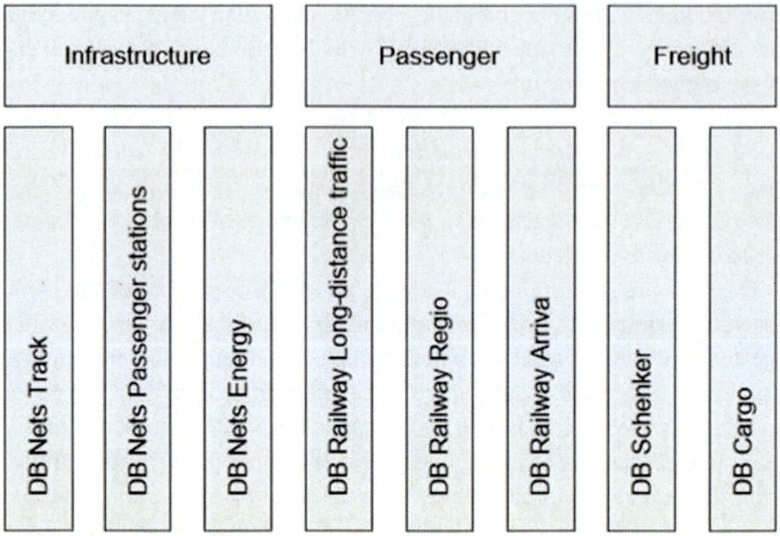

Fig. 16.3 Business units of Deutsche Bahn AG (Source: Based on Deutsche Bahn AG, 2017, p. 1 f)

- In the DB Nets Passenger Stations business unit, DB Station & Service AG is responsible for the operation of passenger stations as traffic stations as well as for the development and marketing of the associated station areas.
- DB Energy supplies traction current and fuels to the railways and all rail transport companies in Germany.
- The DB Railway Long-Distance Traffic business segment has taken over all long-distance passenger traffic and is responsible for its handling. Its aim is to create an attractive range of services and to bind customers to the railway by offering comprehensive traffic and service packages and by coordinating with other modes of traffic.
- In the DB Railway Regio business unit, DB Regio AG has taken over regional passenger traffic from the railways. Since 1996, the federal states have been responsible for all public traffic, including rail traffic. The federal states or specially founded traffic associations decide on the connections to be set up and generally put them out to public tender. In these tenders, DB REGIO AG is usually in competition with other competitors, e.g., state-owned (Hessische Landesbahn), private (Deutsche Eisenbahngesellschaft) or foreign (Mittel-Thurgau-Bahn) railway companies.
- The DB Railway Arriva business unit is responsible for the suburban trains in Berlin and Hamburg as well as bus companies in Germany.
- The DB Schenker business unit manages the Europe-wide activities in rail freight transport. It has a diverse range of services that goes beyond pure transport by rail. For example, door-to-door transports and individual transport solutions are also offered as part of combined transport.

- The DB CARGO business unit offers freight forwarding and logistics services worldwide.

The second major state-owned enterprise—DEUTSCHE LUFTHANSA AG—has been gradually privatized since 1965. In 1997, the German government gave up its last shares. Today Lufthansa is a holding company with a complex structure of subsidiaries, some of which cooperate closely with each other in the provision of services. In order to survive in international competition, Lufthansa established the Star Alliance with other airlines as a strategic alliance. This made it possible for Lufthansa to expand its international services without increasing the size of its own network.

A legal form created specifically for maritime shipping is the *partial ship owner*. In such a single-ship company, several persons join together as co- or part shipowners, each of whom owns a fractional ownership in a ship (ship's part) used for joint account for acquisition by sea.[46] The rights and obligations of the partial ship owner are determined by the ownership shares, which are referred to as "parten". Management is usually entrusted to a so-called correspondent shipowner, who need not be one of the partial ship owner but is often a shipbroker. Externally, the partial ship owner acts under the name of the ship on the basis of majority resolutions of the partial ship owner.

Forms of Organization

The considerations regarding the organization of logistics companies are basically the same for the choice of legal form as for other companies, so that, for example, the same centralization criteria come into question for the formation of the organizational units.[47] When applying the centralization criterion object, there are only two special features. The first special feature is that in a product-oriented organization, the organizational units can be formed according to the various logistical service types rather than according to tangible goods. The second special feature results from the fact that logistics companies often have several regional branches. Consequently, the centralization criterion of region generally plays a greater role for logistics companies than for companies in the shipping industry, where it only becomes important in very large companies.

References

Aberle G (2009) Transportwirtschaft. Einzelwirtschaftliche und gesamtwirtschaftliche Grundlagen. 5., überarb. und erg. Aufl. München

Aberle G, Hedderich A (1993) Diskriminierungsfreier Netzzugang bei den Eisenbahnen – Umsetzung der Bahnstrukturreform. In: Internationales Verkehrswesen 45 1/2, S. 15-26

[46] See Schneck, 2006, p. 146 f.

[47] For an example of the organizational structure of a freight forwarding and warehousing company, see Schumacher, 1988, p. 144.

Arnold D u.a. (Hrsg) (2008) Handbuch Logistik. 3., neu bearb. Aufl. Berlin Heidelberg
Bartsch R (1994) Erinnerungswertes Transportmittel – Alternative Binnenschiffahrt. In: Hossner R (Hrsg) Jahrbuch der Logistik 1994. 8. Jg. Düsseldorf
Beder H (1998) Der Luftfrachtverkehr. In: Isermann H (Hrsg) Logistik: Gestaltung von Logistiksystemen. 2., überarb. und erw. Aufl. Landsberg a. L., S. 125-138
Berg C C (1999) City-Logistik Das Münchner Modell. Ottobrunn
Biebig P, Althof W, Wagener N (2008) Seeverkehrswirtschaft. Kompendium. 4., bearbeitete und aktualisierte Aufl. München
Bjelicic B (1990) Internationaler Unternehmenswettbewerb im gewerblichen Güterverkehr. München
Bottler S (2008) Bündeln durch enge Gassen. In: Deutsche Logistik-Zeitung. 62 70, S. 6
Brandenburg H, Gutermuth J, Oelfke D (2006) Güterverkehr-Spedition-Logistik. Leistungserstellung in Spedition und Logistik. 37. Aufl. Troisdorf
Browne M (2010) Urban Freight Transport and Logistics. In: Schönberger R, Elbert R (Hrsg) Dimensionen der Logistik. Funktionen, Institutionen und Handlungsebenen. Wiesbaden, S. 139-154
Bundesministerium für Verkehr, Bau und Stadtentwicklung (BMVBS) (2008) Masterplan Güterverkehr und Logistik. Berlin
Deutsche Bahn AG (2017): Daten & Fakten 2016. Berlin
Deutscher Speditions- und Logistikverband e. V. (DSLV) (2005): Zahlen, Daten, Fakten aus Spedition und Logistik. Bonn
Eberhart K (1995) Güterverkehrsmanagement: City-Logistik – Kooperationen von Speditionen. In: Internationales Verkehrswesen 47 3, S. 116-122
Eßig M (2007) Sourcing-Strategien von Logistikdienstleistern. In: Stölzle W u.a. (Hrsg) Handbuch Kontraktlogistik: Management komplexer Logistikdienstleistungen. Weinheim, S. 425-446
Florian M (1995) Lean Logistik durch virtuelle Unternehmen. Was könnnen Speditionen aus den "Revolutionen" in Business Bestsellern lernen? In: Logistik + Arbeit o.Jg. 6, S. 44-61
Gareis K (2002) Das Konzept Industriepark aus dynamischer Sicht. Theoretische Fundierung – empirische Ergebnisse – Gestaltungsempfehlungen. Wiesbaden
Glaser J (1993) Güterverkehrszentren (GVZ). Konzepte zwischen Euphorie und Skepsis. In: Läpple D (Hrsg) Güterverkehr, Logistik und Umwelt. Berlin, S. 207-253
Gudehus T (2010) Logistik: Grundlagen, Strategien, Anwendungen. 4., aktual. Aufl. Berlin u.a.
Haubold V, Stahl D (1994) Implikationen für die Speditionsbranche: Lean Production in der Industrie. In: Internationales Verkehrswesen 46 6, S. 317-325
Hatzfeld U, Hesse M (1994) Mangel an Informationen und Stimulanzien: Stadtlogistik – Interessen "statt Logistik"? In: Internationales Verkehrswesen 46 11, S. 646-653
IKB Deutsche Industriebank AG (2007) Transport und Logistik. In: IKB Branchenbericht 12.2007
Jünemann R, Schmidt T (1999) Materialflußsysteme. Systemtechnische Grundlagen. 2. Aufl. Berlin u.a.
Kaupp M (1998) City-Logistik als kooperatives Güterverkehrs-Management. Wiesbaden
Klaus P, Kille C (2008) Die Top 100 der Logistik – Marktgrößen, Marktsegmente und Marktführer in der Logistikdienstleistungswirtschaft, Ausgabe 2008/2009, Hamburg
Kowalski T (1999) Aus konzerninternem Netzwerk zum Logistik-Dienstleister mutiert. In: Logistik im Unternehmen 13 11/12, S. 42-43
Kummer S (2006) Internationales Transport- und Logistikmanagement. Stuttgart
Mosler G F (1993) Strukturveränderungen in der Luftfracht durch Ersatzverkehr: Auswirkungen des "Trucking" auf Airlines und Flughäfen. In: Internationales Verkehrswe-sen 45 9, S. 507-512
Pfohl H-Chr (1993) Logistische Dienstleistungen im Zusammenwirken von Industrie, Handel und Verkehr. In: Simon H (Hrsg) Industrielle Dienstleistungen. Stuttgart, S. 109-132

Pfohl H-Chr, Wagner St (2015) Eine alte neue Idee. In: Verkehrsrundschau Sonderheft. Who is who – Logistik, S. 42-45

Pompl W (2008) Luftverkehr: Eine ökonomische und politische Einführung. 5., überarb. Aufl. Berlin u.a.

Schieck A (2008) Internationale Logistik Objekte, Prozesse und Infrastrukturen grenzüberschreitender Güterströme. München u.a.

Schneck O (2006) Betriebswirtschaft konkret: alles, was Sie wissen müssen. 1. Aufl., Weinheim

Schumacher W (1988) Die Entwicklung der betriebswirtschaftslichen Logistik und ihr Einfluß auf das zukünftige Leistungsbild des deutschen Speditions- und Lagereigewerbes. Köln

Stabenau H (1994) Verkehrsbetriebslehre. Betriebswirtschaftliche Grundlagen für eine langfristig orientierte Unternehmenspolitik in strukturell sich ändernden Verkehrsmärkten. 3. Aufl. Düsseldorf

Statistisches Bundesamt (2017) Pressemitteilung Nr. 057 vom 17.02.2017. Wiesbaden

Vahrenkamp R (2007) Logistik. Management und Strategien. 6., überarb. und erw. Aufl. München

Wittenbrink P (1995) City-Logistik. In: Baumgarten H u.a. (Hrsg) RKW-Handbuch Logistik. Berlin, 21. Lfg. V/95, Kennziffer 8720, S. 1-33.

Inter-organizational Logistics Systems 17

17.1 Logistical Interfaces and Inter-organizational Relationships

Logistical Interfaces

Interfaces can generally be defined as system boundaries. Logistical interfaces are therefore boundaries between a particular logistics system and other logistics systems or with other types of systems (e.g. procurement, production or sales systems). Since the flow of goods and information through a company or between companies always crosses several technical, organizational or legal systems, interfaces play a major role in the logistics conception. After all, the core of the logistics conception is to design the interfaces between the logistics subsystems in such a way that logistics problems can be solved on the basis of systems thinking. Figure 17.1 provides an overview of logistics interfaces, which are classified according to the type of transmission object (goods or information) and the type of interface (type of system boundary).

First- and second-order logistical interfaces are internal company interfaces. The first-order interfaces were explicitly referred to in the treatment of the activity-specific subsystems of logistics.[1] The second-order interfaces were the focus of the explanations of the phase-specific subsystems of logistics as well as in the discussion of the integration of logistics into a multidimensional organizational structure.[2] In the following discussion no longer focuses on these intra-organizational interfaces, but rather on the inter-company or inter-organizational interfaces, which are referred to as *third-order interfaces. These* interfaces have already been addressed in the discussion of linking logistical information systems in the context of order processing[3] and in the treatment of logistical units and

[1] See Part II.

[2] See Parts III and IV, Sect. 14.4.

[3] See Part II, Sect. 4.4.

H.-C. Pfohl, *Logistics Systems*,
https://doi.org/10.1007/978-3-662-64349-5_17

Type of interface	Transfer object				
	Planning information	Dispositive information	Execution information	Service information	Physical objects
Logistical Interface 1st order	Integrated planning of logistical subsystems and elements	i.e. between: Order processing and stock management, program planning and warehousing etc.	i.e. between: Warehousing and stock management, transport and order processing etc.	i.e. between: Logistics planning and executing bodies	Alignment of: Product, packaging, standard loads, handling-, storage and transport system
Logistical Interface 2nd order	Alignment of logistics planning with other company plans (financial, investment, personnel planning etc.)	i.e. between: Sales and program planning, purchasing and procurement etc.	i.e. between: Order processing and sales, production and transportation etc.	Alignment of purchasing-, sales- and production capacities through logistics	i.e. between: The logistical elements and the production system
Logistical Interface 3rd order	Cross-company alignment of logistical systems between cooperating companies	i.e. between: Order processing supplier company and disposition customer company	i.e. between: Shipping supplier company and procurement customer company	i.e. between: Market research customer and program planning supplier (communication of registered market shifts)	i.e. between: Shipping supplier and receiving customer, exchanged standard loads and the respective production systems etc.

Fig. 17.1 Classification and examples of logistics interfaces (Source: Feierabend, 1980, p. 59)

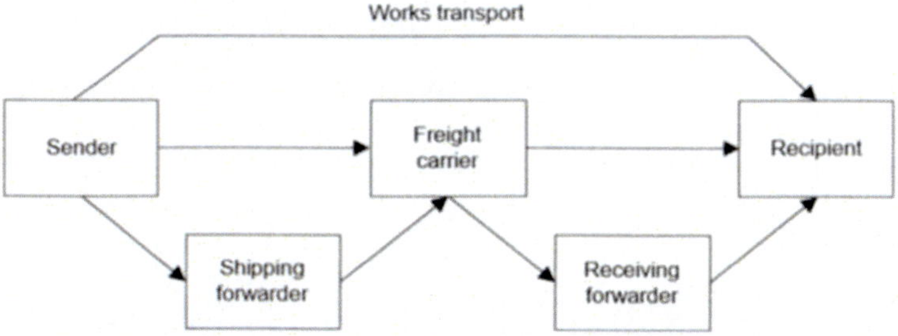

Fig. 17.2 Inter-organizational interfaces in a transport chain (Source: based on Federal Ministry of Research and Technology, 1980, p. 2)

transportation.[4] The variety of inter-organizational interfaces is illustrated in Fig. 17.2 using the example of the transport chain.

Inter-organizational Relations

The performance and coordination of tasks in the inter-organizational flow of goods and information can be associated with a variety of problems at the third-order interfaces described above. How these problems are handled at the interfaces depends on the nature of the respective inter-organizational relationships, which are shown in Fig. 17.3 in the context of the overall relationship structure of a company.

According to this, a distinction can be made between inter-organizational relationships with *mainly non-conflicting interests* and those with *mainly conflicting interests* of involved parties. In the case of the latter, a distinction is made between competition and conflict: in the case of competition, there is a general mutual rivalry between two organizations which is not tied to individual concrete decision-making situations, whereas conflict always relates to a concrete decision-making situation. In the case of inter-organizational relationships with predominantly aligned interests, a distinction is made between the normal business relations, the cooperation, the corporation and the merger as the degree of interdependence increases, whereby the latter two types can be classified as different manifestations of concentration.[5] The normal business relationship is characterised by the fact that, in contrast to cooperation, one of the parties involved can be replaced at any time at short notice. The interdependence is even stronger in the case of the corporation than in the case of the cooperation. This is because the corporation companies retain their legal independence, but lose their economic independence. The strongest degree is achieved in the case of a merger, since in this case legal independence is also lost and the inter-organizational relationships between the merging companies become intra-organizational relationships.

[4]See Part II, Sects. 7.3, 8.2 and 8.4.

[5]See Krass, 1984, pp. 66ff.; Freichel, 1992, pp. 54ff.

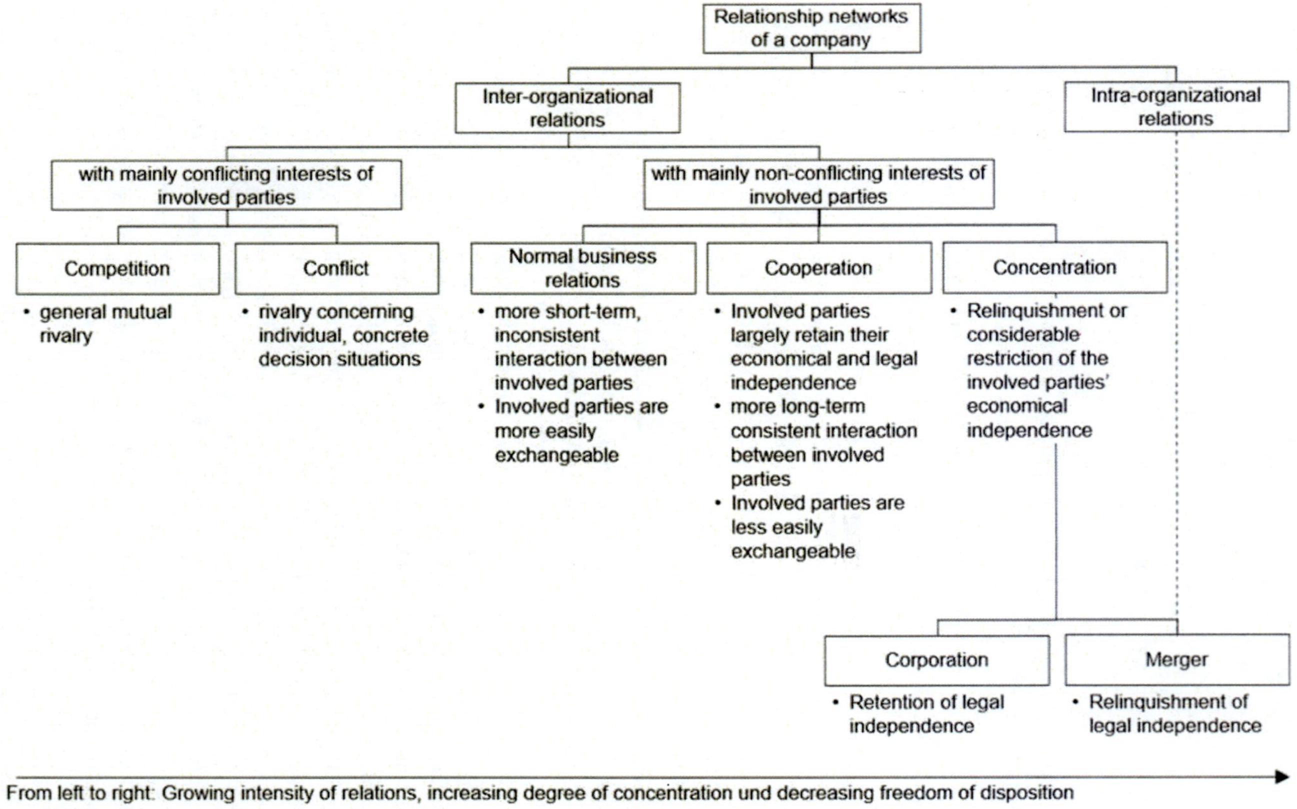

Fig. 17.3 Relationship structure of a company (Source: with modifications taken from Freichel, 1992, p. 54)

In the following, we will discuss *cooperation* as an inter-organizational relationship that gives rise to *mesologistics* systems.[6] Constitutive features of a cooperation are the joint, goal-oriented activity and the economic and legal independence of the participating companies.[7] By *joint action* is meant that conscious agreements are made between two cooperation partners. These agreements may relate to common objectives, to decisions on the manner in which tasks are to be performed or to the actual performance of activities. The decisive factor is that the intensity of the agreement makes a short-term exchange of the cooperation partners and thus a transition to the normal business relationship impossible (e.g. because of investments made, adaptation of technical systems, adaptation of organizational structures). It follows that the *long-term nature* of the cooperation is another characteristic of a cooperation. Although the freedom of decision can be restricted by mutual agreement, the *economic independence of* the cooperation members is preserved in cooperations to the extent that each of the companies can decide autonomously on its withdrawal from the cooperation. The *legal independence of* the participating companies allows a differentiation of the cooperation from the merger.

Conflicts in Cooperations

Interorganizational relationships with predominantly like-directed and predominantly opposed interests are not mutually exclusive. Like intra-organizational relations in institutions, inter-organizational relations between institutions are characterized by the existence of conflicts. However, conflicts do not necessarily prevent cooperation between institutions in the logistics channel. Between two institutions cooperating in the logistics channel, there will always be both like-directed and opposite-directed relationships. While it is obvious that the existence of like-directed relationships facilitates the emergence of cooperation and must prevail over opposite-directed relationships in existing cooperation. But the issue is not to eliminate the conflicts resulting from the counter-directed relations. As in the case of intra-organizational conflicts, the issue in a cooperation is to regulate or handle the conflicts through appropriate conflict management.

Conflicts are caused by goal, role, power and communication relationships.[8] *Goal conflicts* arise when the achievement of the goal of one cooperation partner impairs the achievement of the goal of another cooperation partner. *Role conflicts* occur when the ideas about the role to be performed by an institution in the logistics channel—e.g. performance of logistics functions by a freight forwarder in the role of a classic freight forwarder in accordance with the minimum legal requirements or in the role of a logistics company offering a range of logistics service packages up to and including logistics consulting—do not coincide. Such conflicts arise particularly because, as a result of logistics thinking, traditional ideas about what role a member of the logistics channel has to play in the

[6] See Fig. 1.5 in Part I, Sect. 1.4.

[7] See Krass, 1984, pp. 70ff; Kleer, 1991, pp. 60ff.

[8] See Pfohl, 1987, pp. 18ff; Specht/Fritz, 2005, pp. 439ff.

performance of logistics tasks are changing. *Power conflicts* result from the fact that one member of the logistics channel, because of his power, can force other logistics channel members to make decisions that they otherwise would not have made. The ability to exercise power and the ability to mobilize power resources and support from other logistics channel members are important characteristics of firms that act as leaders in logistics channels and extend their span of logistical control.[9] *Information conflict* results from members of the logistics channel making decisions based on different information. The differences in informativeness may be due to a lack of communication in the logistics channel, but also to a deliberate withholding of information by a firm.

17.2 Cooperation at Different Levels

Cooperation in the logistics channel is nothing new. Rationalization and performance considerations have already led to diverse forms of cooperation in the past. While some forms have proven themselves over the long term and have thus almost become a matter of course, other forms have had very different successes, which suggests a higher degree of instability or dependence on the respective internal and external influences of the company. Some reservations about cooperation that used to exist in the past, for example because of a feared loss of flexibility,[10] have now given way to an understanding of the necessity of cooperation. The growing importance of inter-organizational relationships results from the increasing tendency in companies to concentrate on core competencies and to outsource an increasing share of tasks to external service providers and suppliers.[11] As a result, the number of interfaces between companies is increasing. This requires improved coordination between the companies in the supply chain,[12] which presupposes correspondingly cooperative behavior so that the advantages of this form of logistics can be realized.[13] Business relationships are becoming closer and more long-term.

Cooperation in the logistics channel can be carried out at inter-company and supra-company level.[14] In the case of *supra-company* cooperation, logistics tasks are coordinated in a jointly supported institution or outsourced to it. Either the institution is newly founded

[9] On channel leadership in distribution channels, see Kirsch et al., 1973, pp. 367ff.; Ihde, 2001, p. 50; Specht/Fritz, 2005, pp. 457 f.

[10] On the expected and perceived disadvantages of cooperation from the point of view of shippers and their effects on the willingness of companies to enter into cooperation, see Kleer, 1991, pp. 172 f.

[11] For supply chain management, see Sect. 17.3.

[12] See Baumgarten/Wolff, 1999, pp. 48ff; Frunzke, 2004, pp. 28 f. See also the trend towards a reduction in vertical integration in Part III, Sect. 9.2.

[13] See Pfohl, 1994, p. 216.

[14] In the context of the levels of inter-organizational relationship structures, in German language the term operations is usually used (here as inter- or supra-operations), even if the description of the constitutive characteristics of the cooperation refers to economically independent companies.

for this purpose or a participation in an already existing inter-company institution takes place. The level of *inter-company* cooperation is characterized by direct relationships between two or more participating institutions. In this case, outsourcing of logistics tasks leads to direct service exchange relationships between the cooperation partners. However, the division into inter- and intra-company cooperation is also criticized with the remark that a joint institution, e.g., in the form of a joint venture, can have both inter- and intra-company character. It is therefore proposed to speak of inter-organizational relationship structures with or without a central body in order to characterise different relationship levels.[15]

Supra-Company Cooperation

Supra-company cooperation in the logistics channel takes place in the form of commercial and political cooperation. The road transport cooperatives, which form the classic form of supra-company economic institutions, as well as the transport industry's support institutions, belong to the *commercial cooperation.* The 18 road transport cooperatives[16] have about 30 motorway depots which, with their supply facilities for personnel and vehicles, have become central points for national and international road haulage. Furthermore, they maintain loading space distribution centers, which arrange loading goods and loading space for the affiliated member companies, especially for return journeys. Other activities include freight billing or auditing.[17] Support facilities include centralized accountancy offices, which use IT to handle the bookkeeping of logistics companies, special insurance and business advisory services.

In the meantime, supra-company cooperation in the form of certain types of logistics service networks has become much more important.[18] In this case, several forwarding companies cooperate by forming a joint venture (often on the basis of franchise agreements), among other things in order to be able to realize a nationwide, Germany-wide or Europe-wide range of logistics services. Such cooperations can be found in the furniture transport industry, in parcel services or in general cargo transport, to name just a few examples. As the European markets continue to grow together and the demand for cross-border transport services increases as a result, the need for cooperation is likely to continue to grow in the future. Small and medium-sized companies in particular will only be able to compete with service providers operating throughout Europe if they have international partners, especially as the larger providers in this market segment are currently actively seeking to expand their business internationally, usually by buying existing

[15] See Freichel, 1992, p. 65.

[16] Members (cooperatives) are companies in the road haulage and road haulage industry. The majority of road transport cooperatives are members of the Bundes- Zentralgenossenschaft Straßenverkehr (Federal Central Road Transport Cooperative), Frankfurt a. M., Germany.

[17] See Brandenburg/Gutermuth/Oelfke, 2016, p. 133.

[18] For a detailed discussion of the organization of logistics services networks, see Freichel, 1992, pp. 65ff.

companies.[19] Another example of supra-company cooperation is certain forms of the city logistics concept, in which a joint venture is founded by the partners involved, which then carries out and coordinates inner-city freight transport.[20]

Political cooperation includes the trade associations and federations of the transport industry, but also of other sectors of the economy, as well as their technical committees. It is in particular the technical committees through which cooperation takes place in the standardisation of documents, packaging and means of transport, for example, which were mentioned in the discussion of the interlinking of logistical information systems, logistical units, modular packaging and combined transport.[21] A typical example of this type of cooperation, however, is the development of uniform article numbering systems, which greatly facilitate the exchange of information throughout the marketing channel. For this purpose, the nationally uniform article numbering system for the food trade (ban-L) was developed in Germany more than 30 years ago, followed by the European Article Number System (EAN) in the mid-1970s. In the meantime, this article numbering system is used in over 100 countries worldwide.[22]

Inter-company Cooperation

Inter-company cooperation in the logistics channel takes place in different directions: horizontally, vertically and diagonally. Figure 17.4 shows these three relationship directions and the corresponding interorganizational relationships between the companies involved in each cooperation. *Horizontal cooperation* in the logistics channel initially concerns cooperation between logistics companies that provide services at the same logistics channel level (e.g. freight forwarding companies).[23] Well-known examples of this are groupage consortiums, correspondence relationships in groupage traffic (shipping and receiving forwarders), handling consortiums and meeting transports. Other forms of horizontal cooperation are, for example, the joint purchase of spare parts or operating resources, cooperation in logistics consulting or joint, alternating dispatching and route planning within the framework of certain city logistics projects. Horizontal cooperation also takes place between shippers at the same distribution channel level.[24] An example of this is the cooperation between competing wholesalers for wine and spirits in the area of distribution logistics.

---

[19]For an overview of international cooperation activities in the freight forwarding industry, see Arnold et al. 2008, pp. 16ff. and pp. 981ff.

[20]See Wittenbrink, 1995, p. 10. On the city logistics concept, see also Sect. 16.4.

[21]See Part II, Sects. 4.4, 7.3, 7.4 and 8.4.

[22]The activities in connection with the development of these numbering systems are carried out in Germany by GS1 GERMANY GMBH (formerly Centrale für Coorganisation).

[23]According to this definition, the supra-company cooperation described above in the form of a logistics service network is also to some extent a horizontal cooperation.

[24]For various cooperative distribution logistics concepts see Arnold et al., 2008, p. 999.

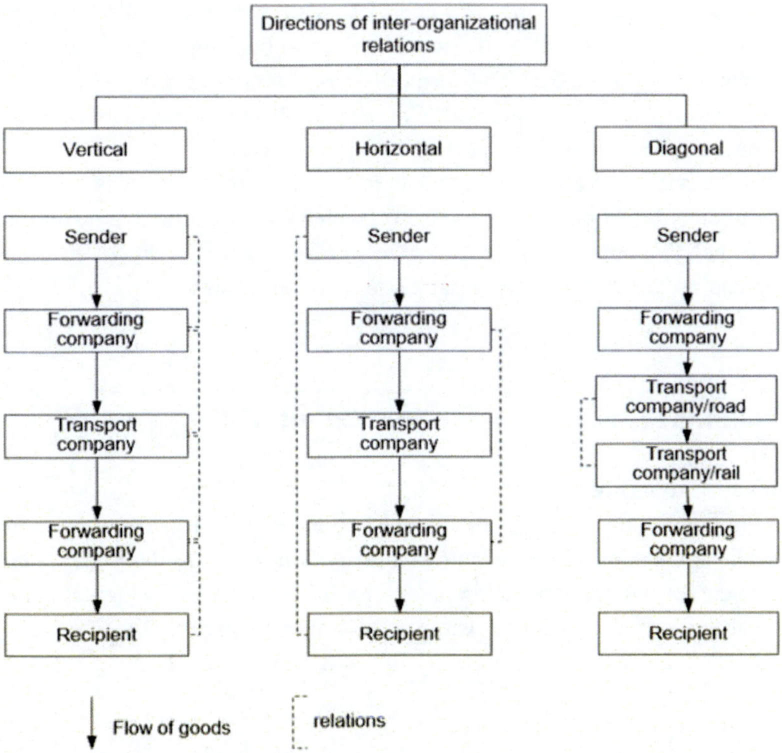

Fig. 17.4 Directions of inter-organizational relationships (Source: Freichel, 1992, p. 61)

Vertical cooperation in the logistics channel occurs between logistics companies and shippers offering services at different logistics channel levels, e.g. freight forwarders and transport companies, and among shippers at different marketing channel levels, e.g. manufacturers and retailers. This is usually a consolidation of relationships that had already existed in the form of normal business relationships between the parties. The consolidation usually consists of a longer-term contractual commitment on the one hand and a more comprehensive range of services on the other. An example of vertical cooperation between logistics companies is the cooperation between medium-sized or large freight forwarders and contractually bound transport companies.

Examples of vertical cooperation between shippers can be found in the joint operation of distribution centres by manufacturers and retailers in a wide variety of industries. There are many forms of vertical cooperation, particularly in the area of procurement logistics in the automotive industry, both between suppliers and logistics companies and between suppliers and buyers or even between all three members of the logistics channel.[25]

[25]In the automotive industry, vertical, cooperative logistics concepts such as just-in-time delivery or vendor managed inventory are now standard. On cooperation between shippers and logistics

Diagonal cooperation in the logistics channel—also sometimes called complementary cooperation—includes cooperation between different transport techniques as well as between different transport areas. The former is based on the idea of the transport chain, which is realised in intermodal transport. Cooperation between transport areas exists, for example, between port companies and road or rail transport.

After this overview of various approaches to cooperation in the logistics channel, the following section will first outline the factors on which the willingness of institutions in the logistics channel to cooperate depends. Afterwards, it will be shown to what extent cooperation can be reflected in inter-organizational relationships.

17.3 Willingness to Cooperate and Extent of Cooperation

Willingness to Cooperate

Two essential conditions for a successful realization of inter-organizational cooperation result from the application of the *incentive-contribution theory to* the logistics channel.[26] According to the incentive-contribution theory, each institution in the logistics channel has to make contributions and receives certain incentives in return. The logistics channel system is in an incentive-contribution equilibrium when the following two conditions are met:

- Logistics channel members value the incentives received more highly than the contributions to be made, according to their individual values.
- The contributions of the individual logistics channel members enable the logistics mesosystem to provide them with the necessary incentives.

Equilibrium capacity is thus influenced by the efficiency of the logistical mesosystem and by the distribution of the success it ensures. How the individual institution perceives the share attributable to it as an incentive is determined by its subjective assessment of the distribution of the cooperation success. Whether such cooperation success is achieved at all depends on the success in finding cooperation partners whose strengths and weaknesses complement each other. Ultimately, cooperation is about making better use of the strengths of the partners and jointly compensating for their weaknesses.

companies in the automotive industry, see Schob et al., 2007, pp. 619ff. A tendency to outsource logistical task packages can also be observed in other sectors, e.g. in the chemical industry, see Hardt/Karsch, 2007, pp. 639ff; or in the textile industry, see Nothardt et al., 2007, pp. 679ff; Pfohl/Shen, 2008, pp. 20 f.

[26] For the following, see Pfohl, 1987, pp. 19 f.; Koppelmann, 2004, pp. 59ff. and the literature listed there. On the willingness to cooperate in the logistics channel, see Krass, 1984, pp. 126ff.; Kleer, 1991, pp. 100ff.

However, the willingness to cooperate is not only determined by the *utilitarian conditions* captured in incentive-contribution theory, but ultimately by the extent to which the goals of the individual institutions are fulfilled by cooperative membership; *structural conditions* are also influential. Thus, the more frequent and stronger the interactions between institutions in the logistics channel, the greater will be the willingness to cooperate, ceteris paribus. An institution involved in several logistics channels is generally less interested in cooperation than an institution that is a member of only one channel. The greater the risk an institution bears, the greater its interest in cooperation to reduce risk will normally be. The willingness to cooperate will also depend on the competitive relationships within a logistics channel and the competitive relationships with other logistics channels. Since developing cooperation without exercising power solely on the basis of consensus with other stakeholders is often illusory in reality, willingness to cooperate will also be influenced by the existence of a leader in the logistics channel (channel leader) who initiates inter-organizational cooperation as a power promoter and drives its realization. Last but not least, normative factors expressed in beliefs and value attitudes have a major influence. The existence of a corresponding spirit of cooperation among the participants is sometimes regarded as a decisive prerequisite for cooperation to come about. This spirit of cooperation should be all the more pronounced the greater the extent of cooperation.

Extent of Cooperation

The extent of cooperation depends on the intensity and scope of cooperation. The *intensity of cooperation* differs for the basic possibilities of cooperation, namely coordination and task transfer (task outsourcing). The intensity of cooperation in the case of *coordination of logistics tasks*, for which the standardisation efforts in the logistics sector provide numerous examples, is lower than in the case of task transfer. In the case of the *transfer of logistics tasks* from a shipper to a logistics company, the dependence of the cooperation intensity is shown in Fig. 17.5.

Accordingly, the intensity of cooperation is determined by the width of cooperation, measured by the number and type of (outsourced) logistics tasks transferred, and the depth of cooperation, measured by the type and number of phases of task performance. In the case of the width of cooperation, the plausible consideration is that, due to the existing logistics interdependencies, the strength of the intervention in the logistics area of a company and thus the required intensity of cooperation increases in the sequence of transport, packaging, warehouse, stockkeeping (inventory management) and order processing. In the case of the depth of cooperation, the plausible assumption is that, due to the existing influence on logistical decision-making premises, the intensity of cooperation increases in the sequence of execution, control and planning.[27]

[27] For a detailed justification, see Krass, 1984, pp. 109ff.

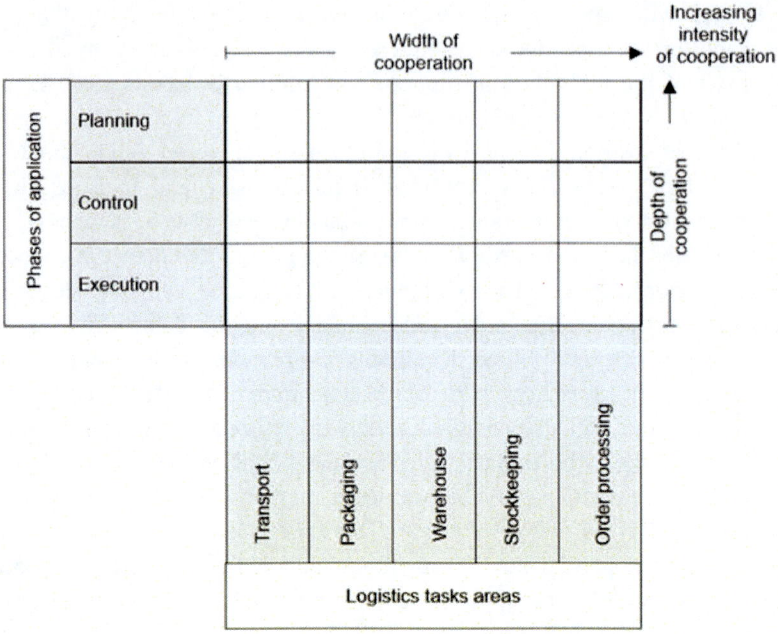

Fig. 17.5 Dependence of the intensity of cooperation in the transfer of logistics tasks from a shipper to a logistics company (Source: with minor modifications taken from Krass, 1984, p. 133.)

The results of empirical studies of shippers who had entered into vertical cooperations with logistics companies confirm these considerations. Thus, the function of inventory management is outsourced significantly less often than the performance of transport. Figures 17.6 and 17.7 illustrate this using the example of outsourced tasks. The results shown there also confirm the considerations on the increase of the cooperation depth in the described order. In almost all cooperations, the activities of the logistical core processes of transport, handling and warehousing form the focus of the cooperation. Tasks of planning and control, such as supply chain network design, are less frequently transferred to the cooperation partner. Tasks such as material planning and order management are only transferred to service providers to a lesser extent.[28]

The consideration of cooperation intensity based on the dimensions of cooperation width and depth can be supplemented by the dimension of logistics channel echelon, as it seems quite plausible that the outsourcing of a central warehouse to a logistics company entails a higher intensity of cooperation than when, for example, transshipment or delivery warehouses are outsourced. In addition to the type of warehouse, importance is also attached to the number of warehouse levels transferred.[29]

[28] A study conducted in the USA shows that planning, monitoring and control tasks are also increasingly being outsourced there. See Capgemini/Langley, 2017.

[29] See Kleer, 1991, p. 127.

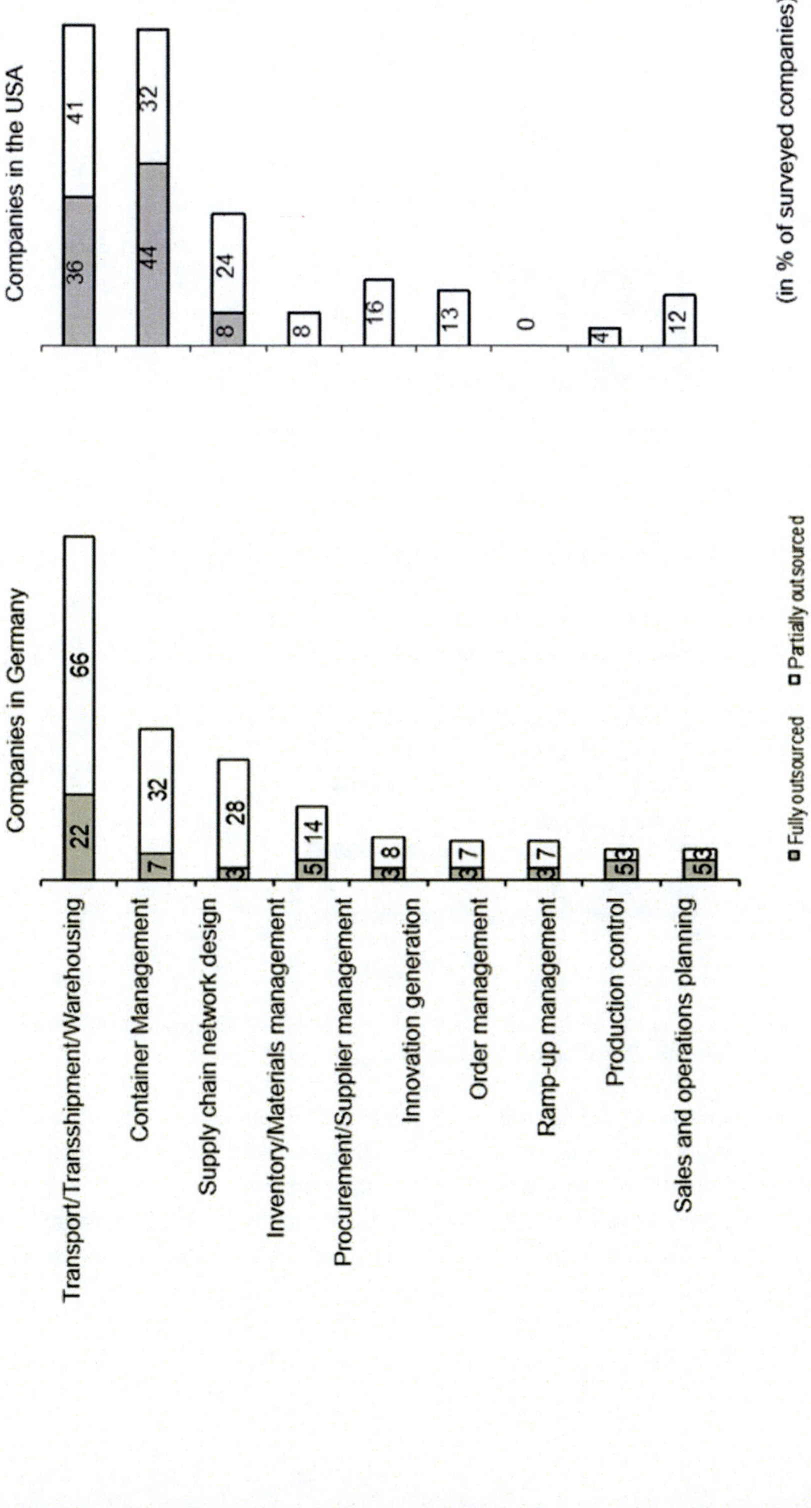

Fig. 17.6 Outsourcing of logistics functions. Results of a survey of 1189 companies in 2008 in Germany, China and the USA (Source: Straube/Pfohl, 2008, p. 25ff)

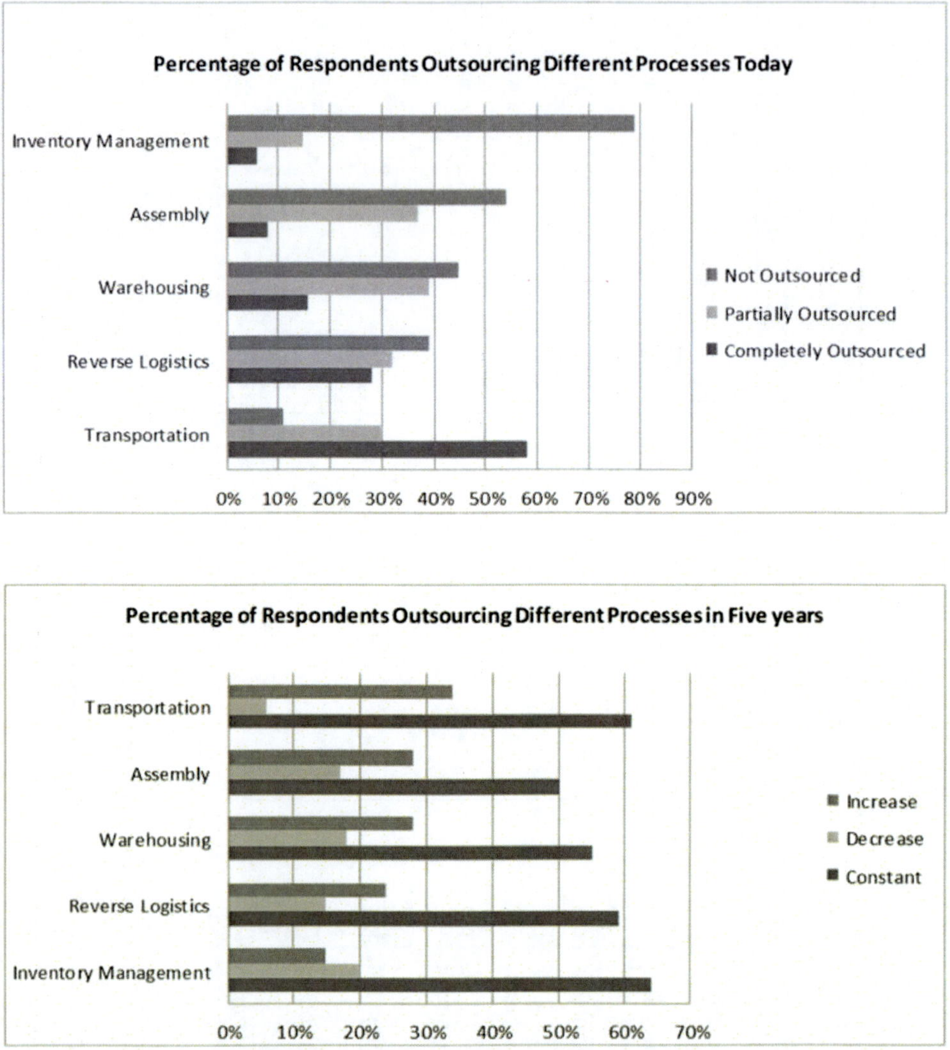

Fig. 17.7 Outsourcing of logistics functions. Results of a survey of 1757 logistics experts in 2013 in Asia, Brazil, Europe and the USA (Source: Hanfield et al., 2013, p. 57)

The *scope of cooperation* is measured by the share of a shipper's total flow of goods that the shipper transfers to the logistics company.[30] If the shipper transfers all inbound and outbound flows of goods to a logistics company, this represents a maximum scope of cooperation. If only parts of the flow of goods are transferred to a logistics company, there are several plausible classification criteria that can be used to determine the scope of

[30] See Krass, 1984, pp. 112ff.

cooperation. Examples of this are regional delimitations in which tasks are assigned to a logistics company, the restriction to certain types of goods, to certain types of transport, to certain shipment sizes, to the time of occurrence of the flow of goods or to the flow of goods entering or leaving a company (procurement and distribution logistics).

A survey carried out in 1987 among 25 shippers in Germany showed that a restriction of the scope according to the criterion "geographical area" was most frequently encountered, followed by restrictions of the scope of cooperation according to the type of goods or the size of the consignment.[31] Nowadays, many large logistics service providers (3PL) offer services throughout Europe or worldwide due to the new structuring of the logistics market (interlinking and acquisition).

Which extent of cooperation is sought by a company in a specific case depends on the utilitarian and structural conditions at hand, which must be subjected to analysis. In general, however, some effects can be named that can be derived from plausible considerations. Before these are explained in the next section, two forms of inter-organizational cooperation in which the extent of cooperation is of great importance will be discussed, namely supply chain management and company networks.

Supply Chain Management[32]

The term *supply chain* refers to the chain of connections between manufacturers and their suppliers, wholesalers and retailers acting as sales intermediaries, logistics companies acting as sales supporters and the end customers. Supply chain management is based on the flow or process orientation that is characteristic of the logistics concept. The focus of the consideration of object flows is on the integration of the second and third order logistical interfaces characterized in Fig. 17.1. Beyond goods and logistical information as *objects,* the integrated view is extended to include acquisitive information, financial resources, and rights. Thus, it is taken into account that the logistics channel is only one part of the *marketing channel*, which describes the path of transactions of a company with its procurement and sales markets.[33] The other part of the marketing channel is the acquisition channel, which includes the contracting flow (flow of rights), the sales or procurement promoting flow (flow of acquisitive information), and the financial resources flow (payment flow). The supply chain management concept calls for intensive consideration of the relationship between the logistics channel and the acquisition channel, which requires a high degree of cooperation across divisional and corporate boundaries. The transformation of interfaces into seams should take place in *all three interface orders*. The object flows are then not only integrated in each case, but mutually coordinated.

[31] See Kleer, 1991, pp. 196 f.

[32] See Pfohl, 2000, pp. 4ff.

[33] For other definitions of supply chain management, which assign considerably more objects to supply chain management, see Gomm, 2008, pp. 31ff. Often the entire value chain can then no longer be distinguished from the supply chain.

At the third order interface level, the focus is on the inter-organizational (across organizations) structure of the supply chain. At this level, it is necessary to create connecting interfaces between the institutions and to make holistic decisions about the members and their arrangement in the supply chain. The roles of the members of the supply chain and their relationships to each other must be identified, the structure of the supply chain must be analyzed and the various relationships in the supply chain must be managed.[34]

At the level of second order interfaces, the existence of interfaces between the logistics functions and the adjacent functional areas such as research and development, procurement, production and marketing is of particular importance.

The second order interface level captures not only the horizontal integration of different functional areas, but also the vertical integration of different planning levels. Here, the focus of the supply chain management concept is on decisions within the framework of strategic and tactical planning and the creation of interfaces to the other information processing levels of operational planning and control. In this context, the planning and control instruments used at the various levels must be coordinated with each other.

The relationships outlined suggest an increasing importance of *inter-organizational*, *inter-functional* and *inter-instrumental integration tasks* in the implementation of the supply chain management concept.[35] The focus is on cost, time and quality targets for the entire supply chain measured against the benefits for the end customer. Accordingly, competition no longer takes place between individual companies, but between supply chains.

Finally, the two object flows that play a special role in the supply chain in connection with logistics, namely the flows of financial resources and rights, should be briefly discussed. Parts of finance and cash management are included in the supply chain management concept in the form of the *flow of financial funds.*[36] Invoicing, which is directly related to cash flows, is a component of order processing. In addition to inventory, accounts receivable are a major component of current assets, where capital is tied up throughout the supply chain. Disruptions in cash flow have a negative impact on the entire supply chain. Thus, reducing net working capital, which consists of accounts receivable, inventory, and accounts payable, is a prerequisite for successful supply chain management. The *flow of rights* through the supply chain determines which institution can dispose of the goods and information and how. This flow is shaped by contracts or rights of use.

[34] See Cooper/Lambert/Pagh, 1998, pp. 7ff. On the management of network relationships, see Weber/Kummer, 1998, pp. 350ff.; Pfohl, 2004, pp. 12ff. Trumpfheller/Hofmann, 2004, pp. 72ff.; Pfohl, 2016, pp. 330ff.

[35] See the consequences of logistics thinking in Part I, Sect. 2.6.

[36] For financial aspects of logistics (supply chain finance), see Pfohl, 2016, pp. 228ff.

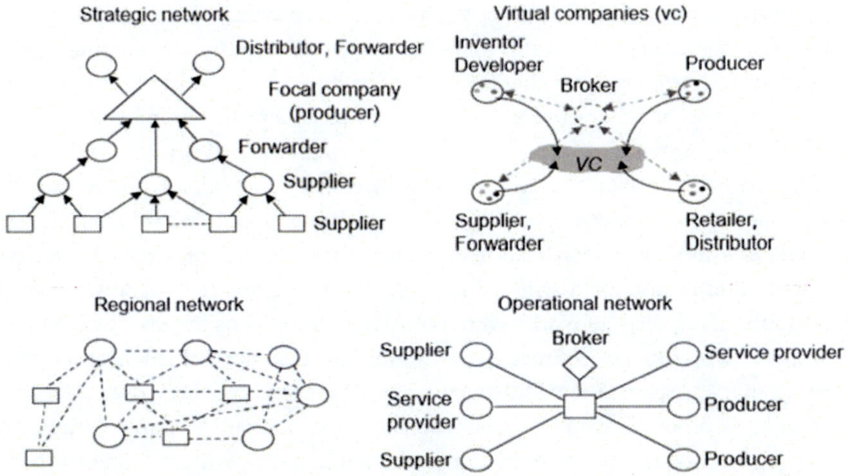

Fig. 17.8 Different types of company networks (Source: Pfohl/Buse, 1998, p. 51)

Company Networks[37]

Four types of company networks can be distinguished (see Fig. 17.8), namely, the strategic network, the virtual enterprise, the regional network and the operational network. The extent of cooperation varies greatly among these network types.

Strategic networks are strategically led by a focal company, which is often an end-product manufacturer or trading company with corresponding proximity to the end customer, are relatively stable, often have investments in network-specific resources, and aim to jointly achieve competitive advantages. The focal firm determines the organization of the network to a significant extent. The other network companies are often also closely bound to the focal company by contract, but also offer their services to other buyers outside the network in order to maintain their independence and own competitiveness. The purpose of the network is typically to serve a relatively predictable, comparatively stable market. Well-known examples are supply networks in the automotive industry.

Virtual companies are created by independent companies working together on the basis of a common understanding of business in order to take advantage of a business opportunity that arises. The basic idea behind this type of network is to bring together partners who each have individual core competencies that can be combined synergistically. Project-like cooperation, mutual trust, the arbitrary spatial distribution of the companies, the intensive use of information and communication technology, the renunciation of detailed contracts and specific investments as well as the uniform appearance towards the customers are regarded as characteristic features of this type. The fields of application for this type of network are primarily those value-added processes which, such as software production, the media industry and information services, are based to a considerable extent on information

[37] See Pfohl/Buse, 1998, pp. 50ff. and the literature cited there.

technology infrastructure. Other fields of application are low-tech value creation processes with very short product life cycles (clothing, toys) or rapidly developing high-tech industries (microelectronics, biotechnology).

Regional networks are characterized by a case-by-case repeated cooperation of many—mostly small—companies that are located in close spatial proximity to each other. The companies usually have latent relationships with a large number of potential partners, which are sometimes activated depending on the order situation. Personal contacts, similar corporate cultures and high specialisation of the companies are characteristic of this type of network. In northern Italy, the textile industry and the manufacture of spectacles are well-known examples. Regional networks with intensive relationships are also called clusters.

Operational networks are characterized by the fact that the participating companies, supported by an interorganizational information system, can access services at short notice, but especially free production and logistics capacities (with regard to storage, transport and packaging) of the partners, for example for peak load balancing in production. Characteristic is the processing of largely standardized transactions involving individual value-added activities and the trading of capacities instead of physical goods. The joint use of pooled resources, for example the joint operation of a warehouse, is also conceivable. An inter-organizational information system forms the basis for the use of market-analog mechanisms to coordinate the exchange process. In this respect, this type has essential characteristics of an electronic market, but is characterised by a higher degree of social organization.

17.4 Effects of Cooperation in the Logistics Channel

Cooperation Advantages

The effects of cooperation include, first of all, the following basic advantages that are attributed to cooperation in the logistics channel[38]:

- *Unnecessary duplication* of logistical activities (redundancy of logistical activities)—e.g. extensive inventories for a product at several stages in the logistics channel—can be avoided.
- *Coordination* of the logistical decisions of the companies working together in the logistics channel is made possible.
- A *counterweight to concentration tendencies* is formed and thus the growth of economic power of individual companies is counteracted. However, this only applies if the cooperation is not intended as a preliminary stage for a later concentration.
- *Technological possibilities* for rationalizing the flow of goods that are already known can often only be realised within the framework of large logistics systems. Such can be

[38] See Pfohl, 1975, p. 286 and the literature cited there.

created precisely through cooperation. The opinion is therefore often expressed that the actual, far-reaching innovations in the logistics sector today no longer consist so much in technological change as in inter-organizational change.

These fundamental advantages of the formation of cooperations in the logistics channel and thus the emergence of inter-organizational logistics systems can be seen as an approach to achieving competitive advantages. According to this view, the focus is no longer on the individual company, but on the *entire value chain* or—in terms of logistics—the cooperation of companies in the logistics channel from the manufacturer of the raw materials to the end customer. The aim is to bring about cost reductions or service improvements through the described advantages of cooperation, which make it possible to achieve a competitive advantage for the entire value chain. It is important that all cooperation partners participate in the successes of the entire value chain and that advantages are not achieved at the expense of individual companies within the cooperation.[39]

Impact on Objectives

In addition to these basic cooperation advantages, the effects of cooperation on the logistics channel can be discussed on the basis of important objectives.[40]

Division of labor: If one assumes that a further increase in the division of labor is to be expected in many sectors, both nationally and internationally, then the exchange of goods will continue to increase. This places high demands on the macrologistics and mesologistics systems. In the future, it will no longer be possible to afford weak points at the interfaces between the companies' micro-logistic systems. Cooperation is a good way to eliminate such weak points. The goal of division of labor can thus be supported by logistics cooperation. The division of labor is to be applied in particular to logistics itself. This then results in an increasing importance of logistics companies in the flow of goods. But it also leads to new forms of cooperation between suppliers and buyers on the supply market, which are no longer regulated by price alone but by the exchange of planning information. The benefits that can be achieved by extending logistics analysis beyond the inter-organizational interfaces to the entire inter-organizational logistics system are expressed in the cost curves in Fig. 17.9. There it is shown which different stages of the analysis can be run through and to what extent only through the inter-organizational logistics cost and profit analysis can a design alternative of the logistics system be found which offers advantages for all cooperation partners.

Rationalization through the use of technology: Many hardware and software technologies, such as the use of pallets and containers or computerized order processing,

[39] In this context, the term win-win situation is also used, see Arnold et al., 2008, p. 274 and p. 1069. For a detailed account of the relationship between the formation of inter-organizational logistics systems and the achievement of competitive advantages, see Pfohl, 1994, pp. 216ff.

[40] See Feierabend, 1980, pp. 112ff.

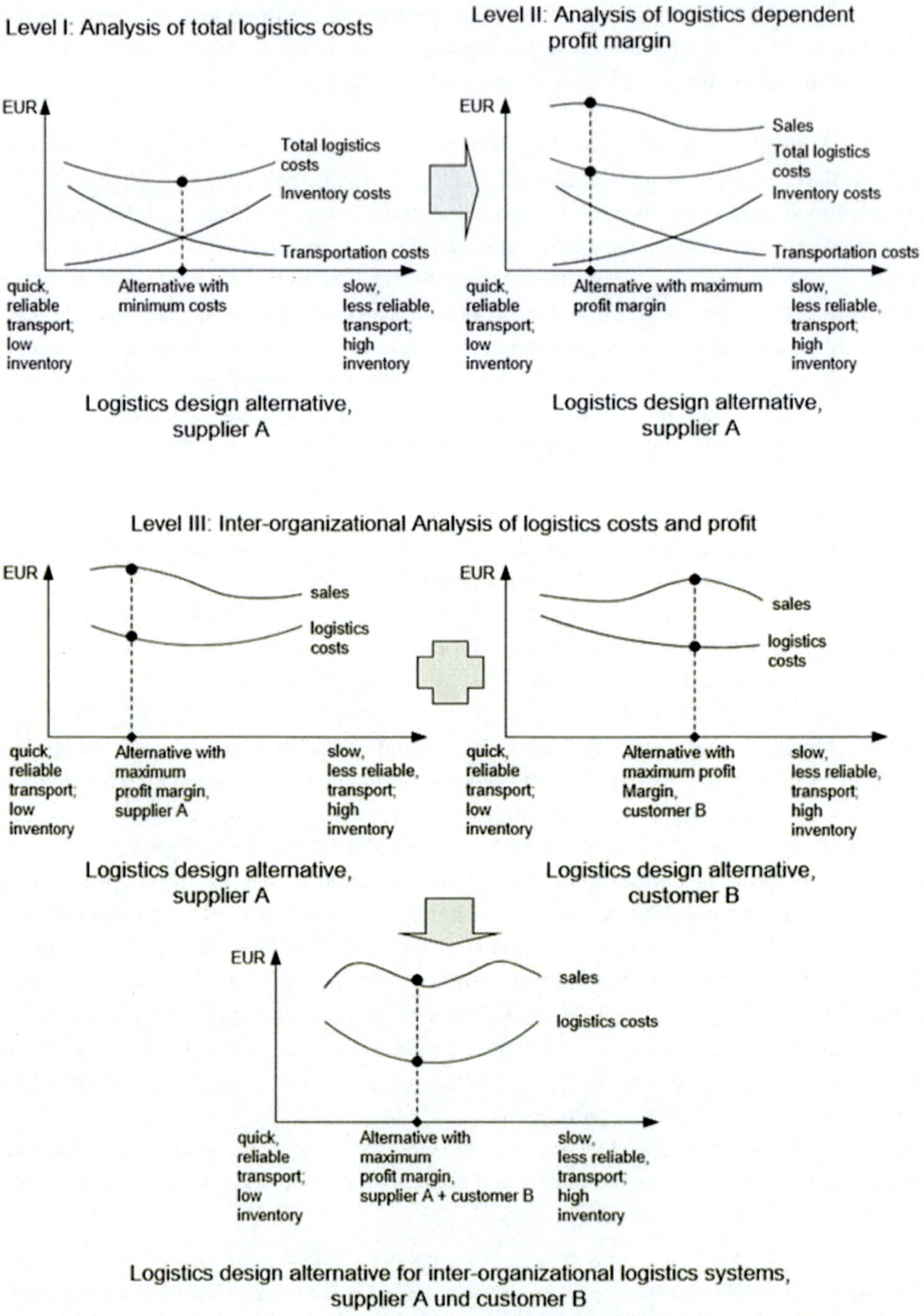

Fig. 17.9 Development stages of logistics analysis as a basis for logistics decisions in the design of inter-organizational logistics systems (Source: Glaskowsky et al., 1992, p. 648. Translation by the author)

require coordination of the companies' micrologistics systems. A substantial part of the costs caused by the flow of goods can only be influenced by mutual agreement between the companies involved in the logistics channel. The reduced autonomous influenceability of these costs can be called the "dilemma of inter-organizational logistics planning". This dilemma, as well as the fact that mechanization and automation often place certain demands on the size of logistical systems anyway, highlights the positive effect of cooperation on the use of technology.

Safety: Cooperation can obviously make a significant contribution to achieving the safety objective. For example, improved information exchange at the inter-company interfaces in the flow of goods reduces the uncertainty of many logistics decisions. In addition, cooperative relationships strengthen general business relationships in the long term, which generally reduces uncertainty in a company's environment.

Capital commitment: The level of capital commitment in a company is decisively influenced by the level of inventories. In addition to the inventories at the points where the flow of goods is interrupted, the inventories during the movement, the so-called pipeline inventories or transit inventories, must also be taken into account. The capital tied up in both types of inventory can be reduced through cooperation, for example by reducing the number of points at which goods are stored in the logistics channel or by speeding up the flow of goods through faster order processing, handling and transportation.[41]

Delivery or supply service: Cooperations can be pursued not only for cost reasons, but also for performance reasons. For example, accelerating the flow of goods not only reduces capital commitment costs, but also shortens delivery times (procurement time) and thus improves an important element of the delivery and supply service. In addition, improved communication within the framework of the cooperation increases the knowledge of the frequently diverging service ideas among the parties involved in the flow of goods.

Independence/competition: Cooperation has two different effects on independence and competition. On the one hand, cooperation reduces independence in economic sub-areas and competition—at least in the case of market-oriented cooperation. On the other hand, cooperation enables a company to reduce its costs and improve its performance. By reducing its independence in economic sub-areas as a result of cooperation, a company can therefore secure its legal and, incidentally, also its economic independence. This counteracts concentration and preserves competition.

Flexibility: The goal of flexibility includes the ability of a company to adapt to changing environmental conditions. With regard to this objective, two different correlations can again be identified. The technical-physical coordination of the logistics systems of the cooperation partners in the flow of goods does not have to, but can lead to a lower flexibility of the individual company. Coordination in the flow of information, on the other hand, will increase flexibility.

[41] See Part II, Sect. 5.2.

Striving for power/prestige: The connection between cooperation and the striving for power of a company and the striving for prestige (reputation) cannot be characterized in general terms—not even in terms of tendencies. It will depend on the objects to which the striving for power extends and the standards by which prestige is measured.

In summary, it can be said with regard to this discussion of objectives that a logistics cooperation can have a positive effect on many corporate objectives. However, this positive effect can only be achieved if—as was already emphasized in the discussion of conflicts—suitable *conflict management* is installed for the conflicts between the cooperation partners, which can never be eliminated. Last but not least, the achievement of positive effects is also influenced by how the macro logistics infrastructure is developed. This will be discussed in the following chapter.

References

Arnold D u.a. (Hrsg) (2008) Handbuch Logistik. 3. neu bearb. Aufl. Berlin Heidelberg

Baumgarten H, Wolff S (1999) The Next Wave of Logistics - Global Supply Chain efficiency. Berlin/Boston

Brandenburg H, Gutermuth J, Oelfke D (2016) Güterverkehr-Spedition-Logistik. Leistungserstellung in Spedition und Logistik. 42., Aufl. Troisdorf

Bundesministerium für Forschung und Technologie (1980) Informatik im Güterverkehr. Endbericht. Karlsruhe

Capgemini, Langley J (2017) 21st Annual Third-Party Logistics Study: The State of Logistics Outsourcing

Cooper M C, Lambert D M, Pagh J D (1998) Supply Chain Management: Implementation Issues and Research Opportunities. In: The International Journal of Logistics Management 8 2, S. 1-19

Feierabend R (1980) Beitrag zur Abstimmung und Gestaltung unternehmungsübergreifender Schnittstellen. Bremen

Freichel S L K (1992: Organisation von Logistikservice-Netzwerken. Theoretische Konzeption und empirische Fallstudien. Berlin

Frunzke H (2004) Von der Kompetenz im strategischen Management zur Netzkompetenz. In: Pfohl H-Chr (Hrsg) Netzkompetenz in Supply Chains: Grundlage und Umetzung. Wiesbaden, S. 13-41

Glaskowsky Jr N A, Hudson D R, Ivie R M (1992) Business Logistics. 3. Aufl. Fort Worth u.a.

Gomm M (2008) Supply Chain Finanzierung. Optimierung der Finanzflüsse in Wertschöpfungsketten. Berlin

Hardt A, Karsch C (2007) Kontraktlogistik in der Chemieindustrie. In: Stölzle u.a. (Hrsg) Handbuch Kontraktlogistik. Management komplexer Logistikdienstleistungen. Weinheim

Handfield R, Straube F, Pfohl H-Chr, Wieland A (2013) Trends und Strategien in Logistik und Supply Chain Management. Hamburg

Ihde G B (2001) Transport, Verkehr, Logistik. Gesamtwirtschaftliche Aspekte und einzelwirtschaftliche Handhabung. 3., völlig überarb. u. erw. Aufl. München

Kirsch W u.a. (1973) Betriebswirtschaftliche Logistik. Systeme, Entscheidungen, Methoden. Wiesbaden

Kleer M (1991) Gestaltung von Kooperationen zwischen Industrie- und Logistikunternehmen. Ergebnisse theoretischer und empirischer Untersuchungen. Berlin

Koppelmann U (2004) Beschaffungsmarketing. 4., neu bearb. Aufl. Berlin u.a.

Krass R (1984) Kooperation zwischen Verlader und Spedition. Konzept zur Effizienzsteigerung der Marketing-Logistik. Darmstadt

Nothardt F, Schmitter F, Trede T (2007) Logistik in der Bekleidungsbranche – quo vadis? In: Stölzle u.a. (Hrsg) Handbuch Kontraktlogistik. Management komplexer Logistikdienstleistungen. Weinheim

Pfohl H-Chr (1975) Interorganisatorische Zusammenarbeit bei der Warenverteilung im Absatzkanal. Ein Beispiel für kooperatives Marketing. In: Jahrbuch der Absatz- und Verbrauchsforschung 21 3, S. 284-306

Pfohl H-Chr (1987) Stichwort "Vertikales Marketing". In: Poth L G (Hrsg) Marketing. 2. Aufl. Neuwied, Kennziffer 36, S. 1-40

Pfohl H-Chr (1994) Interorganisatorische Probleme in der Logistikkette. In: Pfohl H-Chr (Hrsg) Management der Logistikkette: Kostensenkung – Leistungssteigerung – Erfolgspotential. Berlin, S. 201-251

Pfohl H-Chr (2000) Supply Chain Management: Konzept, Trends, Strategien. In: Pfohl H-Chr (Hrsg): Supply Chain Management: Logistik plus? Logistikkette – Marketingkette – Finanzkette. Berlin, S. 1-43

Pfohl H-Chr (2004) Grundlagen der Kooperation in logistischen Netzwerken. In: Pfohl H-Chr (Hrsg) Erfolgsfaktor Kooperation in der Logistik. Berlin, S. 1-36

Pfohl H-Chr (2016) Logistikmanagement. Funktionen und Instrumente. 3., neu bearb. und aktual. Aufl. Berlin/Heidelberg

Pfohl H-Chr, Buse H-P (1998) Marketing-Logistik in Unternehmensnetzwerken. In: Thexis 15 1, S. 50-58

Pfohl H-Chr, Shen X (2008) Apparel Supply Chain between Europe and China. A Guide to Apparel Sourcing and Distribution in China. Darmstadt

Schob U, Halsband E, Anders F (2007) Kontraktlogistik in der Automobilindustrie. In: Stölzle u.a. (Hrsg) Handbuch Kontraktlogistik. Management komplexer Logistikdienstleistungen. Weinheim

Specht G, Fritz W (2005) Distributionsmanagement. 4., vollst. überarb. und erw. Aufl. Stuttgart

Straube F, Pfohl H-Chr (2008) Trends und Strategien in der Logistik - Globale Netzwerke im Wandel. Umwelt, Sicherheit, Internationalisierung, Menschen. Hamburg

Trumpfheller M, Hofmann E (2004) Supply Chain Relationship Management. In: Pfohl H-Chr (Hrsg) Netzkompetenz in Supply Chains: Grundlage und Umsetzung. Wiesbaden, S. 67-91

Weber J, Kummer S (1998) Logistikmanagement. Stuttgart

Wittenbrink P (1995) City-Logistik. In: Baumgarten H u.a. (Hrsg) RKW-Handbuch Logistik. Berlin, 21. Lfg. V/95, Kennziffer 8720, S. 1-33

Part V
Macroeconomic and International Aspects of Logistics Systems

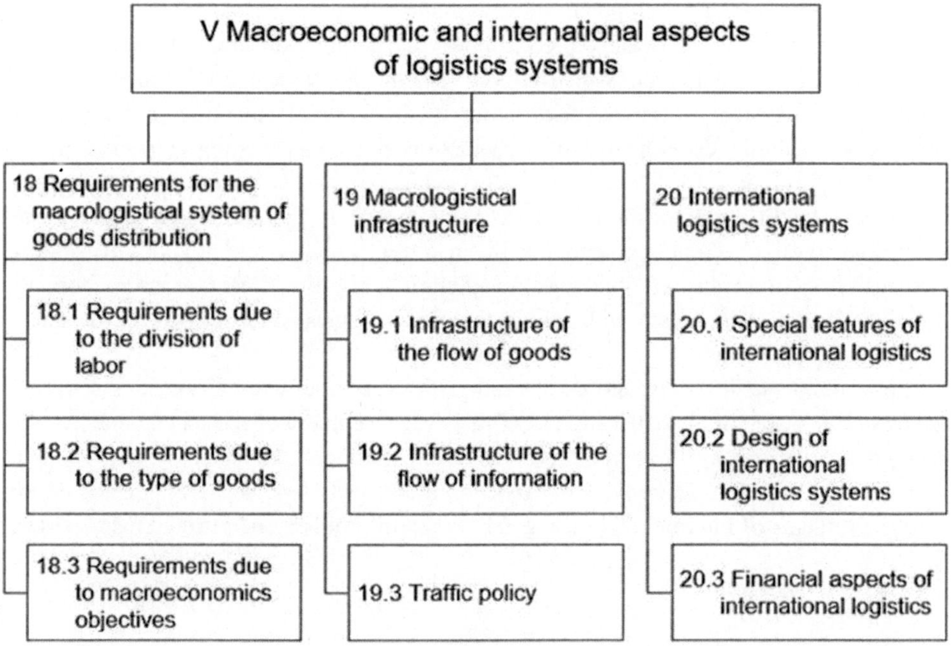

The systems of micro- and mesologistics form the institutional elements of the system of macrologistics. However, a complete explanation and design of processes in the macrologistics system additionally requires a consideration of the relationships between the elements. These can be temporary or permanent, contractual or tangible or intangible relationships between the elements. A useful level of abstraction for such a consideration is achieved by examining the structure-forming components of the macrologistical system. These include the components of the material infrastructure, which primarily comprises the transport and communication networks, whose components have the comparatively most comprehensive and lasting influence on logistical processes.

Fundamentally, it is first necessary to consider the macroeconomic data and interrelationships that form the framework within which logistics tasks can be performed. Due to the close interrelationships that exist between the development of the economy and the macrologistical system of goods distribution, which can be described by a characteristic sequence of phases with certain regularities of production and market development, the macrologistical system of goods distribution is one of the most important features that characterize the level of development of an economy. This is because the flows of goods and information in less developed economies place quite different demands on the macrologistical system of goods distribution than in highly developed economies. The macrologistical system of goods distribution in the Federal Republic of Germany is comparatively highly developed and is characterised by the existing infrastructure for goods and information flows and the market organisation of the domestic goods transport market, which still regulates these flows to some extent. In the discussion of macroeconomic objectives, it becomes apparent that the macrologistical system of goods distribution also has to fulfil non-transport objectives. In addition, it becomes clear that macroeconomic objectives may well compete with microeconomic logistics objectives in the design of macrologistical systems.

In connection with the system of macrologistics, the international logistics system is also treated as a specific logistics system. This is because many of the special features that distinguish international logistics systems from national logistics systems are of macrologistical origin[1]. Moreover, macroeconomic aspects tend to be of greater importance in problems of international management than in problems of national management.

[1]For a ranking of different countries in terms of the quality of their macrologistics (logistics technology and institutional framework conditions for logistics processes), which is significant for the competition between different countries and regions, see World Bank, 2016.

Requirements for the Macrological System of Goods Distribution

18

Logistics problems can be considered at the three levels of aggregation of economic facts usually distinguished in economics. In the previous chapters, logistics problems were discussed at the micro- and mesologistics level. They dealt with economic issues of a company or several companies cooperating in the logistics channel. At the macrologistical level, the focus is on economics issues. They are taken up in the following insofar as they are also of importance for answering business management questions.

18.1 Requirements Due to the Division of Labor

With the provision and use of goods distribution services, a necessary link is formed between the systems of goods provision and goods use. The demands on this macrologistical system of goods distribution grow with increasing *inter-firm division of labor*, above all because of the associated spatial separation of the resulting institutions (enterprises or households). The level of development of an economy is expressed, among other things, in the extent and type of inter-firm division of labor. A distinction can be made here between quantity and kind division, which have different meanings for logistics.[1]

The *kind division* leads to a specialization of the plants or enterprises. As a result of specialisation, larger production volumes are generated in the individual plants or enterprises. Rationalization advantages can be achieved above all in the form of scale degression and learning effects in the system of goods provision. However, the lower production costs compared to non-specialization are offset by higher logistics costs in the goods distribution system.

[1] See Ihde, 2001, pp. 66ff.

H.-C. Pfohl, *Logistics Systems*,
https://doi.org/10.1007/978-3-662-64349-5_18

Quantity division does not result in specialization advantages that can be reflected in lower production costs. Quantity division, on the other hand, seeks to derive production advantages from an optimal size of operation. Accordingly, it may be advantageous to supply a market from several decentralized production locations instead of from one centralized one. But even if the production costs still speak in favor of a central production location, decentralized production locations can be more advantageous when logistics costs are taken into account. In the case of centralized production, the increase in logistics costs resulting from the increasing average procurement and dispatch distances and the increasing turnaround time of the goods in the goods distribution system may be greater than the decrease in production costs.

In principle, it can be stated that in an equilibrium system with recurring production, transfer and consumption processes that take place in an economy based on the division of labor, the flows of goods and information as well as logistics costs are determined by the interdependence of market relations and market sizes. This spatial equilibrium is reflected in a balance between the perception of advantages of the spatial division of labor on the one hand and the acceptance of logistics costs, especially transport costs, on the other.[2] If the system of goods distribution in an economy is characterized by a good macrologistical infrastructure, as is generally the case in highly developed economies, *centralized production* tends to be favored. This is because the additional logistical costs of centralized production do not increase as much with a well-developed macrologistical infrastructure as they do with a poorly developed macrologistical infrastructure. From a marketing point of view, too, such centralization only makes sense in the presence of a well-developed macrologistical infrastructure. This is the only way to achieve the short delivery times required for competitive reasons when supplying the markets.

18.2 Requirements Due to the Type of Goods

Shifting Shares of Particular Types of Goods

However, the level of development of an economy is reflected not only in the extent and nature of the division of labor, but also in the type of goods that flow through the macrologistical system of goods distribution.[3] In the course of the development of an economy, there are large shifts in the shares of the various types of goods. The saturation of the markets for basic needs, which can be observed in highly developed economies, causes bulk goods (e.g. agricultural products, coal or steel) to lose shares in the total volume of the distribution of goods. In contrast, piece goods are gaining shares because the demand for high-value goods, which represent specialized solutions to problems, is rising. Since at the

[2] See Willeke, 1992, pp. 140 f.

[3] See Ihde, 2001, pp. 58ff. For qualitative changes in demand on the freight transport market, see Part I, Sect. 3.3.

same time, for economic reasons, the production processes for processing the raw materials are being relocated to the places of origin, a *dematerialization in the structure of goods* can be observed. This results in the fact that due to the progressive development of an economy by economic growth less and less quantities of goods are produced for the macrologistical system. This has been observed so far in studies on the relationship between the volume of transport (freight transport volume) and the gross national product in various countries. The *transport elasticity*, defined as the quotient of the relative change in transport volume and the relative change in real gross national product (GNP), becomes smaller in the course of an economy's development and is generally smaller than 1 in highly developed economies.[4]

If the quantities of goods to be handled by the macrologistical system as a whole increase less sharply than the real gross national product, the institutions involved in the distribution of goods nevertheless face high demands due to the shifts in the structure of goods. With regard to the demands that the goods to be transported place on transport, one also speaks of transport affinities. These transport affinities are compared with the transport values (e.g. mass efficiency, speed, safety)[5] of the means of transport, with which the efficiency of the means of transport is recorded. This comparison shows that bulk goods generally have a high affinity for rail or inland waterway transport, while road transport is preferred for semi-finished and finished goods.

The aforementioned dematerialization of goods makes it clear that the transport affinities of goods change in the course of the development of an economy and that other modes of transport are therefore preferred. For example, the macrologistical infrastructure in a highly developed economy has to ensure the distribution of high-value goods with high delivery service requirements in particular, which leads to a shift in transport volumes in favour of road freight transport. This shift is referred to as the *freight structure effect*. Along with the logistics effect, the integration effect and the interface effect, it is one of the causes of the disproportionate development of modes of transport, which is expressed in a preference for road freight transport over the other modes of transport and which is rooted in their respective system characteristics.[6] The *logistics effect* is the result of the logistical reorientation of industrial and commercial enterprises, which, e.g. by reducing production depths or because of cross-company flow orientation, place changed demands on freight transport that road freight transport, but not other modes of transport, can meet. The *integration effect* describes the increase in cross-border transport volumes resulting from

---

[4] Here, the definition of transport elasticity is based on the quotient of the change in transport volume [t] and the change in GNP. In addition to this definition, the literature also defines transport elasticity as the quotient of the change in transport performance [tkm] and the change in GNP, see Aberle, 2009, pp. 27 f. Transport performance as the product of transported mass [t] (transport volume) and distance [km] increases in contrast to transport volume in highly developed economies. Accordingly, in this definition of transport elasticity, the quotient is also greater than 1.

[5] See Part II, Sect. 8.3.

[6] See Aberle, 1994, pp. 6ff.; Ihde, 2001, pp. 59ff.

the common European market and the opening of the Eastern European states. In order to cope with these transport flows, continuous transport chains are required, the realization of which is comparatively problematic, e.g. in rail transport, as several national railway companies have to be involved. Finally, the *interface effect is* one of the causes of the shift in modal shares. It affects all transport operations that require a change of transport mode. Here, road freight transport has more favorable system characteristics than the other modes, for example by offering door-to-door transport.

The Scientific Advisory Council to the Federal Minister of Transport[7] points out two further tendencies in addition to the tendencies towards faster and more frequent transports, which go hand in hand with the development of the national economy and through which storage and production costs are substituted by transport costs. One is the tendency towards heavier and larger indivisible transport objects, the importance of which results from the fact that Germany is one of the most important manufacturers and exporters of industrial and transport equipment. As equipment units (e.g. boilers, transformers) or equipment components (e.g. roller stands, bridge girders) take on ever greater dimensions and weights, the need for suitable means of transport increases, the use of which, however, requires a basic infrastructural network with correspondingly wide carriageways, wide curve radii and large clearance heights. The other trend is towards large-volume transport containers for goods with a high stowage coefficient (low unit volume weight). This results from the increasing volume of bulky goods and the transition to ever lighter packaging and necessitates the use of large-capacity vehicles or large-capacity swap bodies.

Life Cycle of Logistics Services

The change in the needs to be satisfied or problems to be solved in the course of the development of an economy in the system of goods use leads to the described change in the volume of goods to be handled by the system of goods distribution. The services to be rendered in the system of goods distribution thus change with the level of development of an economy. This can also be illustrated by the idea of the life cycle of a product or service, according to which they are subject to the "law of becoming and passing away". They are born, grow, grow old and die.[8] Figure 18.1 shows the ideal life cycle for selected logistics services. The positioning of logistical services according to their life cycle phase is likely to approximate Germany's current situation in macroeconomic terms.

[7]For the following, see Scientific Advisory Council to the Federal Ministry of Transport—Transport Economics Group, 1987.

[8]See Meffert et al., 2012, pp. 68ff.

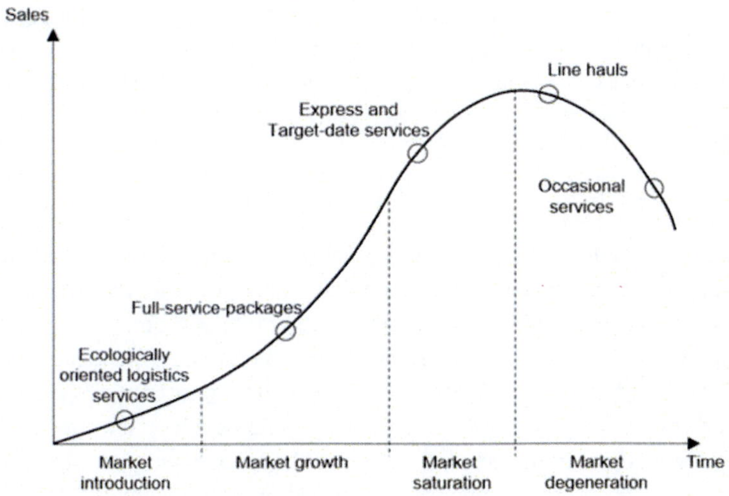

Fig. 18.1 Life cycle phases of selected logistical services in a highly developed economy (Source: based on Göpfert/Wehberg, 1995, p. 107; Ihde, 2001, p. 47)

18.3 Requirements Based on Macroeconomic Objectives

Transport Policy Measures and Macroeconomic Objectives

Transport policy measures include all measures taken by the state to regulate transport within an economy or between economies.[9] First of all, they include *price regulations*, which encompass tariff formation processes and tariff forms as well as the control of government-regulated prices. *Market access* and *capacity regulations,* which include regulations on maximum permissible weights and measures in freight transport, are also part of transport policy measures. The *constraints on supply* that exist for some modes of transport include, in particular, operating and transport obligations (compulsory contracting) and compulsory tariffs, which restrict the freedom of action of the transport industry. *Investment control* concerns investment in the expansion and maintenance of publicly owned transport routes and transport hubs for shipping, air transport, rail transport and road transport, through which intermodal competition is influenced. *Taxes* and *subsidies* have a regulatory effect in the transport industry when they are used in a selectively discriminatory way.

The objectives pursued by these transport policy measures are very heterogeneous and partly compete with each other. It should be noted in particular that transport policy

[9] See Aberle, 2009, pp. 99ff. See also the measures on freight transport management, which includes all measures that do not concern the modification or expansion of infrastructure, in Rühl/Boltze, 2017, pp. 171ff.

measures are taken not only to achieve transport objectives, but also to achieve non-transport objectives. When examining the effect of transport policy measures on microeconomic objectives, it must be taken into account that there may be divergences between microeconomic and macroeconomic objectives.

Non-transport Objectives

Due to its function of linking the economic systems of goods provision (production) and goods use (consumption), there are many relationships between the system of goods distribution and these two systems. Developments in the goods distribution system therefore often have an impact on developments in the other two economic subsystems and vice versa. Transport policy measures are therefore seen as having the potential to trigger design processes in these subsystems and thus to reshape existing economic, spatial, ecological and social structures. They can influence the economic development prospects of entire regions or individual locations, spatially shift environmental burdens through transport infrastructure projects and open up employment and growth potential.[10] Transport policy measures can therefore also pursue economic and social policy objectives, e.g. social policy objectives through social tariffs in passenger transport or environmental policy objectives within the framework of traffic noise legislation.

If there are *conflicts of objectives* between non-transport and transport objectives, there is a danger that, if transport policy measures are taken to achieve non-transport objectives, the transport economy will be negatively affected to such an extent that the positive contributions to objectives in the non-transport sectors will not be able to compensate for the disadvantages in the transport economy.[11] In this case, the question would have to be asked whether it would not make more sense to focus solely on transport objectives when implementing transport policy measures in order to optimally organize the transport economy in this way. Negative target contributions in all other economic or social sectors could then be offset by direct measures, e.g. fare refunds in the social policy field.

If, on the other hand, the positive effects of an intervention in the national economy outweigh the positive effects of other areas of society, it seems sensible to intervene in the national economy by means of transport policy measures. This approach is pursued in particular when ecological objectives are taken into account. The aim is to reduce environmental impacts caused by infrastructure projects and transport processes. Environmental impacts of transport occur both on the input side, e.g. through soil sealing or the consumption of energy sources, and on the output side in the form of noise, material or gaseous emissions or pollution through accidents. Transport policy approaches to reducing these environmental impacts include, for example, exhaust emission regulations for trucks or the introduction of motorway tolls, one of the aims of which is to shift more traffic to modes of

[10] See Baum, 1991, p. 13.

[11] See Ihde, 2001, p. 110.

transport that are considered to be more environmentally friendly, such as rail freight transport.

Transport Objectives

Among the transport objectives,[12] the *provision of transport services to the population* is at the top of the list. Achieving this objective in a market economy is unproblematic if a sufficient and acceptable transport supply can be achieved via the market mechanism. In this case, transport policy measures can be limited to shaping the framework conditions for the transport industry in such a way that intramodal (within one mode of transport) competition on the one hand and intermodal competition (between different modes of transport) on the other are not impaired.[13] Transport policy must ensure the provision of public transport services if transport needs that are recognized as urgent in society (e.g. serving peripheral areas and small towns or regions) are not satisfied by private-sector companies via the market mechanism.

A further transport objective is the *equality of transport modes* and *freedom of action for transport providers and users*. This objective is an essential foundation of the European Union's common transport policy. Equality between modes of transport means that they can assert and freely develop their inherent advantages in the transport sector. This concept is based on the idea of a division of labor that exploits the specific advantages of each mode of transport. The objective of freedom of action in the transport industry means that decisions on the supply of and demand for transport services are in principle subject to the free formation of will.

From the two transport objectives mentioned in the first place follows a third transport objective, namely to make the necessary *expansion and replacement investments* in the transport routes and nodes to be maintained by the public sector.

Finally, as in other areas of the economy, the *principle of economic efficiency* must be observed in the transport sector when scarce resources are used. Since this principle also underlies microeconomic decisions, it might be assumed that in this case there is agreement (complementarity of objectives) between microeconomic and macroeconomic objectives. However, this is only partially the case due to the existence of external effects.

External Effects

Measures to achieve microeconomic objectives can also serve to achieve macroeconomic objectives if, for example, "through organizational and investment measures

- the number of empty runs of the transport vessels is reduced,

[12] See Kirsch et al., 1973, pp. 127ff.; Fichert/Grandjot, 2007, pp. 147ff. For the objectives and areas of action of the Common Transport Policy of the European Communities, see Commission of the European Communities, 2006.

[13] On the types of competition, see Claussen, 1979, pp. 76ff.

- the time, weight and/or volume utilization of the transport vessels is raised,
- labor- and capital-saving continuous transport chains are formed,
- harmonization of packaging and loading equipment (pallets, containers) is carried through,
- warehousing by standardizing and adapting the storage equipment to the transport equipment is facilitated, and
- the organization and disposition of multi-stage goods manipulation (external and internal transport, pre-delivery, intermediate and delivery storage, operational production processes) are improved through the use of technologically advanced information and communication systems."[14]

However, there may also be conflicts between microeconomic and macroeconomic objectives.[15] So-called external effects are effects of microeconomic measures on third parties that are not taken into account in microeconomic decisions. They can occur both as external advantages (external economies) and as external disadvantages (external diseconomies). One speaks of *external advantages* or social benefits when third parties (households, companies, the state) benefit from the measures of a cause without the cause receiving a corresponding countervalue. *External disadvantages* or social costs occur when third parties are adversely affected by the actions of a cause without the cause receiving a corresponding countervalue. Externalities are of great importance in the burdened with industry, as the following examples show:

- "Transport operations always use infrastructure services (roads, waterways, ports, airports, rail transport services, transshipment points for general, bulk and containerised goods). To the extent that the users of these infrastructure capacities are not called upon to finance the commitment of resources, they receive input factors with no or distorted cost charges, with the result that their decisions on the choice of transport mode (modal split), transport routes and transport distance, and possibly also with regard to investment in vehicles, are flawed in macroeconomic terms. This brings us to the so-called "route cost problem" or rather the "route calculation problem".
- The conducting of transport services pollutes the environment through harmful substance and noise emissions from vehicles; the construction of transport routes often results in environmentally problematic fragmentation and land-use change processes. The numerous internalization efforts of these negative environmental effects are intended to influence the decision-making in both transport infrastructure investments and individual transport operations, and thus in freight flow logistics."[16]

[14] Aberle, 1983, p. 4.

[15] See Claussen, 1979, pp. 90ff; Ihde, 2001, pp. 71ff; Aberle, 2009, pp. 574ff.

[16] Aberle, 1983, p. 6.

If a mode of transport is allowed to incur social costs in relation to its competitors, or if it is not compensated for a social benefit it has caused, *competition is distorted*. Externalities thus lead not only to conflicts between microeconomic and macroeconomic objectives, but also to conflicts between the microeconomic objectives of different modes of transport.

In general, the internalization of external costs is regarded as a prerequisite for a true-cost calculation and thus also for an efficient management of transport according to market principles.[17] There are basically two phases for quantifying external effects:

- "the determination of the quantity structures and
- the transformation of quantity values into the economic category of money units, that is, the valuation of quantity structures (monetization)."[18]

However, there are uncertainties about the *valuation and attribution of external effects*. For this reason, there are various calculations of the benefits and costs of individual modes of transport with widely differing results, especially with regard to road transport.[19] Four main reasons can be given for these discrepancies. First, there are often gaps in knowledge and uncertainties about the emissions or impacts caused, which are then bridged with uncertain or arbitrary assumptions (e.g. correlations between certain noise levels and the resulting damage to health). Second, there are considerable uncertainties in the monetary valuation of consequential effects (e.g. in the valuation of odour nuisance). Third, the determination of causality in the application of the polluter-pays principle leads to considerable problems (e.g. state and municipal traffic route planning as a possible cause of traffic congestion). Fourth, the valuation of externalities is subject to political decisions.

Carbon Footprint (CF) is often discussed to capture the externalities. Carbon footprint is the total amount of carbon dioxide (CO_2) and other greenhouse gas emissions along the supply chain or life cycle of a product. Carbon footprint is quantified by indicators such as global warming potential (GWP). ISO 14040-14044 provide requirements for calculating the carbon footprint.[20]

References

Aberle G (1983) Gesamtwirtschaftliche Aspekte der Unternehmenslogistik. In: Baumgarten H u.a. (Hrsg) RKW-Handbuch Logistik. Berlin, 5. Lfg. I/83, Kennziffer 0850, S. 1-26

Aberle G (1994) Makrologistische Rahmenbedingungen für den Aufbau von Logistikketten. In: Pfohl H-Chr (Hrsg) Management der Logistikkette. Berlin, S. 1-32

[17] See Willeke, 1993, pp. 219ff.

[18] Aberle, 2009, p. 612.

[19] See Aberle, 1994, p. 20; Aberle, 2009, pp. 612ff.

[20] See European Commission, 2007, pp. 1 f.

Aberle G (2009) Transportwirtschaft. Einzelwirtschaftliche und gesamtwirtschaftliche Grundlagen. 5., überarb. und erg. Aufl. München

Baum H (1991) Infrastrukturpolitik als Mittel zur Steuerung des Verkehrsträgerwettbewerbs. In: Zeitschrift für Verkehrswissenschaft 62 1, S. 6-19

Claussen Th (1979) Grundlagen der Güterverkehrsökonomie. Hamburg

European Commission (2007) Carbon Footprint – what it is and how to measure it. http://lca.jrc.ec.europa.eu. Zugriff am: 31.03.2009

Fichert F, Grandjot H (2007) Akteure, Ziele und Instrumente. In: Schöller O, Canzler W, Knie A (Hrsg) Handbuch Verkehrspolitik. S. 138-160

Ihde G B (2001) Transport, Verkehr, Logistik. Gesamtwirtschaftliche Aspekte und einzelwirtschaftliche Handhabung. 3., völlig überarb. u. erw. Aufl. München

Kirsch W u.a.m (1973) Betriebswirtschaftliche Logistik. Systeme, Entscheidungen, Methoden. Wiesbaden

Kommission der Europäischen Gemeinschaften (2006) Für ein mobiles Europa – Nachhaltige Mobilität für unseren Kontinent. Halbzeitbilanz zum Verkehrsweißbuch der Europäischen Kommission von 2001. Mitteilung der Kommission an den Rat und an das europäische Parlament, KOM 314, Brüssel

Meffert H, Burmann Chr, Kirchgeorg M (2012) Marketing: Grundlagen marktorientierter Unternehmensführung: Grundlagen – Instrumente – Praxisbeispiele. 11., überarb. und erw. Aufl. Wiesbaden

Rühl F, Boltze M (2017) Freight Transport Demand Management: Influencing the Freight Transport Demand Within Traffic Management. In: Abele E, Boltze M, Pfohl H-Chr (Hrsg) Dynamic and Seamless Integration of Production, Logistics and Traffic. Fundamentals of Interdisciplinary Decision Support. Switzerland, S. 163-184

Willeke R (1992) Nutzen des Verkehrs und der verschiedenen Verkehrsmittel. In: Zeitschrift für Verkehrswissenschaft 63 3, S. 137-152

Willeke R (1993) Zur Frage der externen Kosten und Nutzen des motorisierten Straßenverkehrs. In: Zeitschrift für Verkehrswissenschaft 65 4, S. 215-236

Wissenschaftlicher Beirat beim Bundesminister für Verkehr – Gruppe Verkehrswirtschaft (1987) Verkehrsinfrastruktur als Voraussetzung für die gesellschaftliche und wirtschaftliche Entwicklung in der Bundesrepublik Deutschland. In: Zeitschrift für Verkehrswissenschaft 58 3, S. 131-153

World Bank (2016) Connecting to Complete 2016. Trade Logistics in the Global Economy. The Logistics Performance Index and Its Indicators. Washington

Macrological Infrastructure 19

19.1 Infrastructure of the Flow of Goods

Basics

Whether the requirements discussed in the previous section can be met by the macrologistical system of an economic area depends to a large extent on the existing logistical infrastructure. The term *logistical infrastructure* should be understood in the material sense[1] as the network of an economic area, e.g. an economy, in which goods and information can flow between companies and households. In the following, the infrastructure of the flow of goods is dealt with first.

More precisely, the infrastructure of an economic area does not consist of one network, but is made up of a *large number of sub-networks* which can be distinguished, for example, according to the different modes of transport. In this sense, a maritime network, an inland waterway network, a road transport network, a rail transport network, an air transport network and a pipeline transport network can be delineated.

Each of these subnetworks consists of *transport routes* and *transport nodes*. Examples of transport routes are roads or navigable rivers and canals. Transport routes are also used for passenger transport, which can lead to competition between freight and passenger transport for usage opportunities. Transport nodes are the links in transport networks. Their importance grows with the significance that transhipment processes have for the individual mode of transport. Thus, nodes in the road transport network, e.g. transhipment facilities of road transport companies, can be comparatively simple in design. Nodes in maritime transport (seaports), on the other hand, are extremely complex structures.

[1] On the distinction between tangible (material) and intangible (immaterial) infrastructure, see Ihde, 2001, p. 135.

H.-C. Pfohl, *Logistics Systems*,
https://doi.org/10.1007/978-3-662-64349-5_19

Transport hubs fulfil a *dual function*. On the one hand, they serve as connecting elements in the network of one mode of transport and can thus also link the networks of different economic areas. On the other hand, they also represent interfaces between the networks of different modes of transport. Examples of this are the seaports, which connect the maritime transport network with the inland transport networks of the hinterland. Other examples are the combined transport terminals, which link the transport networks of the modes involved by means of appropriate transhipment facilities.

As a rule, statements on logistics infrastructure are related to an economy as the *relevant economic area*. However, the definition of the relevant economic area also depends on the mode of transport under consideration. In the case of maritime transport, for example, a national delimitation makes little sense. The relevant economic area is the world economy or at least a trade lane between two geographical regions. From a German perspective, on the other hand, the relevant economic area for inland waterway transport or rail transport is certainly Europe. Frequently, however, the infrastructure of a region or even an urban area will also be of interest.

An important parameter of the logistical infrastructure of the flow of goods with regard to transport routes is the *network density*.[2] The network density is determined as the quotient of the length of the transport routes and the area of an economic region. The network density can be used as a comparative parameter for the infrastructure equipment of different economic areas or also for comparison within an economic area. Another important parameter is the length of transport routes per inhabitant. However, these parameters do not allow any statement to be made about the *quality of* transport routes. This requires quality criteria specific to the mode of transport, such as the degree of electrification of railway lines or the proportion of paved roads in the overall network.

The level of potential benefit from an existing logistics infrastructure depends on its *load*. In the Federal Republic of Germany, for example, there are sometimes considerable bottlenecks due to the high volume of traffic, despite the fact that the infrastructure is good by international standards.[3] However, the quality of the logistics infrastructure also depends on the number, distribution and performance of the transport hubs. For example, the availability of sufficiently large seaports and airports is the primary factor determining the quality of maritime and air transport infrastructures.

In the following, the transport routes and transport hubs will first be considered according to the individual modes of transport. Subsequently, the interface function of transport nodes will be discussed in connection with the combination of transport networks.

Maritime

Maritime shipping uses an essential part of the infrastructure of *international logistics*: the seas and the seaports. In 2016, 296.5 million t of goods were shipped from German

---

[2]For further parameters see Ihde, 2001, pp. 111ff.

[3]See Aberle, 1998, pp. 110 f. See Sect. 19.4.

seaports, of which 116.8 million t went to ports outside Germany. 171.1 million t were received from abroad.[4]

While freight rates for chartered shipping can be freely negotiated, liner shipping freight rates are set in cartels known as *shipping conferences*. Most of these conferences, of which there are about 350 today, relate to maritime shipping between two geographical regions. However, the emergence of new and different types of service means that these conferences no longer have the strong position they did in the past. These new competitors do not use their own vessels (*non-vessel operators*) and pursue the idea of through-freight with their range of services. They either use chartered shipping space or so-called landbridge services. One well-known example is the land bridge between Europe and the Far East in the form of the Trans-Siberian Railway.

The growth in the size of ships observed in the development of shipping, combined with the trend towards containerization, is leading to high fixed cost burdens in maritime shipping. This results in the need to utilize the loading capacity as much as possible during a voyage and to increase the ships' speed of rotation. This in turn leads some providers to offer so-called *one-port services*. The aim here is to reduce the number of ports to be called at and to call at only one port per country or per economic region in a round-the-world service (linking the most important traffic flows by means of global liner services).[5]

This further increases the already great importance of individual ports. This is because the logistical infrastructure of maritime shipping is primarily determined by the arrangement of the *seaports as hubs*, since in principle the entire seas are available as transport areas. In 2007, for example, Hamburg alone accounted for 120.3 million tonnes of the 253.7 million tonnes of goods handled by the German North Sea ports.[6] Hamburg also accounted for 46% of the 15.2 million TEU[7] containers handled in German seaports in 2007, and Bremen-Bremerhaven for 20%.[8] In the same year, 461.2 million tonnes were handled in Rotterdam, 40% of which, however, were petroleum and petroleum products.[9] When there are only very few ports of call, greater demands are placed on them in terms of handling goods and supplying the hinterland with appropriate feeder services. This increases the importance of the hinterland location of seaports compared to their sea or coastal location. The seaport must be connected to the hinterland with appropriate infrastructure through feeder services (coastal shipping), inland shipping, rail and road transport. This infrastructure plays an important role in the competition between seaports.

---

[4] See Federal Statistical Office, 2017a, p. 1.

[5] See Schieck, 2008, pp. 192ff.

[6] See Federal Ministry of Transport, Building and Urban Affairs, 2008, p. 71.

[7] TEU: Twenty foot equivalent unit. Common conversion of transports in container traffic, which are carried out with different container units, to a container with a length of 20 feet.

[8] See Federal Ministry of Transport, Building and Urban Affairs, 2008, p. 75.

[9] See Federal Ministry of Transport, Building and Urban Affairs, 2008, p. 317. However, in contrast to the figure stated for Hamburg, this figure includes the tare weights of transport vehicles, containers, trailers, etc.

The choice from the available range of services in maritime shipping is restricted by *protectionist measures*. Many countries want to secure for themselves a certain share of the transport volume of maritime shipping, above all in the form of so-called local content provisions. At the centre of these considerations on the granting of privileges in the chartering of national fleets for conference liner shipping is the quota system in accordance with the Unctad Code. According to this, 40% of the bilateral freight transport volume between two trading partners is to be allocated to the national fleets involved, while 20% is to be reserved for the free market. However, the OECD countries waive this rule among themselves, which is why the code is mainly applied in trade with developing countries.[10] With regard to national fleets, the phenomenon of *flagging out of seagoing vessels* is also evident in many industrialized countries, including the Federal Republic of Germany. In 1998, for example, almost 52% of gross tonnage was operated under the flags of so-called flag states, which can result in savings of up to half a million euros per ship and year for modern ships, due to tax advantages as well as labor and social law benefits.[11] However, the progressive flagging out seems to have been halted by the establishment of a *second register* in Germany. As a result of inclusion in the second register, certain labor and social legislation in Germany no longer applies to the foreign crews of these ships.

Inland Navigation

In 2017, the *length of German waterways was* 7290 km.[12] Thus, the *network density* is only 0.02 km/km^2. The average *transport distance* of inland navigation in Germany is 260 km, i.e. inland vessels are mainly used for long-distance transports.[13] A breakdown by *waterway class* for the whole of Germany is available for 1991. According to this, only 1682 km are navigable for large vessels of 1500 t or more.[14] These waterways are almost exclusively located in the "old" federal states, while the "new" federal states are dominated by those of the lower classes. The *number of German inland waterway vessels* has continued to decrease. In 2000, there were only 2569 cargo vessels in the old federal states, whereas in 1980 there were 3812 (only former Federal Republic of Germany) and in 1992 there were 3453. In 2015, there are only 2029.[15] However, the total carrying capacity of vessels has decreased less dramatically due to larger designs. Inland navigation traffic has developed comparatively positively. The transport volume decreased only slightly to 221.3 million tonnes (2016) compared to 221.4 million tonnes (2015).[16]

---

[10]See Ihde, 2001, pp. 149 f.

[11]See Ihde, 2001, p. 148; Vahrenkamp, 2007, p. 320.

[12]See WSV, 2017.

[13]See Federal Ministry of Transport, Building and Urban Affairs, 2008, p. 243.

[14]See Federal Ministry of Transport, 1994, p. 114.

[15]See WSV, 2016, p. 1.

[16]See Federal Statistical Office, 2017b, p. 2.

In Western Europe, inland navigation primarily means navigation on the Rhine. In 2015, around 80% of the transport performance on German inland waterways was generated in the Rhine region.[17] This is promoted by the Mannheim Act of 1868, according to which *navigation on the Rhine* is free for all Rhine riparian states. The dominance of the Rhine area also becomes clear when the goods transhipment of the *inland ports* is broken down by waterway area.[18] In terms of goods transhipment, the largest German inland ports are Duisburg, Cologne, Hamburg, Mannheim and Ludwigshafen. The Rhine navigation gained additional importance with the opening of the 171 km long *Main-Danube Canal* in 1992, which created a continuous European connection from the North Sea to the Black Sea.

The inland waterways still have considerable *capacity reserves*.[19] At present, inland navigation is primarily used for the transport of bulk goods. For the transport of the goods groups coal, petroleum and petroleum products, ores and metal waste, iron, steel and non-ferrous metals as well as stones and earths and agriculture, 47.9% of the transport performance of inland navigation in Germany is provided.[20] With total transport volumes remaining almost constant, container traffic on German waterways has increased in recent years from 748,600 TEU in 1995 to over 2.4 million TEU in 2015.[21] Inland navigation could therefore contribute to relieving the infrastructure of other modes of transport, in particular rail and road transport, primarily through the further expansion of container transport chains.

Air Traffic

As in the case of maritime transport, air transport infrastructure provides an important basis for international logistics. To facilitate cross-border air traffic, the *ICAO agreement*[22] was concluded at state level to guarantee the "five freedoms of the air". [23] These are transit, landing and transport rights of the participating countries in another contracting state. While a multilateral regulation of all five freedoms could be achieved for occasional traffic, there are only transit agreements for scheduled traffic, which allow the overflying of another state and landing for non-commercial purposes (technical freedoms). Commercial landing and transport rights (commercial freedoms) were regulated bilaterally, contrary to the original intention.

In air transport, the *INTERNATIONAL AIR TRANSPORT ASSOCIATION (IATA)* continues to play an important role as a regulatory factor, although its influence has

[17] See Bundesverband der Deutschen Binnenschifffahrt e.V., 2017.

[18] See Federal Statistical Office, 1999, p. 314.

[19] See Aberle, 1998, p. 110.

[20] See Bundesverband der Deutschen Binnenschifffahrt e.V.., 2017.

[21] See Bundesverband der Deutschen Binnenschifffahrt e.V., 2017.

[22] ICAO: INTERNATIONAL CIVIL AVIATION ORGANIZATION.

[23] See Ihde, 2001, pp. 181ff.

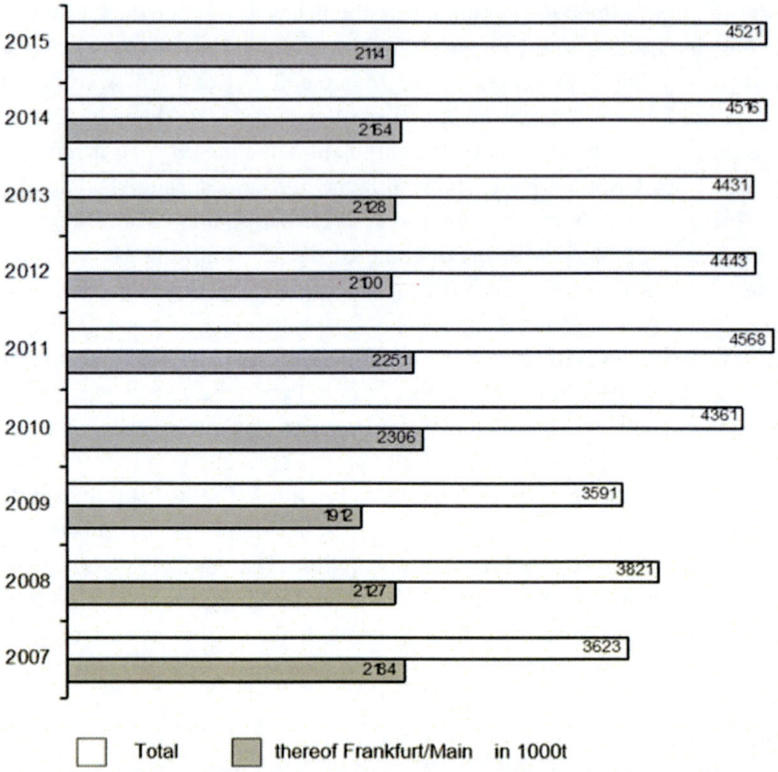

Fig. 19.1 Development of freight and mail volumes at German commercial airports (Source: Federal Ministry of Transport, Building and Urban Affairs, 2016, p. 91)

declined in the context of the deregulation of air transport. Unlike ICAO, the members of IATA are not the states, but the airlines. IATA is a price and conditions cartel for commercial air transport, dealing with territorial agreements, quotation, pooling, monetary settlement problems, organizational cooperation and standardization of documents.

Although, in principle, air traffic—analogous to maritime shipping—has unlimited space available for executing transports, the movements of commercial aviation are limited to the controlled routes. The logistical infrastructure of air transport is largely determined by the location of *airports* and their capacity. Although air freight accounts for a small share of total transport volume, it has significantly increased the load on German commercial airports. Figure 19.1 shows the 25% *increase in mail and freight* volumes at German commercial airports. In addition, the number of passengers also increased by 17% during the same period.[24] Accordingly, there are bottlenecks in the allocation of take-off and landing possibilities (slots).

[24]See Statista, 2017, pp. 14 f.

	D	GR	E	F	I	NL	GB	H	RO
Route length in 1.000 km	41,3	1,8	12,8	29,1	16,6	2,8	20,0	7,5	10,8
thereof electrified in %	55,7	4,6	58,9	50,4	69,4	73,4	25,1	35,2	36,9
Network density in km/1.000 km²	115,7	13,6	25,4	43,1	55,1	67,4	81,7	80,6	45,3

Fig. 19.2 Route lengths, network densities and quality characteristics of the railways of selected European countries in 2000 (Source: Federal Ministry of Transport, Building and Urban Affairs, 2008, p. 53; Bundesverband Güterkraftverkehr Logistik und Entsorgung, 2001, p.78)

The importance of an airport as a logistics hub depends, among other things, on the immediate airport surroundings, in which the necessary infrastructure must exist for the organization of pre- and post-carriage. In addition to this immediate airport hinterland, the catchment area of an airport is also important for the volume of traffic. This is comparable to the hinterland of seaports. The airports of the individual countries or economic regions are hierarchically related in that smaller airports provide feeder and distribution services for larger airports. For example, Frankfurt/Main airport is at the top of the hierarchy of European airports for freight distribution and, as a transit station, has an important hub or turntable function in international air freight traffic. Due to their technical facilities and organization, transit airports must be able to handle cargo in the shortest possible time. If, as a result of the increase in air traffic, the traditional major hubs become congested, this may lead to the development of secondary or bypass hubs.[25]

Rail Transport

In 2014, a *rail network of* 33,500 km operated by Deutsche Bahn AG and 5300 by other operators was available for rail traffic in the Federal Republic of Germany. However, only 25% of this rail network, the so-called operationally optimal network, handles over 75% of the traffic.[26] So while a large part of the rail network is hardly used for freight transport, other parts are fully utilized.

An important *quality feature* of the rail network is the proportion of electrified and multi-track lines. The lengths, network densities and quality characteristics of the railway lines of selected European countries are shown in Fig. 19.2. Compared to other European countries, the Federal Republic of Germany has a well-developed rail network. The network density is significantly higher than in other countries. However, Germany tends to occupy a middle position in terms of the proportion of electrified lines.

Different problems arise in the use of the rail network in wagonload traffic and in less than wagonload traffic (general cargo and express goods traffic). While the use of *wagonload traffic* depends heavily on the connection of consignors and consignees to the rail

[25] See Trumpfheller, 2006, pp. 33ff.

[26] See Ihde, 2001, p. 159.

network by sidings, *less than wagonload traffic* places particular demands on the stations serving as breakup and concentration points. Due to economic considerations, Deutsche Bahn AG has reduced the number of general cargo stations in the Federal Republic of Germany to less than 250.[27] This trend towards concentrating the handling of general cargo on a smaller number of hubs in the rail network requires increased cooperation between the railways and the road haulage industry in order to be able to guarantee surface service for general cargo transport.

General cargo traffic is to be handled via approx. 40 *freight centers* that can be served by direct connections. For this purpose, the entry from or the distribution to the area will be realized by increased truck connections. As the following remarks on the linking of the logistics infrastructure of different modes of transport will show, it is planned to increasingly integrate these freight centers, as well as the transhipment terminals of combined transport, into *freight transport centres.*

Road Traffic

The infrastructure of road freight transport is primarily formed by the *transport routes,* although, as will be shown, nodes are of great importance for linking up with the networks of other modes of transport. The lorry can make use of a much more closely meshed network of routes than the inland waterway vessel and the railway. This gives the road transport infrastructure a leading role in opening up the area.

In 1999, 230,700 km of *interurban roads* were available to trucks in the Federal Republic of Germany. In addition, there were approx. 413,000 km (as of 1992) of local roads. The *average network density of* interurban roads in 1999 was 0.646 km/km^2.[28] Figure 19.3 shows the lengths and densities of the road networks of selected European countries.

Compared with its European neighbours, Germany has a very well developed *freeway network.* Only smaller countries with a high population density (Netherlands, Belgium) have a much better network density. In less well-developed countries, on the other hand, the freeway infrastructure is consistently poor (e.g. Greece, Ireland, Romania). An important quality indicator is also the proportion of roads covered, which in Fig. 19.3 relates to the entire road network, i.e. including local roads.

However, the infrastructure equipment must be seen in relation to the volume of traffic—freight and passenger traffic. In 2007, for example, the *average daily traffic volume* (DTV) on the freeways of the old federal states was 49,200 motor vehicles.[29] However, this average value still understates the actual situation. This is shown by the following examples

[27] See Ihde, 2001, pp. 169f.; Vahrenkamp, 2007, pp. 304ff.

[28] Discrepancies between the total length of the interurban road network given and the network length resulting from Fig. 19.3 are due to the fact that the so-called district roads with a length of 91,100 km have also been taken into account here, See Federal Ministry of Transport, 2001, p. 111.

[29] See Federal Ministry of Transport, Building and Urban Affairs, 2008, p. 107. The measured quantity DTV indicates the number of passages in both directions at a counting point.

	D	GR	E	F	I	NL	GB	H	RO
Freeway in 1.000 km	12,4	0,9	12,0	10,8	6,6	2,5	3,5	0,5	0,2
Network density in km/1.000 km²	34,7	6,8	23,8	16,0	21,9	60,2	14,3	5,4	0,8
Federal highways in 1.000 km	50,0	10,2	24,9	25,2	46,0	6,7	46,7	30,5	14,8
Network density in km/1.000 km²	140,0	77,3	49,3	37,3	152,7	161,3	190,8	327,8	62,1
Regional roads in 1.000 km	178,1	30,9	139,7	365,0	119,9	57,5	114,7	53,3	36,0
Network density in km/1.000 km²	498,7	234,2	276,8	540,9	397,9	1.384,6	468,5	572,9	151,0
Share fortified roads of total road network in %	100	k. A.	99,0	100	100	k. A.	100	43,9	30,2

Fig. 19.3 Route lengths, network densities and quality characteristics of road transport routes in selected European countries 2005; GB, RO as of 2004; I, NL, E, H as of 2003 (Source: Bundesverband Güterkraftverkehr Logistik und Entsorgung, 2001, p. 78)

of DTV at freeway junctions: on the A3 Köln-Heumar junction 236,900 motor vehicles, including 29,035 HGVs; on the A5 Nordwestkreuz Frankfurt 251,800 motor vehicles, including 42,335 HGVs; on the A831 Stuttgart-Vaihingen 127,400 motor vehicles, including 4713 HGVs.[30]

Combination of Transport Networks

Traditionally, seaports, airports, inland ports and combined transport terminals have fulfilled the *function of linking the sub-networks of the various modes of transport*. A prime example of this is the seaport, which links the infrastructure of maritime shipping with that of its hinterland, i.e. inland waterways, railways and the road network. The road network plays a key role in the overall infrastructure, as its high network density enables it to guarantee accessibility to all companies and households in an economic area.

Freight transport centers (GVZ), as the main nodes of the macrological infrastructure, occupy a central place in the infrastructure discussion.[31] Essential elements of GVZs are

[30] See Federal Highway Research Institute, 2007, pp. 7ff.

[31] See Nobel, 2004; see Part IV, Sect. 16.4.

transhipment facilities, e.g. container terminals, which realize the system transition between the modes of transport. Ideally, all modes of transport should be brought together as far as possible, whereby the combination of road and rail is regarded as constitutive and the connection of inland shipping as desirable. In addition, the interface between local and long-distance transport is an essential feature of GVZs. Existing combined road/rail terminals, container terminals, rail freight centers, inland ports and seaports are therefore suitable "nuclei" for GVZs. However, a prerequisite for a GVZ is the existence of a sufficiently large hinterland, which guarantees a correspondingly large volume of cargo.

Pipeline Traffic

Pipeline transport also has a certain significance for the distribution of bulk goods. This primarily involves the transport of *liquids* and *gases* (water, crude oil and petroleum products as well as natural gas). The development of so-called *solid pipelines* is only in its infancy, whereby a distinction must be made between two directions of development. These pipelines are either designed as hydraulic systems in which solid transport goods are floated or they are pneumatic systems in which solid transport goods are moved in closed capsules. Larger growth rates in pipeline traffic can probably no longer be realized in liquid or gas transport, but only in solid transport.

19.2 Infrastructure of the Flow of Information

Trends in Logistics IT and Starting Points for Improvement

The *demands on the flow of information* in logistics are becoming ever greater as a result of the shift in the proportion of goods towards urgent goods that are transported in smaller quantities and also as a result of the global distribution of value-added activities. Despite the almost ubiquitous access to the Internet and the enormous increase in the transmission and processing power of IT systems, there are still some starting points for improvement. A central trend in logistics IT will therefore be to improve the integration capability for vertical and horizontal data exchange between companies.[32] In principle, there are three starting points for expanding a powerful, cross-company information infrastructure[33]:

- Standardization of data exchanged in the value chain.
- Improvement of the technical-organizational possibilities of information exchange between the institutions cooperating in the value chain.
- Security and legal bindingness of the exchanged data.

[32] See Albrecht, 2009, p. 44.

[33] Ihde, 2001, pp. 193ff.

A central prerequisite for the establishment of a cross-company data network is the *standardization of data*. Different country and industry-specific standards lead to similar, co-existing systems for information processing, which, however, do not allow any linkage across the country and industry borders. With the definition of the international, industry-independent EDIFACT standard, an attempt was made to create a uniform structure and format for all data required in the information flow. Due to the complexity of a cross-industry standard, various EDIFACT subsets were defined that are compatible with each other, such as EDIFOR for freight forwarders or EDITRANS for the transport industry. Additional requirements for the exchange of construction data and multimedia documents led to the development of further standards within EDIFACT, such as STEP for product data and ODA/ODIF for documents.

The standardization of data and documents facilitates the exchange of information between the information systems of the institutions cooperating in the value chain. However, in order to set up cross-company information and communication systems, *technical-organizational prerequisites* must be created that go beyond this and that, on the one hand, allow decentralized data collection, reproduction and processing (information availability) and, on the other hand, enable protected access to sensitive data (information confidentiality).[34] Protected here means that only those actors have access to the respective data who should have knowledge of it and that others are denied access. In particular, this is an essential prerequisite for greater integration of digital solutions (for example, through communication between machines, the application of artificial intelligence or fully autonomous logistics units) to improve inter-company communication and transparency about material flows.[35] Such technical-organizational prerequisites relate both to vertical communication in a transport chain from the consignor via the dispatching forwarder, carrier, receiving forwarder to the consignee and back, as well as to horizontal communication between several carriers or forwarders providing logistics services at the same production stage.[36]

Further considerable potential for improvement in IT systems for electronic data exchange exists in the area of *security and legal bindingness of the exchanged data*. In addition to the above-mentioned protection goals of availability and confidentiality in electronic data exchange, the aspects of information authenticity, integrity and bindingness play a decisive role in the handling of cross-company, electronic business processes.[37] Here, companies must use technical procedures to exclude the possibility of misuse in electronic data exchange. Hash and encryption procedures are a first approach, but they

[34] See Ihde, 2001, pp. 193ff. On information availability and confidentiality, See Eckert, 2007, pp. 8ff.

[35] See Pfohl et al., 2015, pp. 40ff.

[36] For an overview of horizontal and vertical communication systems, see Straube, 2004, pp. 144ff.; Mertens 2007, pp. 5ff.

[37] See Eckert 2007, pp. 6ff.

cannot secure authorship beyond doubt in the legal sense. These procedures, which are often incorrectly referred to as electronic signatures, therefore have no legal probative force in legal disputes. For this reason, the model of electronic signatures was developed which fulfils the formal legal requirements valid for paper form—these include document authenticity, conclusion function, visualisability, identity function, authenticity function, disputability, significance, immediacy and transparency. Complex cryptographic procedures are used here, which have all the above-mentioned properties and whose security depends only on a secret key (e.g. a 2048-bit long numerical code with certain properties). With this key, the owner can ensure exactly the required goals of information authenticity, integrity and bindingness. An important function in this model is played by electronic notaries, the so-called trust centers or trusted third parties. They are considered trustworthy by all participants and guarantee the identity of the key owners. In Germany, this function is performed by the Federal Network Agency, for example. The model of electronic signatures is also provided for in the EDIFACT standard.[38]

On the one hand, the approaches to improving *vertical communication* can be traced back to the initiative of individual logistics companies (e.g. Deutsche Bahn AG, seaports, airlines) and associations (e.g. Bundesverband für Spedition und Lagerei, Verband der Automobilindustrie, Verband Deutscher Maschinen- und Anlagenbau). On the other hand, there are pilot projects funded by the Federal Government and the European Union. At the same time, new information service providers are emerging with the participation of leading telecommunications and information technology providers offering open, cross-industry logistics information systems. In particular, the recent rapid and high-frequency development of innovative solutions for inter-company information exchange offers a large market and a variety of alternatives for companies to choose from.

One example is the EURO-LOG application, which provides an EDI communication system for the exchange of documents between the shipper, the logistics service providers and the recipient of the goods.[39] In addition, there is a shipment tracking system for intermodal transports[40] and a vehicle tracking system using digital mobile radio networks and satellite systems. These functions make it possible to reach the various systems of the logistics service providers via a general, cross-modal interface.

Considerations for improving *horizontal communication* focus on increasing the capacity utilization of the existing macrologistical infrastructure. For example, capacity utilization by weight in commercial road freight transport is only 66.6% on average, which

[38]EDIFACT is defined in the ISO 9735 standard. Part ISO 9735-9 describes the use of electronic signatures.

[39]See Gromball, 1992.

[40]At this point, the results of the research cooperation "CairGolution", which developed a module for the "tracking" of air freight loads, should also be mentioned. In principle, these can be used for all multimodal transport systems. A telematics module was developed here: See Pfohl et al., 2016, pp. 94ff.; Seidler et al., 2015, pp. 113ff.

indicates a large number of empty runs.[41] Improved information on available loading capacities and return loads could increase the utilization of existing capacities. The use of innovative IT platforms that control the direct transport of partial truck loads across freight forwarders would be helpful in this respect.[42] Electronic markets can significantly simplify the market transactions required for this. One example of this is Transpotel, a European freight and loading space exchange for road haulage. With the Docitel information system, the European railways also have a platform for making their loading space available via electronic systems.[43]

Overall, a development from centralized organizations to the *decentralized use of IT systems* can be observed in the area of information processing. However, this results in higher demands on the underlying communication infrastructure and technology via which the exchange or retrieval of information is carried out. Two decentralized types of IT infrastructure systems will therefore be discussed in more detail below: Cloud technology and Blockchain technology. After the respective explanation, their possible uses in logistics practice will also be discussed.

Cloud technology is a star-shaped IT infrastructure system. It allows member organizations to access computer resources via the Internet that are physically stored on a central database at any time. The respective organizational unit in the cloud network accesses this data either via a browser function or there is a direct connection between the "cloud" database and the company's internal IT applications.[44] A distinction is also made between the "Infrastructure-as-a-Service" (IaaS), the "Platform-as-a-Service" (PaaS) and the "Software-as-a-Service" (SaaS) model. While the IaaS model enables virtual access to hardware services, the PaaS model involves access to development and test platforms, for example for software.

The latter SaaS model enables the use of application software. The main advantages of all models are that, firstly, the service can be used on a consumption-oriented basis and, secondly, the hardware and software do not have to be maintained.[45] Thus, in star IT infrastructure systems, a central actor or an intermediary takes over the management and maintenance of the data flows. An example from order processing has already been described: the so-called "Supplier Information Management" systems (SIM).[46] This is a SaaS model in which a technology company is responsible for designing the cloud processes and software, for data security, and for data maintenance and completeness. A supplier can (e.g. after being approached by the technology company) centrally maintain its own master data (based on which the invoice is also paid after order fulfilment), to which

[41] See German Freight Forwarding and Logistics Association, 2005, p. 12.

[42] See Tummel, 2015, pp. 83ff.

[43] See Schmid, 1993, p. 473.

[44] See Bedner, 2013, p. 22.

[45] See Wang, 2016, pp. 2833ff.

[46] See Part II, Sect. 4.3.

the buyer company then has direct access. This eliminates information pathologies and increases the efficiency of order fulfillment processes.[47] In addition to order fulfillment, cloud solutions are also used for other purposes in inter-company relationships. For example, for the consolidation of transport data in a central supply chain unit in the network.[48]

Blockchain technology is an IT infrastructure system[49] developed by an anonymous person at the beginning of the twenty-first century and is often considered a disruptive technology innovation, as the role of central actors in the network is no longer needed.[50] Basically, it allows a secure and completely decentralized management of rights of availability. It is a distributive network that, unlike cloud technology, does not have a data focus. This is because the task of storing and updating data is shared by all network actors—no network actor can present a different accounting at a given time. Under a consensus mechanism, all actors document transactions in a predetermined time cycle. Because the existing, stored state of information in this network cannot be changed from a technological perspective, the blockchain brings together two previously incompatible advantages. There is no need for a central actor to aggregate the data and vouch for its accuracy. Nevertheless, the consensus mechanism enables a very high level of trustworthiness of the existing database, which is guaranteed by the network as a whole and not by a central actor.[51]

The following four possible applications of blockchain technology in logistics are discussed in the relevant literature and are partly implemented in operational practice[52]:

- *Replacement of EDI technology*: Instead of a purely bilateral information relationship, data could also be exchanged within an entire network. Data transfer is much more secure and cost-effective with blockchain technology. But questions about data security in particular (for sensitive information) will initially hold back the development of blockchain as a substitute for EDI technology.
- *Tracking and Tracing of Goods:* If, for example, a direct technological link between the flow of goods and information is established by using RFID technology and connected

[47] See Yahsi, 2017, pp. 171ff.

[48] A market analysis of existing supply chain software leads to the following list of vendors: PTC, Kinaxis, Tableau, Adexa, Intellicus, Llamasoft, River Logic, OM Part- ners, Synchron, Blue Ridge, Arena, Qlik, etc. These vendors develop solutions for connecting the flow of materials with the flow of information, for "Sales and Operations" (S&OP) projects, and for short- and long-term planning and decision-making processes.

[49] See Nakamoto, 2008.

[50] On blockchain technology and its potential contribution to the fourth industrial revolution, see Tapscott/Tapscott, 2016, pp. 8ff; Sundararajan, 2016, pp. 58ff.

[51] For a very detailed discussion of blockchain technology and how a transaction is created through the consensus mechanism, see Yahsi, 2017, pp. 161–171.

[52] For the possible applications in operational practice mentioned below, see Cecere, 2017, pp. 60–61.

to blockchain technology, then, on the one hand, the location and state of each individual good can be stored in the blockchain. On the other hand, so-called smart contracts (algorithms that decide on the basis of events in the blockchain infrastructure according to predefined rules, thus creating contracts) can decide on the next steps completely autonomously on the basis of location data.
- *Transparency about flows of goods*: If the data on incoming and outgoing goods is continuously documented in the blockchain by all network actors, then on the one hand there is greater transparency about the flows of goods themselves, but also about the actors involved in the value chain. A large automotive group at the end of a transport chain would in this way be able to gain an overview of its suppliers from the second to the last supplier level.
- *Cooperative financing of the supply chain (deep-tier financing)*: Blockchain technology, whose actual intention was to enable a financial system without central intermediaries, could be used for the management of financial flows in a value chain. On the one hand, financial values could be exchanged within a supply chain in a matter of seconds and with almost no transaction costs. In that case, this would even be possible without the involvement of a third party—for example a bank. On the other hand, it is possible to revolutionize the letter of credit financing process[53] as a trade financing option. This would, firstly, extremely reduce the administrative effort and, secondly, make it possible to finance suppliers across several stages of the value chain (with a regularly high refinancing interest rate) at the best possible interest rate in the supply chain.[54]

External Communications Infrastructure and Technology

In the macrologistical system, the development of the external communication infrastructure or technology is of primary interest, which in the B2B area serves communication between companies or enterprises and in the B2C area between companies and end consumers. While Deutsche Telekom used to dominate the market for communications infrastructure in the Federal Republic of Germany, there are now various companies offering different solutions for companies and also end consumers. The *development direction of communications technology* can be described by four interrelated trends:

- The growing availability of broadband mobile services enables communication from and to any location, independent of stationary connections.
- Networks with very high transmission speeds allow the transmission of large amounts of data and the operation of multimedia applications via the Internet.

[53] See international logistics (Fig. 20.9).

[54] For an example of the use of blockchain in the context of a supply chain finance concept, see Yahsi, 2017, pp. 297ff. See on deep-tier financing Pfohl/Yahsi (2017).

- International standards and the interconnection of national networks reduce the importance of national borders. This is particularly evident in the Internet, which today is a powerful and cost-effective network for data transfer that knows no national borders.
- Value-added services and electronic services are based on public or private networks and extend these with additional service offerings. Examples are the assumption of payment processing by clearing houses or information offers by online services.

Some important networks and services for data communication are explained below[55]:
Nets:

- *Public networks*: Form of data exchange based on publicly recognized, usually international standards, usually the Internet.
- *Private networks*: Private networks are limited to one or a few participating organizations. They are generally local networks located in a building or closed area (e.g. company premises). Virtual private networks (VPN), which use the infrastructure of public networks and prevent access to the transmitted data through encryption, are used to connect several locations.
- *Radio networks*: In the case of radio networks, a distinction must be made between, on the one hand, the national GSM (Global System for Mobile Communications) and UMTS (Universal Mobile Telecommunications System) networks of the mobile phone providers and, on the other hand, the local radio networks of companies and private individuals based on the WLAN (Wireless Local Area Network) standards. GSM and UMTS are second- and third-generation mobile communications networks that comply with uniform global standards. UMTS, for example, enables transmission speeds of up to 7.2 Mbit/s using the HSDPA (High Speed Downlink Packet Access) method. Wireless networks according to the WLAN standard IEEE 802.11, on the other hand, can be set up at any location with the desired expansion and, with transmission rates of over 100 Mbit/s, represent a very good infrastructure for private networks. In addition, there are some special radio networks that are used, for example, to bridge long distances (e.g. WIMAX, GNSS) or for decentralized communication between technical devices (e.g. Bluetooth).[56]
- *Wired networks*: The most common standard for wired networks connecting to the Internet is DSL (Digital Subscriber Line). DSL refers to a set of standards that utilize unused frequencies in telephone lines for high-speed data transmission, providing a simple means of broadband Internet connectivity. Depending on the standard, transmission rates of up to 210 Mbit/s can be achieved on the basis of DSL (VDSL). In addition

[55] Network refers to the interconnection of electronic systems that form the infrastructure for the provision of (network) services. Services make it possible to transport information of a certain type via such networks. On data networks See Schneider/Werner 2007, pp. 308ff.

[56] See Schenk, 2015, pp. 263ff.

to DSL, there is also the possibility of being connected to the Internet via TV cable or the power grid. Companies also have the option of renting their own fiber optic lines for very high transmission rates.

Communication Services:

- *Telephone services*: The separation of system logic and service in the intelligent network enables flexible tariffing, location- and time-dependent traffic routing to target connections, and automatic call distribution and rerouting. Examples include Service 130, Service 190 for private information providers, Telebox, Teleconference, Televotum and Teledialog.
- *Fax (telecopier)*: Copies are not made at the location of the original, but at geographically distant recipients. Terminal devices are special fax machines or computers equipped with fax cards. Fax is a purely analogue process which, due to rapid technological development, continues to lose importance in business practice.
- *Mobile satellite systems*: Mobile satellite networks are mobile radio networks for voice and data communication that can be received worldwide (with the exception of certain regions, depending on the network). Mobile satellite radio systems are mostly installed on ships, aircraft and trucks. There are also mobile devices, but they differ significantly in size and weight from ordinary mobile phones. These devices do not require base stations for communication, but communicate directly with each other via satellites orbiting the earth or with connections in normal telephone networks.
- *Satellite navigation systems*: Besides satellite communication, satellite navigation (e.g. GPS—Global Positioning System and GLONASS—Global Navigation Satellite System) has the greater importance. With suitable navigation receivers, it is possible to determine the location of mobile units with an accuracy of about 10 m. If satellites are also supported by differential GPS correction data (DGPS) or an odometry measurement, the accuracy of the location can be increased to up to 0.5 m.[57] In conjunction with map material stored on data carriers, navigation systems are widely used to support road traffic. On ships and in airplanes these systems have been standard for a long time. The combination of different systems mentioned here, e.g. mobile radio and mobile satellite radio for continuous communication as well as satellite navigation for location determination, enables global tracking and tracing of the transport units equipped with them, e.g. on the basis of transport containers or means of transport.
- *Terrestrial tracking systems*: Terrestrial positioning systems use stationary reading stations that read features for identification on logistical objects and forward them to (usually) central IT systems. The higher the density of the installation of reading stations, the higher the locating accuracy. Therefore, localization with centimeter accuracy can only be achieved if up to a thousand measurements are taken per second.

[57] See Schenk, 2015, p. 266.

19.3 Transport Policy

Transport infrastructure is of paramount importance to an economy. Although it cannot necessarily be assumed that the infrastructure must grow at the same rate as the economy,[58] an inadequate transport network would hinder economic development. Particularly in view of an increasing division of labor in economic life, which usually entails the transport of goods between the companies involved, transport infrastructure and the conditions and costs under which it can be used are becoming increasingly important.[59] At the same time, transport places a burden on the environment, so that both the extent and the nature of the use of transport networks need to be managed in terms of environmental impact.

Development of the Infrastructure Load

The macrologistical infrastructure of the Federal Republic of Germany shows bottlenecks. There has long been talk of a *transport infrastructure crisis.*[60] There have even been warnings of "traffic gridlock" and, even worse, "traffic collapse".[61] If one looks at the development of traffic to date and the available forecasts, these fears are understandable. This is most obvious in the case of road traffic.

In the case of *road freight transport*, the transport performance in long-distance road freight transport has been increasing continuously since 1945. For example, it rose from 80 billion tkm (1980) to 140.0 billion tkm (1992, old federal states only).[62] In 2014, it was 172.5 billion tkm.[63] Short-distance road freight transport has also grown strongly. It grew by about 20% between 1980 and 1992 alone. In addition, the road infrastructure is burdened by passenger car traffic. The stock of passenger cars increased from 42.3 million (1999) to 43.8 million (2014).[64] This resulted in an increase in total mileage of about 13% for the period 1999–2014, with little change in average mileage [km] per passenger car per year over this period.[65] Between 1999 and 2014, the average mileage [km] per passenger car of all passenger cars registered in Germany increased by 0.6%.[66] This indicates that the increase in mileage per passenger car has declined in recent years. The main causes of the problems in road traffic are the local and temporal (rush hour) peak loads already mentioned in Sect. 2.1.

---

[58] See Rommerskirchen, 1999, pp. 231 f.; Baum, 1995, pp. 14 ff.

[59] See Merath, 1995, p. 284.

[60] See Hamm, 1987, pp. 423ff; Wissenschaftlicher Beirat beim Bundesministerium für Verkehr, 1987, pp. 136ff; Willeke, 1989.

[61] See Arnold et al., 2008, p. 1017.

[62] See Federal Ministry of Transport, 1994, pp. 228 f.

[63] See Federal Ministry of Transport, Building and Urban Affairs, 2016, pp. 244 f.

[64] See Federal Ministry of Transport, Building and Urban Affairs, 2016, pp. 132 f.

[65] See Federal Ministry of Transport, Building and Urban Affairs, 2016, pp. 132 f.

[66] See Federal Ministry of Transport, Building and Urban Affairs, 2016, pp. 152 f.

In the case of *rail freight transport*, the transport performance increased by about 18% from 1999 to 2014. The transport performance in passenger transport has increased steadily in the period from 1998 to 2014—about 40%. The number of passengers has also increased in recent years.

In the case of *air transport*, both passenger transport performance and the number of passengers almost doubled between 1998 and 2014. In contrast to rail transport, the absolute growth per year was almost constant during this period.[67] At the same time, the volume of air freight has also increased by around 123% in the same period.[68]

Real *gross fixed capital spending* in transport infrastructure increased only slightly. For example, gross fixed capital spending in roads and bridges at 1998 prices decreased from EUR 10.8 billion to EUR 11.7 billion (2014). Gross fixed capital spending in railways at 1998 prices increased from EUR 5.1 billion to EUR 6.6 billion (2014).[69]

The development shown is also reflected in the *forecasts of future traffic volumes.*[70] For the *Federal Transport Infrastructure Plan 2030* (BVWP), the development of traffic up to the year 2030 was estimated. These estimates were regularly revised and the plans adjusted to the new forecasts but also to changes in transport policy objectives. The BVWP 2030 pursues the goal of seeing the task of transport policy as strengthening the transport system as a whole. Its focal points are:

- "Enabling mobility in passenger traffic,
- Securing the supply of goods, increasing the competitiveness of companies,
- Increasing traffic safety,
- Reduction of emissions of pollutants and greenhouse gases,
- Limitation of the use of nature and landscape,
- Improving the quality of life, including the noise situation in regions and cities."[71]

The *reasons for the development outlined above* and thus for the infrastructural bottleneck situation are manifold. On the one hand, the demand for transport services has increased with the general growth of the economy, although the transport elasticity in Germany has become less than 1 since the mid-1970s. Furthermore, the change in the structure of goods and the increase in qualitative demands on transport services have led to a shift in the modal split in favor of road transport. Finally, undoubtedly misguided regulatory developments also play a role, leading, for example, to a high proportion of

---

[67] The years 2001 and 2002 are an exception, when air traffic declined due to the bursting of the dotcom bubble and the SARS epidemic. Thereafter, the almost linear trend continued. See German Aerospace Center (DLR), 2008, pp. 5ff.

[68] See Federal Ministry of Transport, Building and Urban Affairs, 2016, pp. 90 f.

[69] See Federal Ministry of Transport, Building and Urban Affairs, 2016, p. 22.

[70] See Federal Ministry of Transport and Digital Infrastructure, 2016.

[71] Federal Ministry of Transport and Digital Infrastructure, 2016, pp. 5ff.

empty runs in works transport. Another major cause is the increasing internationalisation that has accompanied the realization of the European internal market and the opening up of Central and Eastern Europe. Particularly in the case of cross-border freight transport, there have been very significant shifts in the past in favor of road transport.

Infrastructure Policy

The field of infrastructure policy can be differentiated into supply-oriented and demand-oriented policies. The *supply policy* aims at the sufficient supply of logistical infrastructure. This is to be achieved in particular by creating new transport routes and transhipment facilities. The object of a transport *demand policy* is to influence transport demand in order to achieve a relief of the existing infrastructure.

Based on the forecasts shown and with the help of an assessment methodology that takes into account economic, ecological and urban development criteria, an *urgent need for infrastructure and connectivity* was identified in the Freight Transport and Logistics Master Plan. In addition to new projects that have been assessed as particularly beneficial, this also includes projects that were already prioritised in the BVWP 2030 and have not yet been implemented or have only been partially implemented.[72] The Federal Transport Infrastructure Plan estimates an investment volume of EUR 264.5 billion for the railways, federal major roads and federal waterways by the end of 2030. Of this, about 40.3% is for the rail network and 55.4% for the federal major roads. Of this, only EUR 25.2 billion is earmarked for new projects. The lion's share is earmarked for ongoing and firmly scheduled projects.[73]

One of the main objectives is to link the modes of transport into an integrated transport system, which is why combined transport will continue to be strongly promoted. The seaports, inland ports and airports play a key role in an integrated transport system as ideal interfaces for linking the modes of transport. One example of this is the hinterland connections of the German seaports.

In the future, however, it is to be expected that the mostly nationally oriented infrastructure measures will increasingly give way to international projects. This will be driven above all by the advancing process of European integration. Cross-border passenger and freight traffic between the participating EU states is growing. In addition, the integration of Central and Eastern European states into the European Union and the increasing economic and political opening towards these countries requires an improvement of the transport infrastructure in the border areas, at the border crossings and also in these countries.[74]

Today, road and rail already form a well-functioning backbone for the development of the eastern German economy. In the BVWP 2030, the construction of local bypasses for the reconstruction of the east is more in focus.

[72] See Federal Ministry of Transport and Digital Infrastructure, 2016, p. 6.

[73] See Federal Ministry of Transport and Digital Infrastructure, 2016, pp. 1ff.

[74] See Aurbach, 1999, pp. 21ff.

In addition to the expansion of infrastructure, *fiscal policy measures to reduce and shift traffic*—as essential components of transport demand policy—are of particular importance. One possibility here is to make certain transports more expensive by means of appropriate taxes and levies, e.g. the so-called eco-tax on petrol and diesel. A prerequisite for this, however, is that the demand for transport has a sufficiently high price elasticity.[75] In addition to the existing taxes—mineral oil tax and motor vehicle tax—in Germany, for example, German and foreign operators of trucks with a gross vehicle weight of at least 7.5 t, which use the federal motorways and part of the federal roads, have had to pay a certain user charge since 31 August 2003 and from 1 July 2018 for all federal roads (the range of toll rates is between 8.1 and 21.8 euro cents).[76] Trucks are divided into six toll categories, and the amount of the toll depends on the number of axles of the vehicles and the pollutant class. Truck tolls are also considered a transport infrastructure financing measure.[77]

References

Aberle G (1998) Verkehrsinfrastrukturpolitik und deren Auswirkung auf die Unternehmenslogistik. In: Isermann H (Hrsg) Logistik. Gestaltung von Logistiksystemen. 2., überarb. und erw. Aufl. Landsberg a. L., S. 109-124

Albrecht W (2009) Von der Logistik bewegt. In: Logistik inside, Jg. 8, Nr. 1, S. 42 – 45

Arnold D u.a. (Hrsg) (2008) Handbuch Logistik. 3. neu bearb. Aufl. Berlin Heidelberg

Aurbach G (1999) Europäische Verkehrspolitik mit Blickrichtung Mittel- und Osteuropa. In: Faller P (Hrsg) Transportwirtschaft im Umbruch: Strukturwandel, Anpassungserfordernisse, Gestaltungsaufgaben. Wien, S. 21-34

Baum H (1995) Entkopplung von Verkehrswachstum und Wirtschaftsentwicklung. In: Zeitschrift für Verkehrswissenschaft 66, S. 13-32

Bedner M (2013) Cloud Computing - Technik, Sicherheit und rechtliche Gestaltung. Zugl.: Kassel, Univ., Diss., 2012. Kassel: Kassel University Press

Bundesanstalt für Straßenwesen (2007) Manuelle Straßenverkehrszählung 2005. Ergebnisse auf Bundesautobahnen. http://www.bast.de/nn_39112/DE/Statistik/Verkehrsdaten/Downloads/zaehlung-2005-BAB-strassen,templateId=raw,property=publicationFile.pdf/zaehlung-2005-BAB-strassen.pdf. Zugriff am: 31.03.2009

Bundesministerium für Verkehr (Hrsg) (1994) Verkehr in Zahlen 1994. Bonn

Bundesministerium für Verkehr (Hrsg) (2001) Verkehr in Zahlen 2001/2002. Hamburg

Bundesministerium für Verkehr, Bau und Stadtentwicklung (BMVBS) (2016) Verkehr in Zahlen 2016/2017. Berlin

Bundesministerium für Verkehr und digitale Infrastruktur (2016) Bundesverkehrswegeplan 2030. Berlin

Bundesverband der Deutschen Binnenschifffahrt e.V. (BDB) (2017) Daten & Fakten 2016/2017. Duisburg

[75] See Strese, 1994, p. 200.

[76] See Toll Collect, 2017.

[77] See Haase, 2005, pp. 163 f.

Bundesverband Güterkraftverkehr Logistik und Entsorgung (BGL) (Hrsg) (2001) Verkehrswirtschaftliche Zahlen 2005+2006. Frankfurt am Main

Cecere L (2017) Seven use cases for hyperledger in the supply chain. CSCMP's Supply Chain Quarterly, S. 60-61

Deutscher Speditions- und Logistikverband e. V. (DSLV) (2005) Zahlen, Daten, Fakten aus Spedition und Logistik. Bonn

Deutsches Zentrum für Luft- und Raumfahrt e. V. (DLR) (2008) Luftverkehrsbericht 2008. Daten und Kommentierungen des deutschen und weltweiten Luftverkehrs. Köln

Eckert C (2007) IT-Sicherheit. Konzept – Verfahren – Protokolle. 5., überarb. Aufl., München u.a.

Gromball P (1992) EURO-LOG: Nutzung neuer Informationstechniken für die umweltgerechte Steuerung des europäischen Warenflusses. In: Informationstechnik 34 3, S. 168-176

Haase R (2005) Verkehrsinfrastruktur in Deutschland und ihre Finanzierung – Verkehrsbeherrschung durch nutzerfinanzierte Verkehrswege. In: Stopka U, Pällmann W (Hrsg.) Für eine neue deutsche Verkehrspolitik. Mobilität braucht Kommunikation. Hamburg

Hamm W (1987) Strukturwandel im Güterverkehr als Grund für verkehrspolitischen Handlungsbedarf. In: Internationales Verkehrswesen 39 6, S. 422-424

Ihde G B (2001) Transport, Verkehr, Logistik. Gesamtwirtschaftliche Aspekte und einzelwirtschaftliche Handhabung. 3., völlig überarb. u. erw. Aufl. München

Merath F (1995) Verkehrswege als Einsatzfaktor effizienter Produktion: Zum Zusammenhang zwischen Produktionsverlagerung und verkehrlichen Wirkungen. In: Zeitschrift für Verkehrswissenschaft 66 4, S. 279-290

Mertens P (2007) Integrierte Informationsverarbeitung 1. Operative Systeme in der Industrie. 16., überarb. Aufl., Wiesbaden

Nakamoto S (2008) Bitcoin - A peer-to-peer electronic cash system. Online verfügbar: https://bitcoin.org/en/bitcoin-paper

Nobel T (2004) Entwicklung der Güterverkehrszentren in Deutschland - eine am methodischen Instrument Benchmarking orientierte Untersuchung. Bremen

Pfohl H-Chr, Yahsi B (2017) The Impact of Deep-Financing on Supply-Chain- Competitiveness. Proceedings of the 3rd International Conference on Production, Logistics and Transportation

Pfohl H-Chr, Yahsi B, Kurnaz T (2015) The Impact of Industry 4.0 on the Supply Chain. Innovations and Strategies for Logistics and Supply Chains. In: Kersten W, Blecker T, Ringle C M (Hrsg) Proceedings of the Hamburg International Conference on Logistics, S. 31-58

Pfohl H-Chr, Kurnaz T, Yahsi B (2016) Entwicklung eines risikobasierten Funktions- und Prozessmodells für eine hybride Sicherheitsdienstleistung zur Erhöhung der Sicherheit in der Luftfrachtlogistikkette. In: Jahrbuch der Logistik 2016. Wuppertal, S. 94-98

Rommerskirchen S (1999) Entkopplung des Wachstums von Wirtschaft und Verkehr? In: Internationales Verkehrswesen 51 6, S. 131-136

Schenk M (Hrsg.) (2015)Produktion und Logistik mit Zukunft. Digital Engineering and Operation. Berlin, Heidelberg

Schieck A (2008) Internationale Logistik. Objekte, Prozesse und Infrastrukturen grenzüberschreitender Güterströme. München u.a.

Schmid B (1993) Elektronische Märkte. In: Wirtschaftsinformatik 35 5, S. 465-480

Schneider U, Werner D (2007) Taschenbuch der Informatik. 5., neu bearb. Aufl., München u.a.

Seidler T, Kurnaz T, Pfohl H-Chr (2015) CairGoLution - Steigerung der Luftfrachtsicherheit durch die Erkennung von Integritätsverletzungen In: 20. Magdeburger Logistiktage "Sichere und nachhaltige Logistik" S. 113-120

Statista (2017) Passagiere auf deutschen Flughäfen. https://de.statista.com/statistik/daten/studie/77928/umfrage/passagiere-aufdeutschen-flughaefen/. Zugriff am 06.09.0217

Statistisches Bundesamt (Hrsg) (1999) Statistisches Jahrbuch 1999 für die Bundesrepublik Deutschland. Wiesbaden

Statistisches Bundesamt (2017a) Pressemitteilung vom 5. April 2017 – 115/17, S. 2

Statistisches Bundesamt (2017b) Pressemitteilung vom 30. März 2017 – 110/17, S. 2

Straube F (2004) e-Logistik. Ganzheitliches Logistikmanagement. Berlin u.a.

Strese D (1994) Dämpfung des Verkehrswachstums durch Transportverteuerungen? In: Internationales Verkehrswesen 46 4, S. 199-203

Sundararajan A (2016) The sharing economy - The end of employment and the rise of crowd-based capital-ism. Massachusetts, MIT Press

Tapscott D, Tapscott A (2016) Blockchain revolution - How the technology behind bitcoin is changing money, business and the world. New York

Toll Collect (2017) Maut-Tarife. https://www.tollcollect.de/de/toll_collect/bezahlen/maut_tarife/maut_tarife.html. Zugriff am 13.09.2017

Trumpfheller M (2006) Strategisches Flughafenmanagement: Positionierung und Wertschöpfung von Flughafentypen. 1. Aufl., Wiesbaden

Tummel Chr (2015) IT-Kooperationsplattform für speditionsübergreifende Direktverkehre von Lkw-Teilladungen. Zugl.: Aachen, Techn. Hochsch., Diss., 2015. Cuvillier: Göttingen

Vahrenkamp R (2007) Logistik. Management und Strategien. 6., überarb. und erw. Aufl. München

Wang B u.a. (2016) SaaS-based enterprise application integration approach and case study. In: The Journal of Supercomputing, Jg. 72, Heft Nr. 7, S. 2833-2847

Willeke R (1989) Infrastrukturkrise des Verkehrs. In: Informationen aus dem Institut für Verkehrswissenschaft an der Universität zu Köln 24 1/2, S. 1-3

Wissenschaftlicher Beirat beim Bundesministerium für Verkehr – Gruppe Verkehrswirtschaft (1987) Verkehrsinfrastruktur als Voraussetzung für die gesellschaftliche und wirtschaftliche Entwicklung in der Bundesrepublik Deutschland. In: Zeitschrift für Verkehrswirtschaft 58 3, S. 131-153

WSV (2016) Entwicklung der deutschen Binnenflotte. Bonn

WSV (2017) Wasserstraßen. https://www.wsv.de/wasserstrassen/ Zugriff am 06.09.2017

Yahsi B (2017) Financial Supply Chain Management. Erfolgsfaktoren der Gestaltung von Finanznetzwerken. Zugl.: Darmstadt, Technische Universität, Diss., 2017. Universitäts- und Landesbibliothek Darmstadt: Darmstadt

20 International Logistics Systems

20.1 Special Features of International Logistics

Basics

With today's ever-increasing *global economic interconnection*, the importance of international logistics is growing, the special features of which result from the fact that the points of delivery and receipt of goods are located in different countries. It is therefore a matter of planning, implementing and controlling cross-border flows of information and goods. The special features compared to national logistics systems can occur very differently in individual cases in the logistics subsystems of order processing, inventory management, warehousing, packaging and transport. The theoretical basis for dealing with the special features of international logistics must necessarily be taken from the fundamentals of logistics conception on the one hand and the fundamentals of international management on the other.

The systems thinking characteristic of the *logistics conception* as well as the resulting total cost thinking and service thinking apply in principle to the solution of both national and international logistics problems. International logistics systems are only *more complex* in the sense that the framework conditions (the surroundings) for logistics processes (e.g. transport, storage) can differ from country to country.[1] When analyzing cost interdependencies, not only the functional logistics costs caused in the logistics subsystems but also the costs caused by political and economic barriers to trade must be taken into account. Also, the delivery service requirements to be provided by the logistics system often vary from country to country.

For *international management*, which deals with the cross-border activities of companies designed for the long term, three main conceptual research areas can be

[1] See Stock/Lambert, 2001, pp. 551ff.

H.-C. Pfohl, *Logistics Systems*,
https://doi.org/10.1007/978-3-662-64349-5_20

named.[2] *Foreign management*—(the firm has a foreign division)—deals with the economic, political-legal, and social peculiarities that are important for doing business in foreign markets. Many of these peculiarities concern the flows of information and goods in logistics systems. However, joint concepts for the exploitation of synergy effects refer at best to measures and programs in the field of marketing policy instruments. The focus of international management in the sense of *multinational management*—(the enterprise is multinational)—is the design of enterprises present in several countries with direct investments. Despite the geographical dispersion of corporate activities, they are to be aligned in a largely uniform manner in order to exploit synergy effects. On the one hand, standardization and centralization postulates indirectly affect the logistics systems, e.g. via a production network, since only they make such a network possible. On the other hand, these postulates affect the logistics systems directly, e.g. in the form of the demand for a central warehouse for distribution in all Western European countries. Finally, for international management in the sense of *global management*—(cosmocorporation)—nationality is no longer a decision criterion. In order to exploit synergy effects and comparative advantages to the greatest possible extent, the aim is to work on transnational common market segments of the world market. For them, a uniform marketing mix—which also includes the delivery service as a marketing instrument—is designed and procurement and production are carried out in a network. All logistics systems are thus affected by global management decisions.

Framework Conditions (Surroundings) for Logistics Processes

The framework conditions for the design of international logistics systems, in contrast to the design of national logistics systems, can be broken down into general and country-specific framework conditions. The *general framework conditions* that characterize cross-border logistics processes primarily include transport distances, means of transport, institutions, documents and information[3]:

Transport distances: The distances to be covered by logistics systems are greater, which, ceteris paribus, results in longer delivery or replenishment times, greater unreliability both in forecasting demand and in meeting replenishment times, and larger inventories. Logistical errors can thus affect both service and costs to a greater degree than in national logistics systems.

Means of transport: In international logistics systems, transport is inevitably often organized as broken intermodal transport, i.e. with the aid of the use of different means of transport. Leaving aside possible land bridges, maritime and air transport are available for intercontinental connections, while inland waterways, road and rail transport are also

[2]See Macharzina/Engelhardt, 1987, pp. 322ff. On internationalisation strategies, see Holtbrügge/Welge, 2010, pp. 93ff.

[3]See Nelson/Toledano, 1978, pp. 2ff.; Bowersox/Sterling, 1982, pp. 20 ff.; pp. 390ff.; Bender, 1985b, pp. 782ff.; Wood, 1989, pp. 101ff. Specifically on international procurement logistics, See Pfohl/Large, 1991, pp. 23ff.; Piontek, 1994, pp. 123ff.

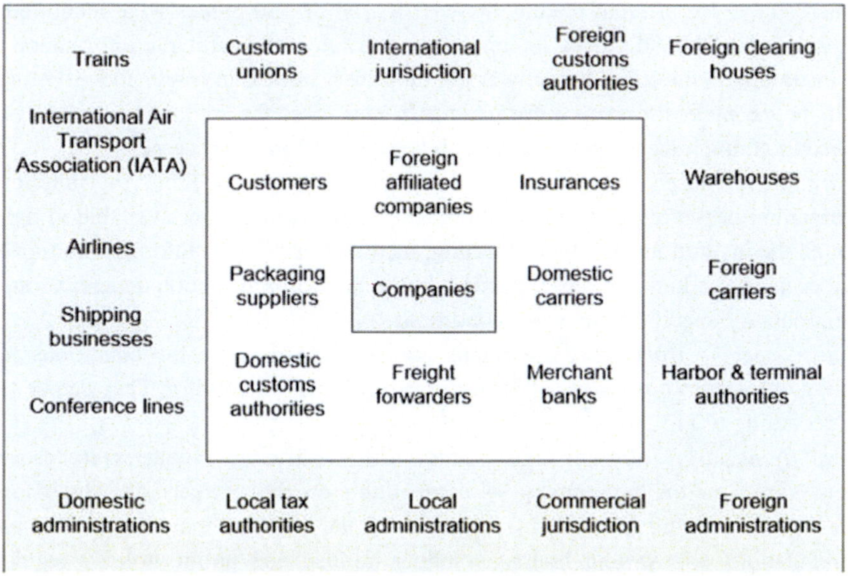

Fig. 20.1 Institutions in international logistics processes (Source: based on Slater, 1980, p. 162)

available for continental connections. In order to set up international transport chains, it is therefore necessary to combine means of transport with very different technical characteristics. Due to their different technical characteristics, the means of transport also have different cost structures (proportion of fixed and variable costs; proportion of capital costs and personnel costs) and performance capabilities (speed, network formation capability, etc.). Knowledge of these is an indispensable prerequisite for a rational decision when purchasing transport services, in particular also for assessing the respective prices.

Institutions: In general, several institutions are involved in the planning, implementation and control of international logistics processes. This applies both to the physical flow of goods and to the flow of information that overlaps it. In addition, in the case of international procurement or distribution, the monetary flow is often closely related to the logistical flow of information and goods, since payment depends, for example, on the presentation of documents accompanying the goods. Figure 20.1 gives an impression of the diversity of institutions.[4] The institutions involved in the procurement or distribution channels may differ from country to country. Clearly, international logistics systems require more coordination, more communication and more control.

Documents: The diversity of institutions involved in international logistics processes results in an even greater diversity of documents exchanged between institutions.[5] In a

[4] See Stock/Lambert, 2001, p. 521.

[5] See Stock/Lambert, 2001, pp. 538 f.

Canadian study, 46 different documents with a total of 360 copies were identified for a typical shipment. The individual documents are tailored to the particular information needs of the institution in question. In many cases, they are historical in origin and differ in form, even if they contain the same information. In any case, the multiplicity of documents complicates communication and leads to higher costs of order processing.[6]

Information: Closely related to the documents is the collection, transmission and transformation of the required information. This information is not available in the same way in all the institutions involved. Missing, incorrect or delayed information impairs the service or increases logistics costs. Barriers in the flow of information depend strongly on different country-specific framework conditions.

Country-specific framework conditions are understood to be the conditions for the execution of logistics processes, which differ from country to country. They can be broken down according to Fig. 20.2.[7]

Legal framework: Different legal frameworks in different countries are caused by different views on the functioning of competition on the markets on which logistics services are offered and demanded (such legal frameworks also include, for example, the legal regulations on working hours, including holiday and public holiday regulations, which influence the time of use of operating resources). Accordingly, regulatory views range from liberal ideas to dirigiste regimentation. In order to facilitate the exchange of goods, bilateral and multilateral agreements aimed at harmonizing national regulations are therefore being concluded alongside national laws.

Administrative framework: They characterize the handling of legal regulations and the organisation of processes in cross-border flows of information and goods and are the result of well-established administrative structures and procedures in the various countries. For this reason, even the same legal regulations can have very different effects on the flow of information and goods because they are handled differently in different countries.

Technical framework: These are also partly the result of different legal regulations, in particular on permissible dimensions and weights of the equipment that can be used. However, they can also be based on differences in the investment policies of the institutions involved in the logistics processes. Examples of this are the different quality of the vehicle fleets of transport companies and the different IT equipment of haulage companies in different countries. In the case of railways, different power systems, track gauges and loading gauge profiles have grown over time.

Infrastructural and geographical framework: First of all, they result from the topographical conditions of the different countries. Furthermore, they are the result of national transport route and communication network policies. Only in more recent times, in the context of larger economic areas such as the EU, has infrastructure become more multinationally oriented. Particular importance is attached to the infrastructural conditions

[6]On cost differences in international logistics, see Schieck, 2008, pp. 71 f.

[7]See Pfohl/Large, 1991, pp. 24ff.; Dülfer/Jöstingmeier, 2008, pp. 213ff.

Ancillary constraints	International road freight traffic e.g.	International rail freight traffic e.g.	International inland freight shipping e.g.
Legal ancillary constraints	• cabotage quotas according to EC-decree • community licenses for cross-border traffic according to EC-decree	• access regulations according to EC-guideline • simplified community transit according to EC-decree	• Mannheim Convention
Administrative ancillary constraints	• control of transported fuel, transport authorization, national freight transport authorization and national social regulations • inspection of goods • working time regulations at customs offices • staffing of customs offices	• change of the freight waggons, motor vehicles and accompanying staff at the borders • inspection of documents and goods • technical inspection of waggons • primacy of certain transports	• checking of the amount of fuel • approval test of ships for the transport of hazardous goods • inspection of material and on-board supplies • working time regulations
Technical ancillary constraints	• permitted dimensions • permitted axle and wheel loads and total weights	• loading gauges • gauge of the tracks • permitted accesses	• permitted dimensions • permitted total weights
Infrastructural and geographical ancillary constraints	• capacities on lorry access and exit routes of border crossings • international main roads • transit duty customs offices in international freight traffic	• capacities in the railway network – alp transits – island positions • marshalling capacities at border stations	• width and depth of waterways • lock capacities
Cultural ancillary constraints	• different languages • value systems	• different languages • holiday regulations	• different languages • value systems

Fig. 20.2 Country-specific framework conditions for international logistics processes with examples of road, rail and inland waterway freight transport (Source: based on Zettelmeyer/Zöllner, 1986, p. 56)

at border crossings, where limited handling capacities hinder international logistics processes.

Cultural framework: The different lifestyles, mentalities, educational systems and languages in different countries are a major barrier to the organization of international logistics systems. Not only does this make communication more difficult, but the different performance and willingness of logistics personnel, caused by different training, living and value standards in different countries, mean that smoothly functioning logistics systems in one country cannot easily be adopted in other countries.

Costs

The different country-specific framework conditions are the reason for the differences in the *share of logistics costs* in the turnover of a product and in their *composition* in different countries. The most comprehensive current comparison of logistics costs in Europe is probably provided by the Top 100 study by Klaus and Kille.[8] For various European countries, it shows the different distribution of logistics costs among the activity-specific subsystems of logistics.

Data from the Davis surveys are also frequently cited in relevant publications. These show that in 2007 logistics costs as a proportion of turnover were higher in the USA (9.74%) than in EU countries (8.39%).[9] However, these results should not be interpreted to mean that companies in certain countries are particularly efficient in their logistics processes. The different values also reflect, above all, different geographic, demographic and cultural *conditions*, the latter of which certainly includes the degree of implementation of the logistics concept.

The costs attributable to cross-border flows of information and goods can be broken down into functional costs and trade barrier costs.[10] The *functional costs* are caused by the activities in the logistics subsystems. These logistics costs will generally be higher for a given product in international logistics systems than in national logistics systems because of the framework conditions described above. In addition, there are the *trade barrier costs*.[11] They are not caused by logistical performance variables, such as distances, weights and times, but by barriers around national and multinational markets. Trade barriers are state-sanctioned—i.e. state-ordered or state-tolerated—interventions in the cross-border exchange of goods and services that discriminate against all foreigners (violation of national parity) or against certain foreigners (violation of most-favored-nation treatment).

Tariff barriers to trade are all types of tariffs. *Non-tariff* trade barriers include first of all direct or indirect protectionist laws. Direct protectionist laws openly influence foreign trade (e.g. import quotas). Indirectly protectionist laws were originally or ostensibly enacted for

[8] See Klaus/Kille, 2008, p. 159.

[9] See Establish, Inc./Herbert W. Davis and Company, 2007, p. 16.

[10] See Cook/Burley, 1985, pp. 27ff.

[11] See Quambusch, 1989.

other purposes, but can be abused in a discriminatory manner through the discretion of administrative authorities (e.g. compulsory labelling for consumer protection). Administrative protectionism includes, in addition to discretionary decisions, arbitrary acts or harassing behavior by the administration (e.g., discriminatory practices in public procurement). Finally, emotional protectionism refers to appeals to national sentiment and calls for boycotts (e.g. appeals to buy domestic products). In the cost analysis of international logistics systems, in addition to the known trade-offs within functional costs, the possible trade-offs between functional costs and trade barrier costs and also the possible trade-offs within trade barrier costs must be taken into account.

Service

Like the level of logistics costs, the *level of service* for delivery service elements can vary in different countries due to country-specific conditions. For example, delivery times in Japan tend to be shorter than in the USA due to the much smaller geographic footprint.[12] The importance of individual *delivery service elements* as factors influencing purchasing decisions can also be assessed differently in different countries.

A survey conducted in Germany, the USA and China in 2008 presents the different logistics objectives of the industrial companies surveyed in the three countries (See Fig. 20.3). Although there is only a slight difference in the basic assessment of objectives in the international comparison, a detailed examination of individual objectives nevertheless reveals differences.

It is typical of international distribution logistics that great importance is also attached to the location of the spare parts warehouse or the availability of spare parts.

20.2 Design of International Logistics Systems

Value Chain as the Basis for Designing International Logistics Systems

Porter classifies internationalization strategies, for which three conceptual research foci were mentioned in the basics of international logistics, according to the three dimensions of value chain, system configuration and system coordination.[13]

Figure 20.4 shows the *value chain* model. The value chain shows how the total value of a product—the amount customers are willing to pay—is made up of the value creation activities and the profit margin. The lower part of the chain lists the primary activities, which deal with physically producing the product and making it available to create value for the customer, distinguishing between upstream and downstream activities. The upper part contains the supporting activities (flanking measures) to maintain the primary

[12]See Stock/Lambert, 1982, p. 6.

[13]Porter, 1989, pp. 25ff.

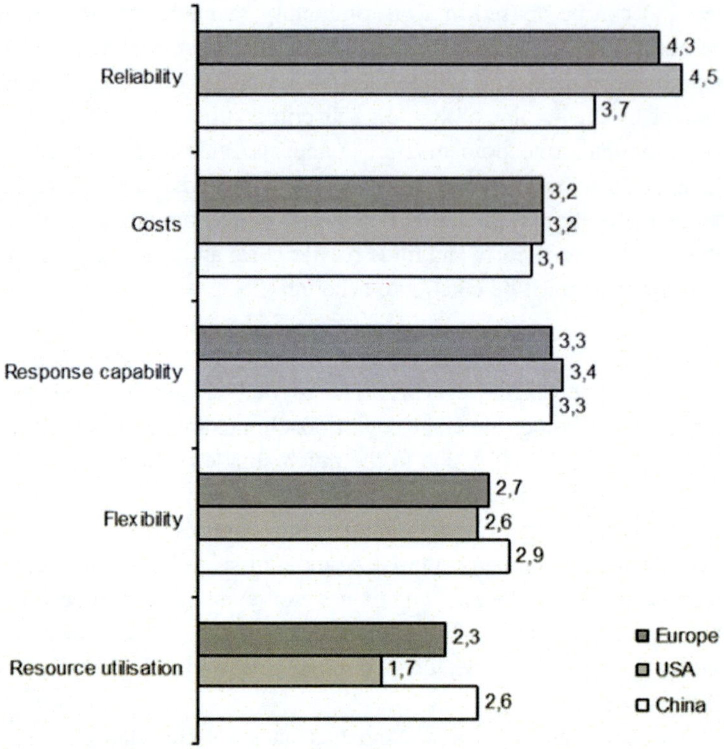

Fig. 20.3 Logistics goals in international comparison. 1 = "low importance" to 5 = "high importance"; mean values from all responses (Source: Straube/Pfohl, 2008, p. 19)

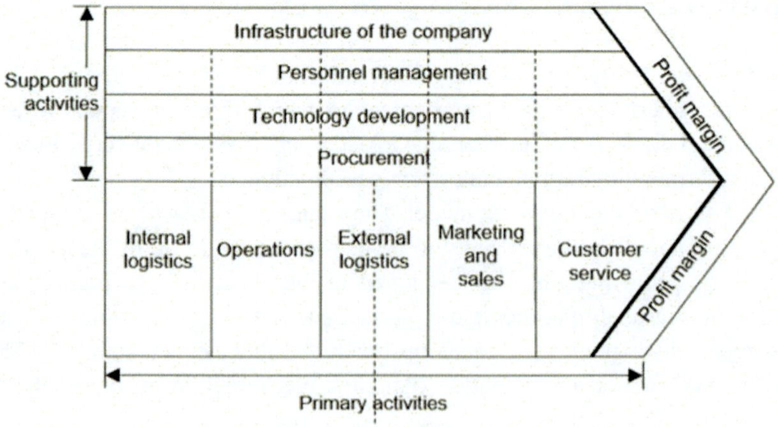

Fig. 20.4 Value chain model (Source: Porter, 2014, p. 64)

activities. PORTER sees the value activities as the building blocks of competitive advantage.

Downstream activities are closely linked to customers and tend to be localized in the countries where the customers are located. This is because in many cases these activities create competitive advantages that are country-specific. Upstream and support activities, on the other hand, tend not to be tied to a specific customer country. This is because the competitive advantages created by these activities arise from the totality of markets in which the firm operates rather than from its presence in each individual country.

The activities in the value chain can be characterized according to the dimensions of *configuration* (centralization–decentralization of activities) and *coordination* (high–low need for coordination). The following factors speak for a centralization of value activities:

- "increasing economies of scale in activities,
- the progressing along the learning curve,
- the comparative cost advantages of concentrating activity in one or a few locations,
- coordination benefits resulting from geographic linkage of related functions (e.g., R&D and manufacturing)."[14]

The first two factors influence decisions about the number of sites at which an activity is carried out. The last two factors influence decisions about the geographic location of these sites.

Just as logistics activities are part of the overall value chain and depend on the other value activities, the international logistics strategy is part of the *internationalization strategy*.[15] Which logistics strategy to choose can therefore only be decided in the context of the overall internationalization strategy.[16] This must be borne in mind when the three conceptual research foci of international management are discussed below with regard to logistics issues. The starting point here is the diagram in Fig. 20.5, which shows the possibilities of international market entry.[17] For the level of activity in international logistics (measured, for example, by the personnel employed in this area), SLATER postulates a course similar to the product life cycle.[18] The basis for this is the thesis of an evolutionary course of the internationalization process. Accordingly, logistics activities increase from indirect export via direct export, reach their maximum with the company's own foreign assembly and certain forms of its own foreign production, and then decrease

[14] Porter, 1989, p. 32.

[15] On internationalisation strategies and logistical requirements, see Baumgarten/Herter, 1999, pp. 828ff.

[16] See Porter, 1989, pp. 50ff.

[17] See Stock/Lambert, 2001, pp. 519ff.; Dülfer/Jöstingmeier, 2008, pp. 173ff.

[18] Slater, 1980, p. 174. See also Holtbrügge/Welge, 2010, pp. 55ff. on product life cycle theory as a theory of internationalization.

Foreign management	Multinational management	Global management
• Indirect export – domestic export companies – foreign import companies	• Direct export – with direct investments in logistics systems abroad	• Worldwide centralization
• Direct export – without direct investments in logistics systems abroad – licensed production abroad	• Assembly abroad – participation / joint venture – own assembly	• Worldwide decentralization
	• Production abroad – participation / joint venture – own Production	• Triad

Fig. 20.5 Opportunities for international market entry

again with other forms of foreign production and global management. The statements regarding the sales channel apply analogously to the procurement channel. Here the forms quasinational procurement, indirect procurement, short term direct procurement, long term direct procurement, multinational procurement and world-wide coordinated procurement can be differentiated.[19]

The International Logistics System in Foreign Management

In the case of foreign management, a company does not set up a special logistics system abroad, but manages the flow of goods and information with its national logistics system or with foreign logistics systems. No international logistics know-how is required for the form of *indirect export*, in which a domestic exporter is involved who handles all cross-border logistics activities. If, on the other hand, a foreign importer is involved, knowledge of the cross-border logistics activities is necessary, although the extent of this can remain comparatively small if the importer's logistics know-how is used or logistics tasks are outsourced to internationally active logistics companies.[20]

In the case of *direct exports* to retailers (who are not specialised importers) or to users, the demands on international logistics know-how increase, even if direct investments are not yet made abroad. This is because, in addition to the cross-border logistics activities of exporting to importers, the logistics activities in the foreign market segments must be

[19] For a more detailed account, see Pfohl/Large, 1991, pp. 26ff.

[20] See Stock/Lambert, 1982, pp. 23ff.; Stock/Lambert, 2001, pp. 535ff.

performed. However, the requirements can also be reduced by outsourcing logistics tasks to logistics companies.[21]

Licensed production abroad can avoid the need to set up an international logistics system. However, in contrast to indirect exporting, more influence is possible on the delivery service to be offered abroad. This is because the delivery service requirements to be met can also be part of the license agreement.[22]

The International Logistics System in Multinational Management

In the case of *direct export with direct investment* in logistics systems abroad, four basic models can be distinguished.[23] In principle, these models can also be implemented without direct investment abroad if the logistics tasks associated with the models are outsourced to logistics companies.

In the *classical system*, the foreign branch of the exporting company operates one or more warehouses in which extensive stocks are maintained. This reduces the frequency of deliveries from the production site to the warehouses. Since transport time is not a major factor, large quantities of goods can be moved by low-cost means of transport, partly because of the grouping (collecting and consolidating) of different shipments. Due to the large transport units, the number of documents can be reduced. Another cost advantage results from the fact that customs duties are not calculated on the customer prices but on the internal company transfer prices. However, these cost advantages are offset by high stockkeeping costs. They result from high inventories during slow-moving goods, capital tied up in a higher stage of the value chain, and high safety stocks that must be maintained to guarantee delivery service. However, inventories maintained in the country of establishment have a positive psychological effect on the customer, who then has less resistance to buying foreign goods in general and to buying goods from distant countries in particular.

The *transit system* differs from the classic system in that no stocks are held in the foreign warehouse. It merely serves as a transshipment warehouse. Although this eliminates the advantages of the classic system, stockkeeping costs are reduced due to the fast means of transport characteristic of this system and the central stockkeeping at the exporting company.

The *regional system* is to be classified between the classical system and the transit system. The stockkeeping is not transferred from the branches back to the country of the exporting company, but to a warehouse (distribution center) that is responsible for a region consisting of several countries. From there, the warehouses of the branches, which function as transshipment warehouses, or, in the case of particularly urgent or large orders, the

[21] On the outsourcing of logistics tasks to international logistics service providers, see Piontek, 1994, pp. 108ff.

[22] See Stock/Lambert, 1982, p. 20; Stock/Lambert, 2001, p. 522.

[23] See Picard, 1982, pp. 28 f.; Sletmo/Picard, 1985, pp. 42 f. For basic structures of logistics systems see Fig. 1.2 in Part I.

customers are supplied directly. An example of this is the centralization of the distribution of goods of an American exporter in one country for the whole of Western Europe. The distribution center is either managed by the branch office of the country in which it is located or it reports directly to the exporting company. In the first case, there is the possibility of a conflict of interest, e.g. in the event of supply bottlenecks, between the requirements of the higher-level branch and the other branches in the region. A particular advantage may arise if the distribution center is established in a free trade zone (e.g. on the site of seaports). Free trade zones are areas from which free trade with all countries is permitted.[24] A warehouse in a free trade zone performs the same functions as a customs bonded warehouse or bonded goods warehouse. When goods are stored in a customs bonded warehouse in the importing country, duties are not payable until the goods are put to further use. In contrast, the bonded goods warehouse in the exporting country is used to store goods that have already been cleared for export but whose subsequent use has not yet been determined. Here, too, duties are only due when the goods are removed from the bonded warehouse. In free trade zones, however, goods can not only be stored and transhipped, but also industrially processed.

In the *direct system*, the foreign branch no longer has anything to do with the physical flow of goods. The foreign customers are supplied directly from the country of origin using fast means of transport. This eliminates any costs for storage and handling abroad. While the classic system has the highest fixed cost components of logistics costs among the basic models, the direct system is characterized by the highest variable cost components. If the customer bears the costs for customs, the advantage of the calculation is lost to the low internal company transfer prices compared to the customer price. In addition, the customer is burdened with administrative problems of customs clearance. Special difficulties also arise when, due to the applicable law for certain products (e.g. pharmaceutical products), quality control must be carried out in the customer country.

In the case of foreign assembly and production, a distinction must be made in each case as to whether this is done under the company's own management or in the form of an equity investment or a joint venture (*ownership strategy*). In the case of participation or joint venture, the investing company can benefit from the possibly existing logistical know-how or the existing logistical network of the foreign company. In all other respects, however, the logistical problems remain the same as in the case of an own-account operation. For this reason, no distinction is made below between the various forms of foreign assembly and production.

Foreign assembly can prove advantageous if, for example, a low wage level can be exploited for the assembly activities abroad or if assembly abroad can be adapted better or faster to the specific customer requirements there than assembly in the home country. Foreign assembly also makes sense if—especially in developing and emerging countries—the import of end products is handled restrictively through high import duties or even to the

[24] On the importance of free trade zones for logistics systems, see Schieck, 2008, p.354.

point of a complete import ban. One example of such foreign assembly is CKD production[25] in the automotive industry.[26] Vehicle components of defined assembly stages are assembled into parts sets and exported to certain countries for assembly. Differentiated according to the regulations of the assembly countries, these parts sets are supplemented there by a spectrum of LC parts.[27] Special requirements are placed on the packaging when supplying CKD assembly plants (maximum protection of goods, minimum use of packaging materials, economically justifiable packaging costs (wages), optimum freight volume or optimum freight costs). Therefore, the packing structure is developed with CAD support in order to find the optimal combination of different parts in one packing module. High requirements for the supply of the CKD assembly plants also result from the necessity to coordinate the delivery of the CKD parts sets from the home country with the delivery of the LC parts from the assembly country.

If the *foreign production* consists of producing in a customer country for this country—e.g. in order to circumvent import restrictions—then this does not result in any new logistics problems for the investing company. These only arise if, within the framework of a *production network* between the various production plants of a company, the advantages of the division of labor are exploited in such a way that parts of a product are only manufactured in one production plant and supplied to other production plants. In order to manage the associated material and information flows, the production sites must be linked by logistics systems, which give rise to corresponding logistics costs.

It depends on the production and logistics cost structure of a product whether centralized versus decentralized parts production is advantageous. Centralization is advantageous in the case of capital-intensive production—i.e. production characterized by high fixed cost components—in which the unit production costs fall sharply as the production volume increases. In contrast, in the case of labor-intensive production—i.e. production characterized by a high proportion of variable costs—unit production costs fall only slightly with increasing production volume. However, production centralization is only worthwhile if the production cost advantages that can be realized as a result are not offset by higher logistics costs. This danger exists above all for parts for which the transport cost components are high, which is the case, for example, if the parts have a low value/weight or value/volume ratio. High trade barrier costs also argue against centralization.

[25] CKD: Completely Knocked Down.

[26] In addition to CKD production, SKD (Semi Knocked Down) and MKD (Multi Knocked Down) production also exist in the automotive industry. The aim of all three forms is to avoid import duties on FBU (Fully Built Up) vehicles. In SKD manufacturing, the vehicle is built up in a plant in the exporting country to the extent permitted by import regulations. In MKD production, only bodyshells are imported. In the highest form, CKD production, the vehicle is completely dismantled, see Schulz/Hesse, 2009, pp. 224 f.

[27] LC: Local Content.

The International Logistics System in Global Management

The need for global management is present in markets characterized by *worldwide homogeneous customer needs* (customer problems). Levitt sees a trend towards such global markets as people learn to articulate the same needs due to the increasingly intensive exchange of information via communication networks spanning the globe.[28] Global markets with homogeneous products can lead to lower logistics unit costs if order processing, packaging, inventory and warehousing costs fall due to standardization effects. However, they can also lead to higher logistics unit costs if communication and transport costs increase when supplying the global markets.

Bender sees five *strategic advantages of* a company operating in global markets[29]:

- *Resource advantages*: Global companies can take advantage of the different availability and cost situation of production factors around the world. Thus, labor-intensive production can be concentrated in low-wage countries, asset-intensive production in countries with favorable financing conditions (low borrowing costs and taxes, investment subsidies, depreciation) and/or favorable operating conditions (no restrictions on capacity utilization, no environmental constraints), and information-intensive production in countries with good information conditions (good information and communication infrastructure, easily accessible technological, economical and social knowledge). For material-intensive products, it depends on their cost structure and weight loss during production whether production is located close to material supply points or demand points.
- *Size advantages*: Global companies can exploit volume and experience cost potentials due to their size (economies of scale).
- *Program advantages*: Global companies can work with a broader and/or deeper production program and achieve synergy effects through the better use of quantitative and qualitative capacities (economies of scope).
- *Trade barrier advantages*: Global companies can avoid trade barrier costs through their presence in the countries concerned.
- *Market presence advantages*: Global companies are able to position themselves on the basis of their worldwide market presence and market knowledge and can adapt to changes in the qualitative and geographical demand structure in good.

Ohmae's *triad strategy* for global companies is based on such presence advantages.[30] According to this strategy, every company should be represented with all its value-creation activities in the USA, Japan and Europe. Only then is one an insider in these important

[28] See Levitt, 1983, pp. 20 f.

[29] See Bender, 1985a, p. 22. See also the considerations made at the beginning of this section on the basis of the value chain.

[30] Ohmae, 1985, pp. 143ff.

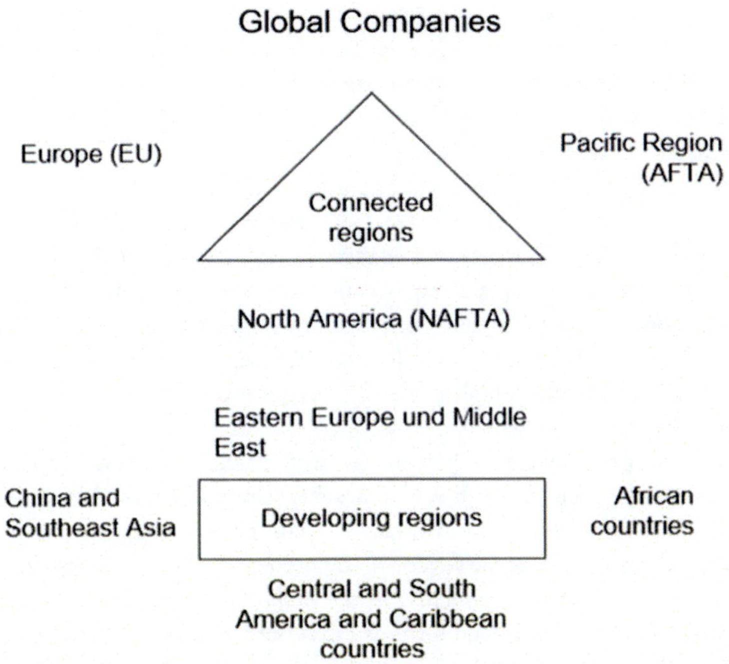

Fig. 20.6 Framework of global logistics requirements (Source: Bowersox, 1994, p. 21)

markets, which have a decisive influence on the existence of global companies. The decision for global management with a centralization, decentralization or triad strategy results in different logistics systems.

Building on the triad strategy, Bowersox characterizes global companies as "*Stateless Enterprises*".[31] These companies are not only represented in the three most developed economic regions of[32] Europe, North America and the Pacific region, but also in markets of economically strongly developing regions, which thus also belong to the framework of global logistics requirements shown in Fig. 20.6. In the management of these companies, more value is placed on the overall development of the company than on individual measures in individual countries or regions. This *equidistant management* is only possible in companies in which managers do not exclusively represent local interests, but whose loyalty is primarily to the company as a whole. These global companies are characterized by the fact that they generally generate less than 40% of their sales in their country of origin.

[31] Bowersox, 1994, pp. 21 f.

[32] Within these regions, these are primarily the customs or economic unions of Europe (EU), North America (NAFTA: North American Free Trade Agreement) and the Western Pacific region (AFTA: Asian Free Trade Agreement).

European Logistics Systems in Transition

The two most comprehensive *changes in Europe* in *the* 1980s and 1990s were the completion of the internal market within the European Union and the opening up of Central and Eastern Europe. Both developments have led to new challenges and to new design approaches for European logistics systems.

The extensive realization of the legal and administrative framework conditions for a *common European market* by January 1, 1993 and the expectation of this change in advance resulted in a comprehensive change in the logistics strategies of Western European companies.[33] Due to the extensive elimination of trade barriers and the progressive harmonization of national framework conditions, a multitude of new—in many cases advantageous—design alternatives arise for the designer of European logistics systems, which, however, increase the complexity of the design task.

The main changes and challenges faced by logistics managers in designing and implementing European logistics systems, as well as the European logistics systems already implemented, were the subject of a study carried out by European Logistics Consultants (ELC) in 1996.[34] More than 300 companies from various industries in Europe, the USA and Asia were surveyed. A comprehensive change in logistics structures was revealed by the significant *reduction in the number of distribution warehouses and production facilities*. There is a trend towards cross-border supply of entire regions from a few locations, although the qualities of logistics service providers in cross-border traffic are often criticized.[35] Various regions have emerged here, representing country clusters in which countries with largely uniform logistical characteristics are grouped together. These are shown as examples in Fig. 20.7. Decisive for the formation of these regions is not only geographical proximity, but also comparable framework conditions with regard to the transport environment, closely interlinked trade structures, uniform customer requirements, the importance of transport times and the value of the products as well as the industry in which they are sold.[36] Customer requirements and adaptation to the development of information and communication technology were assessed as particularly relevant for the development of logistics management.[37]

With the concentration of distribution and production locations and their integration into distribution and production networks, the need for comprehensive *planning and control systems* as well as an adequate organizational structure grows in the sense of multinational

---

[33] See Pfohl, 1993, pp. 64ff.

[34] European Logistics Consultants, 1996.

[35] See European Logistics Consultants, 1996, p. 7.

[36] See A. T. Kearney, 1993, p. 24. In the consumer goods industry, where cultural differences are more important than in the capital goods industry, the Mediterranean countries Portugal, Spain, Italy and Greece are often grouped together in one region.

[37] See European Logistics Consultants, 1996, p. 4.

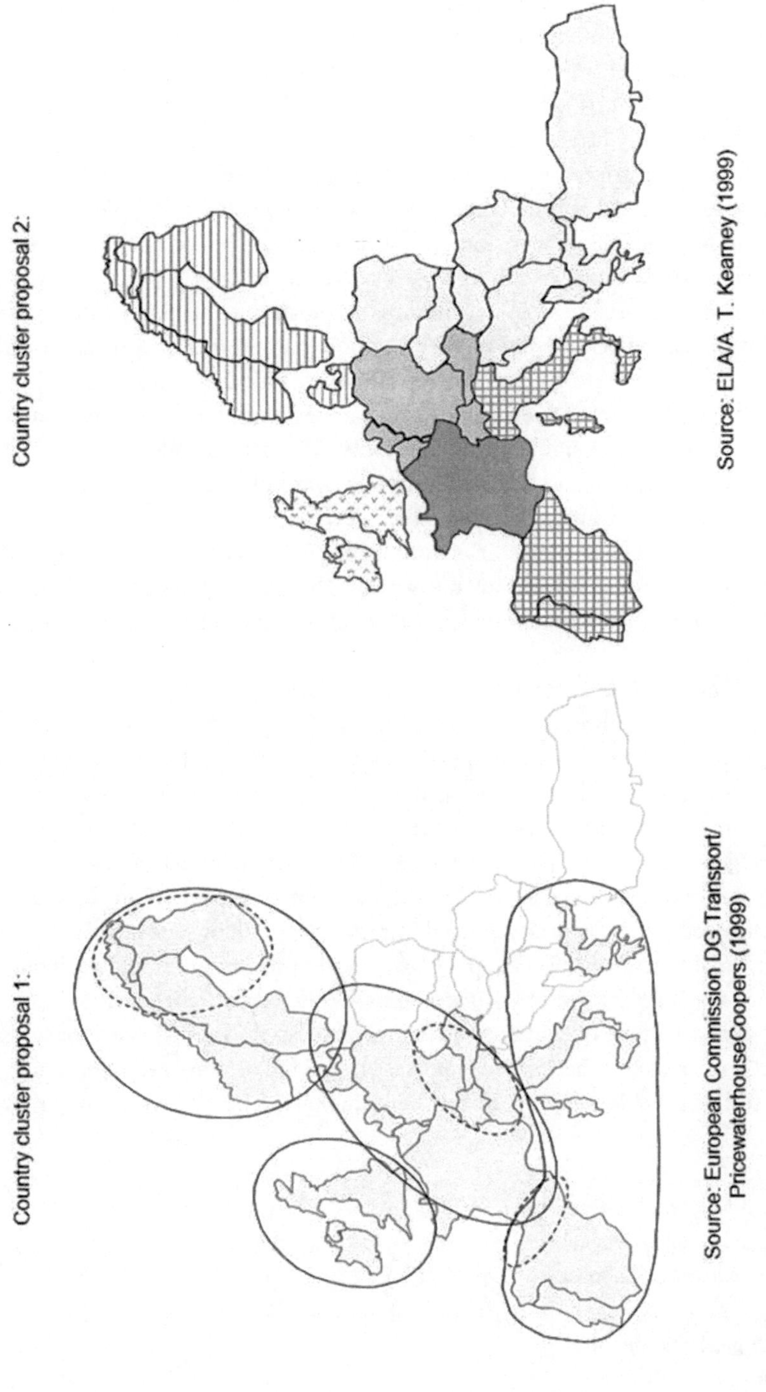

Fig. 20.7 Country clusters in Europe from the logistics perspective

management.[38] Success stories of increasing concentration and integration can therefore not be generalized. Each European logistics system requires its own individual solution, which does not always lie in extreme centralization.

Another trend identified is the progressive concentration on a small number of *logistics service providers.*[39] This is associated with far-reaching changes for such service providers, which not only affect the range of services offered, but above all their efforts to develop into mega carriers by building up Europe-wide service networks.[40] Possible approaches to this were and are the takeover of foreign logistics companies with existing national networks, cooperation with foreign partner forwarding companies or the establishment of their own subsidiaries through direct investment.

While many companies were in the process of adapting their structure to the new possibilities of the European internal market, the political reforms or revolutions initiated the economic *opening of Central and Eastern Europe.* With the increase in trade and the development of production and distribution capacities through direct investments and cooperations, the need to design logistics systems that also include delivery, reception and transfer points in Central and Eastern Europe and provide their services on the basis of the Central and Eastern European infrastructure grew.

The design of logistics systems in Central and Eastern Europe does not differ in principle from that in other regions of the world. This applies to both distribution and procurement systems. Of great importance, however, are the country-specific framework conditions and the basic corporate policy concepts of Western companies on which the commitment is based.[41] One obstacle to the flow of goods to, from and within the countries of Central and Eastern Europe is the comparatively poor state of the infrastructure, especially outside the respective national capitals, although there are considerable differences between the countries.[42] When deciding on locations for distribution centres, a long transport time between the countries in Central and Eastern Europe must be assumed. For general cargo, a national transit time of 24–48 h and cross-border transit times of 3–6 days must be expected.[43] On the other hand, especially in the Eastern and Central European countries, at least the legal framework conditions are largely in place and comparable with those in Western Europe. However, despite the EU's bilateral agreements, especially with the early reform states, there are still numerous trade barriers.

In the initial phase of the opening up of Central and Eastern Europe and the accession of the five new German states, the focus of interest was on the design of *distribution logistics* to ensure fast and reliable supplies to the new markets. The EU enlargement by eight

[38] See O'Laughlin/Cooper/Cabocel, 1993, p. 13.

[39] See O'Laughlin/Cooper/Cabocel, 1993, p. 10.

[40] See O'Laughlin/Cooper/Cabocel, 1993, pp. 77ff.

[41] See Large, 1992.

[42] See Pfohl/Large, 1993, pp. 6ff.

[43] See Vahrenkamp, 2007, p. 134.

countries in 2004 opened up new perspectives. Due to the clear cost advantages associated with production in or sourcing from Central and Eastern Europe, European companies are now increasingly procuring from Central and Eastern European suppliers.[44] As a result, the problem of East-West *procurement logistics* has recently come to the fore. In addition to the problem of setting up the inter-organizational logistics system—often by involving Eastern or Western logistics service providers—Western companies are faced with the task of assessing and, if necessary, securing the efficiency of a potential supplier's internal logistics. Due to the historical development of these countries and the deep transformation crisis of the past 5 years, the establishment of functioning procurement logistics is often associated with comprehensive supplier development and promotion.[45]

Case Study[46]

The example of the toy manufacturer Lego Group can be used to explain the change in European logistics systems. In 2004, the Lego Group was the fourth largest toy manufacturer in the world, but it was making losses in the millions. Besides the failed product strategies, strong competition from video games and cheap products, the high logistics costs and insufficient delivery service are the main reasons. From 2004 to 2008, the toy manufacturer carried out a project to optimize its supply chain.

Before the project, there was a central distribution center linked to the production plants. However, the retail trade was still supplied by around a dozen regional distribution centers. Goods were transferred between different warehouses several times. Each location had its own inventory management, different procedures and ways of working. Almost all logistics activities were carried out by the Lego Group itself.

After the project, the 12 regional distribution centers were consolidated into a single central logistics center in the Czech Republic (see Fig. 20.8). All orders from more than 120 of a total of 135 customer countries worldwide with 14,000 retail customers are processed directly via the warehouse. The Lego Group commissioned Deutsche Post DHL as contract logistics provider for the central warehouse and also for value-added services under a 5-year contract.

As a result of the optimization project, savings of around 25% were achieved in global logistics costs. This corresponds to approximately 150 million euros. The share of logistics costs in sales was reduced from over 12% in 2005 to 9.2% in 2009. At the same time, the delivery service level was increased from around 90% per day in 2004 to as much as 99% per hour in 2009. Furthermore, the consolidation reduced the carbon footprint. In 2009, the transport performance volume kilometers decreased by 25%, which corresponds to a total of five million volume kilometers less per year.

[44] See Baumgarten/Krokowski, 2003, p. 14.

[45] On supplier promotion, see Large, 2013, pp.256ff.

[46] See Pfeiffer, 2009, pp. 10ff.

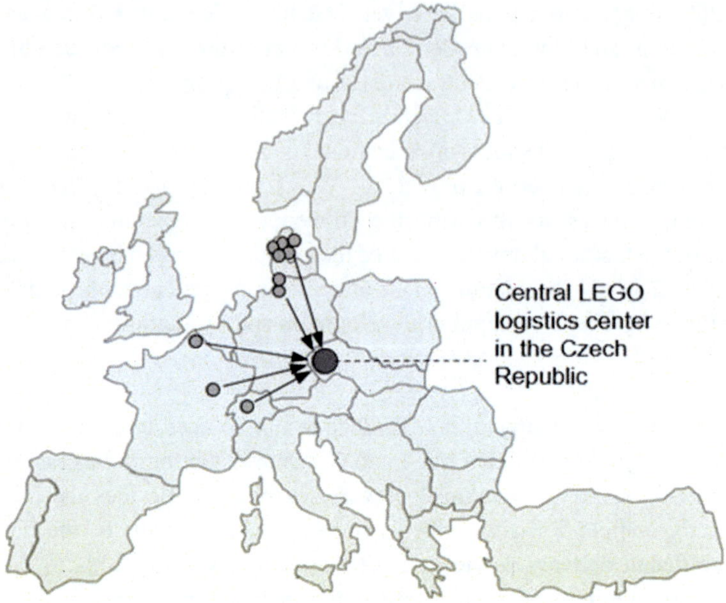

Fig. 20.8 Centralization of LEGO distribution centres

The case study illustrates the two main changes in international logistics systems described in the last section: reduction in the number of distribution warehouses, outsourcing of logistics activities and concentration on a few logistics service providers.

20.3 Financial Aspects of International Logistics

Terms of Delivery and Payment

Due to the lower knowledge of the contractual reliability and creditworthiness of the business partners abroad compared to the domestic market, the longer time for the exchange of information and goods between supplier and customer as well as the different currencies, the risk in international procurement, production and distribution is greater than in the domestic market. The first step is to try to limit this risk by drawing up appropriate contracts with regard to the terms of delivery and payment.[47] These essentially regulate pricing, currency, retention of title, terms of delivery, delivery date, terms of payment, guarantees, warranties, contractual interest, contractual penalties, choice of law, place of jurisdiction, etc.[48]

[47] See the relationship between contracting mix and logistics in Part III, Sect. 11.2.

[48] See Schieck, 2008, p. 151.

As regards the *terms of delivery*, contracts may refer in particular to the Incoterms, an overview of which has already been given in Part III, Chap. 9, as an integral part of the conditions policy and as rules determining the logistical span of control. In addition to the transfer of costs and risks, the terms of delivery also regulate the procurement of various documents (export licence, certificate of origin, consular invoice, etc.) and the payment of ancillary expenses (customs duties, fees, taxes and other charges), which are important in connection with the execution of the order.

Terms of payment regulate the relationship between the flow of goods or information in the logistics system and the flow of payment by specifying the time and conditions of consideration for delivered products and/or services. In contrast to terms of payment without documents, which distinguish between payment before delivery, payment on delivery and payment after delivery, each with different financial risks for the supplier and the customer, terms of payment with documents place special demands on order processing.[49] In this case, payment is made against documents, such as bills of lading, insurance documents, or commercial invoices, with which the power of disposal over the goods is transferred from the seller to the buyer.

In the case of *documentary collection*, when a bank is usually involved in the case of "payment against documents", payment is made by the buyer/importer to the collecting bank (possibly the importer's house bank) against handover of the agreed documents. The documents are previously forwarded to the collecting bank by the seller (exporter) via his house bank. The exporter receives payment in the opposite way. In the case of "documents against acceptance", the exporter issues a bill of exchange to the foreign customer. The collecting bank hands over the documents to the importer when the latter accepts the bill of exchange. The exporter bears the risk of the acceptance payment, unless he requires the importer to provide a bank guarantee.

In the case of a *letter of credit*, the importer (issuer of the letter of credit) instructs his bank (letter of credit bank) to open a letter of credit in favor of the exporter (beneficiary) with his bank (advising bank). In general, a letter of credit is a promise to make a sum of money available to the beneficiary if certain conditions are met. In the case of a documentary letter of credit, the bank's promise to pay is honoured by the presentation of documents relating to the exported goods. In the case of a revocable letter of credit, the letter of credit bank may amend or withdraw the letter of credit, generally at the instigation of the importer, without notifying the exporter. In the case of an irrevocable letter of credit, on the other hand, the mandated bank is obliged to make payment without exception, provided that the prescribed documents are available (See Fig. 20.9).

Financial Engineering

Based on delivery and payment terms, new financial services have been developed by banks under the heading of "financial engineering", by means of which the capital tied up

[49] See Kummer et al., 2010, pp. 155 f.

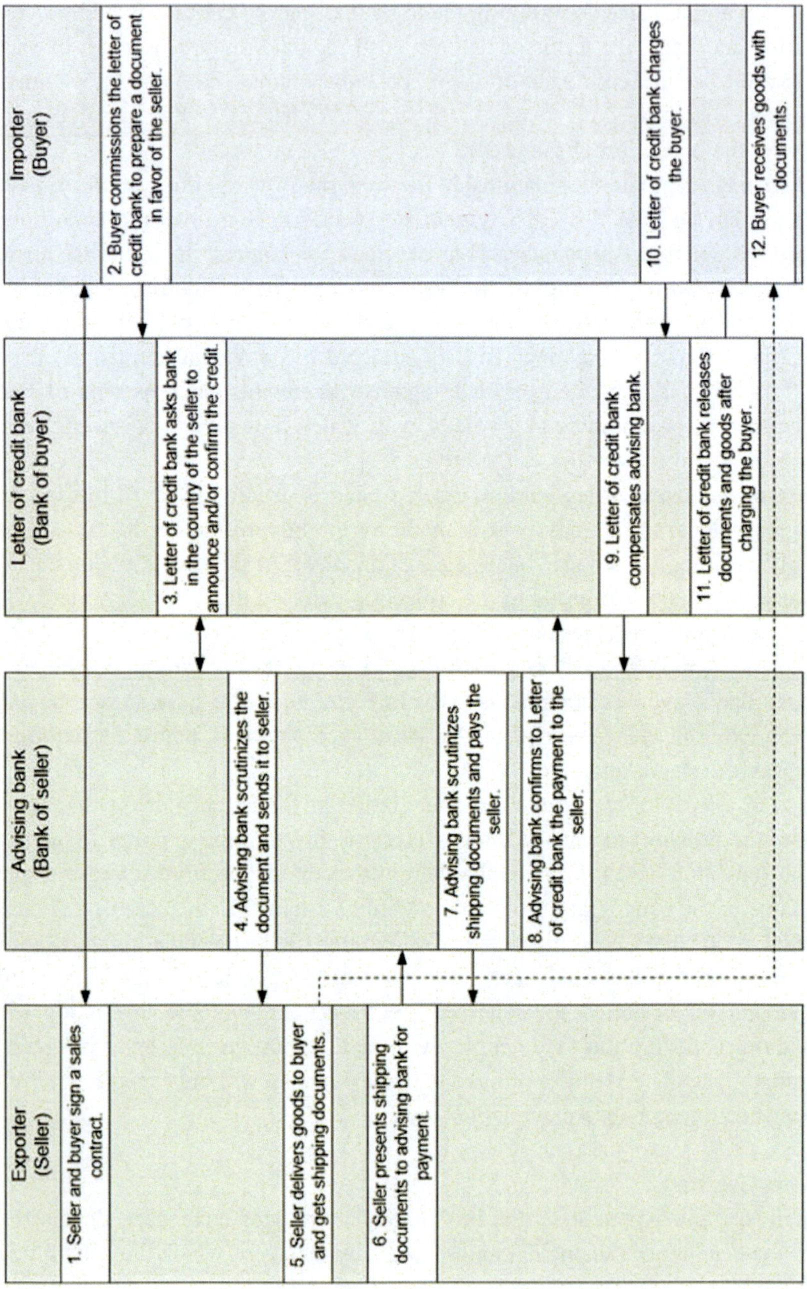

Fig. 20.9 Procedure for a letter of credit (Source: Branch, 2001, p. 136)

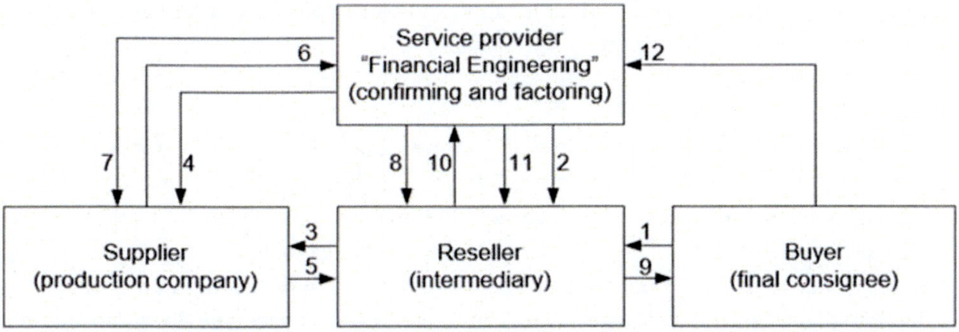

1 Buyer orders goods from reseller
2 Service provider assumes the risk of buyer's inability to pay (acceptance of liability)
3 Reseller orders goods from supplier
4 Service provider assumes the risk of resellers inability to pay (acceptance of liability) and possibly finances the supplier's production costs
5 Supplier delivers goods to reseller
6 Supplier hands over documents to service provider
7 Service provider pays supplier
8 Service provider hands over documents to reseller against payment
9 Reseller delivers goods to buyer
10 Reseller sells claim towards buyer to service provider without recourse
11 Service provider pays purchase price after covering of claim from confirming
12 Buyer pays purchase price to service provider when due

Fig. 20.10 Confirming and factoring in connection with procurement and distribution logistics (Source: based on Hardt, n.d., p. 303)

in the current assets of industrial and commercial enterprises can be reduced. Such financial services, which are based on the financing instruments of letters of credit (order security), storage financing (inventory) and cession credit (trade receivable), are in principle also to be discussed in connection with national logistics systems; however, because of the increased risk and the longer order processing times, they are of particular importance for international logistics.[50]

Figure 20.10 shows such a financial services package to be offered by a service provider from a single source, with the corresponding payment, information and goods flows. The service provider acts as a confirming and factoring company. Confirming companies existed in large numbers in England at the end of the last century. These were large trading houses that, in addition to financing, took care of finding suppliers, loading/shipping, insurance, warehousing, and so on. Today, confirming is the service of risk protection and financing. Factoring refers to the non-recourse sale of receivables to a factoring

[50] See Pfohl et al., 2004, pp. 15ff.; Gomm, 2008, pp. 171 f. and generally on financial services provided by logistics service providers, ibid., pp. 169ff.

company. It is responsible for the administration and collection of the receivable and bears the risk of insolvency of the buyer of the goods.

Currency Impact

The currencies of different countries can affect international logistics in two ways. On the one hand, *exchange rate risk* must be taken into account, both in stockkeeping and in order processing. To manage the exchange rate risk, the financial instruments of exchange rate hedging are available in the first instance.[51] However, the exchange rate risk in terms of inventory valuation can also be reduced by including it in the decision on the location of production plants and warehouses. One can preferentially place the locations in the countries whose currencies are stable or one can also consider the investment in locations as an investment in a currency portfolio. Moreover, reducing the exchange rate risk is also helped by shortening the order processing time.

In addition to exchange rate risk, on the other hand, *foreign exchange management* can affect international logistics. If, for example, foreign exchange is only made available to companies in developing countries for the import of goods to a limited extent, the possibilities for building up inventories in these countries are restricted.

Tax and Customs Influence

As soon as a company is present in several countries with direct investments in accordance with the concepts of multinational or global management, the problem arises that the intra-company cross-border transfer of goods without tax influence is not possible due to national tax jurisdictions.[52] From the point of view of profit distribution to the shareholders of the German parent of a multinational company, therefore, German operating locations can prove to be quite advantageous compared to foreign operating locations due to existing *double taxation* of profits in Germany and abroad. The objective of profit shifting by valuing the cross-border flows of goods between the subsidiaries of a company with corresponding *transfer prices* is limited by the fact that national tax laws require prices that stand up to an arm's length comparison. It often makes sense to deliver the products to the foreign operating locations broken down into their individual parts in order to avoid the higher customs duty on the total product through lower customs duties on the individual parts. In the overall cost calculation, however, it must be taken into account that the transport of the individual parts leads to higher transport costs, which, however, are generally lower than the customs costs saved.

---

[51] See Schieck, 2008, pp. 144ff.

[52] See Holtbrügge/Welge, 2010, pp. 393 f.

References

A T Kearney (1993) Logistics Excellence in Europe. A Study Report Prepared by A. T. Kearney on Behalf of the European Logistics Association (ELA)

Baumgarten H, Herter M (1999) Internationalisierung der Logistik. In: Weber J, Baumgarten H (Hrsg) Handbuch Logistik. Stuttgart, S. 828-845

Baumgarten H, Krokowski W (2003) Trends and Strategies in International Procurement. A Comparison between Asia and Europe. Hamburg

Bender P S (1985a) The Challenge of International Distribution. In: IJoPD 15 4, S. 20-26

Bender P S (1985b) The International Dimension of Physical Distribution Management. In: Robeson J F, House R G (Hrsg) The Distribution Handbook. New York/London, S. 777-814

Bowersox D J (1994) Framing global logistics requirements. In: Brace G (Hrsg) Logistics Technology International. London, S. 19-22

Bowersox D J, Sterling J U (1982) Multinational Logistics. In: Journal of Business Logistics 3 2, S. 15-25

Branch A (2001) International Purchasing and Management. Cengage Learning EMEA

Cook R L, Burley J R (1985) A Framework for Evaluating International Physical Distribution Strategies. In: International Journal for Physical Distribution & Materials Management 15 4, S. 26-38

Dülfer E, Jöstingmeier (2008) Internationales Management in unterschiedlichen Kulturbereichen. 7., vollst. überarb. Aufl. München

Establish Inc, Herbert W Davis and Company (2007) Logistics Cost and Service 2007. Präsentation anläßlich der Jahreskonferenz des Council of Supply Chain Management Professionals (CSCMP)

European Logistics Consultants (1996) Logistics in Europe – The Vision and the Reality: Survey into current developments influencing European manufacturing and logistics

Gomm M (2008) Supply Chain Finanzierung. Optimierung der Finanzflüsse in Wertschöpfungsketten. Berlin

Hardt D (o J) Logistik und neue Wege der Finanzierung. In: Bundesverband Deutscher Volks- und Betriebswirte e. V. (Hrsg) Logistik im Mittelpunkt moderner Betriebsführung. Fachtagung. Minden, S. 298-305

Holtbrügge D, Welge M K (2010) Internationales Management. Theorien, Funktionen, Fallstudien. 5., überarb. Aufl. Stuttgart

Klaus P, Kille C (2008) Die Top 100 der Logistik – Marktgrößen, Marktsegmente und Marktführer in der Logistikdienstleistungswirtschaft, Ausgabe 2008/2009, Hamburg

Kummer S, Schramm H-J, Sudy J (2010) Internationales Transport- und Logistikmanagement. 2. Aufl. Wien

Large R (1992) Gestaltung von Logistiksystemen in Mittel- und Osteuropa. In: Logistik Spektrum 4 4, S. 4-9

Large R (2013) Strategisches Beschaffungsmanagement. Eine praxisorientierte Einführung. Mit Fallstudien. 5., vollst. überarb. Aufl. Wiesbaden

Levitt Th (1983) The Marketing Imagination. New York London

Macharzina K, Engelhard J (1987) Internationales Management. In: DBW 47 3, S. 319-344

Nelson P T, Toledano G (1978) Challenges for International Logistics Management. In: Journal of Business Logistics 1 2, S. 1-21

Ohmae K (1985) Macht der Triade. Die neue Form weltweiten Wettbewerbs. Wiesbaden

O'Laughlin K, Cooper J, Cabocel E (1993) Reconfiguring European Logistics Systems. Oak Brook, IL

Pfeiffer K (2009) Stein für Stein zentralisiert. In: Logistik heute. 31 3, S. 10-13

Pfohl, H.-Chr. (1993): EC Unification and the Outlook for Logistics. In: Journal for Business Logistics 14 1, S. 43-79

Pfohl H-Chr, Elbert R, Hofmann E (2004) Management der „finanziellen" Supply Chain: Charakterisierung – Aufgabenbereiche – Interdependenzen. In: Bundesvereinigung Logistik e.V. (BVL) (Hrsg) Finanzierung – eine neue Dimension der Logistik. Ergebnisse des Arbeitskreises Logistik und Finanzen. Berlin, S. 1-64

Pfohl H-Chr, Large R (1991) Internationale Beschaffung. Einflußfaktor Logistik. In: Beschaffung aktuell 6, S. 22-30

Pfohl H-Chr, Large R (1993) Sourcing from Central and Eastern Europe: Conditions and Implementation. In: International Journal of Physical Distribution and Logistics Management 23 8, S. 5-15

Picard J (1982) Typology of Physical Distribution Systems in Multi-National Corporations. In: International Journal of Physical Distribution & Materials Management 12 6, S. 26-39

Piontek J (1994) Internationale Logistik. Stuttgart

Porter M E (1989) Der Wettbewerb auf globalen Märkten. Ein Rahmenkonzept. In: Porter M E (Hrsg) Globaler Wettbewerb. Strategien der neuen Internationalisierung. Wiesbaden, S. 17-68

Porter M E (2014) Wettbewerbsvorteile. Spitzenleistungen erreichen und behaupten. 8. durchgesehene Aufl. Wiesbaden

Quambusch L (1989) Stichwort "Handelshemmnisse, nicht-tarifäre". In: Macharzina K, Welge M K (Hrsg) Handwörterbuch Export und Internationale Unternehmung. Stuttgart, Sp. 782-799

Schieck A (2008) Internationale Logistik Objekte, Prozesse und Infrastrukturen grenzüberschreitender Güterströme. München u.a.

Schulz R, Hesse F (2009) Das Produktionsnetzwerk des VW-Konzerns und die Versorgung der Überseewerke. In: Göpfert I (Hrsg) Logistik der Zukunft – Logistics for the Future. 5., aktual. und überarbeit. Aufl. Wiesbaden, S. 211-232

Slater A (1980) International Marketing: The Role of Physical Distribution Management. In: International Journal of Physical Distribution & Materials Management 10 4, S. 160-184

Sletmo G K, Picard J (1985) International Distribution Polices and the Role of Air Freight. In: Journal of Business Logistics 6 1, S. 35-52

Stock J R, Lambert D M (1982) International Physical Distribution. A Marketing Perspective. In: International Journal of Physical Distribution & Materials Management 12 2, S. 3-39

Stock J R, Lambert D M (2001) Strategic Logistics Management. 4. Aufl. Boston u.a.

Straube F, Pfohl H-Chr (2008) Trends und Strategien in der Logistik - Globale Netzwerke im Wandel. Umwelt, Sicherheit, Internationalisierung, Menschen. Hamburg

Vahrenkamp R (2007) Logistik. Management und Strategien. 6., überarbeit. und erw. Aufl. München

Wood D F (1989) Towards a Channels Theory of International Logistics. In: Masters J M, Coykendale C L (Hrsg) Logistics Education on Research. A Global Perspective. Proceedings of the 18th annual Transportation and Logistics Educators Conference. The Ohio State University. Columbus, S. 98-113

Zettelmeyer B, Zöllner W (1986) Hemmnisse für Logistik. Randbedingungen des internationalen Güterverkehrs. In: ZfL 7 1/2, S. 55-57

Index

H.-C. Pfohl, *Logistics Systems*,
https://doi.org/10.1007/978-3-662-64349-5